Elevator Traffic Handbook

Elevator Traffic Handbook

Theory and practice

Dr Gina Barney

Spon Press
Taylor & Francis Group

LONDON AND NEW YORK

First published 2003 by Spon Press
11 New Fetter Lane, London EC4P 4EE

Simultaneously published in the USA and Canada
by Spon Press
29 West 35th Street, New York, NY 10001

Spon Press is an imprint of the Taylor & Francis Group

Printed and bound in Great Britain by TJ International, Padstow, Cornwall.

British Library Cataloguing in Publication Data
A catalogue record for this book is available from the British Library.

Library of Congress Cataloging in Publication Data
A catalog record for this title has been requested.

ISBN 0-415-27476-1

Contents

To Josie
without whose love, friendship and support
many things would not have been possible.

Foreword

People who are familiar with Dr Barney's book "Elevator Traffic Analysis Design and Control" (written with S.M. Dos Santos) will be quite surprised when they review Dr Barney's latest − this volume. Where the initial book left off, this one picks up, repeats some of the previous information and goes way beyond.

I was impressed that the title of the book is "Elevator Traffic Handbook" yet most of the text refers to "lifts". Those familiar with the current lift (one must get used to the word "lift" instead of "elevator" when reading this book) and escalator literature can recognise some information and will appreciate its association in Dr Barney's context.

Dr Barney is a good student and an astute observer. She has taken the appropriate information about pedestrian handling and movement and consolidated it into this handbook. It is quite interesting that Dr Barney starts off with pedestrian circulation in her initial chapters. This is one area which few elevator (lift) consultants consider and its proper provision can be an asset or detriment to any vertical transportation scheme. This is particularly true of escalators, which Dr Barney emphasises.

An interesting feature is added after the initial chapters in a series of "case studies" where various traffic situations are presented and solutions offered. I am certain many of the readers will relate to some of these cases and could add others from their own experience.

Beyond that the book settles into the "nitty gritty" of elevator, people handling, theory and design. How elevators serve people, how people react to elevators, the many time factors, the estimation of demand, various traffic patterns − the myriad of aspects that constitute the entire art of elevator traffic design are covered in these ensuing chapters. One may or may not agree with what is presented but the debate goes on as long as anyone involved in the process has any judgement or opinion.

One of the most controversial aspects of vertical transportation is covered in Chapter Six, which is the determination of passenger demand. This relates to the population and the use of various buildings, which is probably the most "blue sky" aspect of the whole design process. Who can predict what will happen in the years or more between the initiation and the completion of any building ? As an observation, some consultants have either over populated or over elevated a building and have been overly pessimistic about the elevator's ability to handle the projected population and have ended up being "safe".

Leaving the theoretical behind, the book proceeds into the various aspects of elevator operation and traffic situations. The mathematics becomes more extensive and various individuals who have had input to the process are quoted and discussed. I was pleased to see the reference to loading based on volume rather than weight, which those familiar with my work will recognise as one of my pleas.

Dr Barney does an extensive job in this middle third of the book. Special elevator situations such as sky lobbies, shuttles and double deck are investigated as well as the more esoteric aspects of elevator shape, door design, lobby size and so on. Tables and case studies are provided to illustrate and consolidate information. If you are looking for something you will find it and the extensive table of contents and comprehensive index will help.

The ensuing chapters are devoted to various operating systems and approaches to traffic handling. Included are discussions of call allocation and fixed and dynamic sectoring. Common and unusual traffic situations are discussed as well as some of the aspects of traffic handling during peak periods.

Modern elevator literature would not be complete without investigation of computer control which Dr Barney does in-depth. I have to confess that much of it goes over my head, being a veteran of the transition from relay control to solid state. I trust that the modern elevator person will have little trouble understanding what Dr Barney presents. Reinforcing the chapters are a number of case studies which enhance the related information.

The final chapters are an in-depth study of elevator traffic handling, which is an expansion of the studies presented in the earlier book. Helping the discussion is a study of simulation which, of course, was a difficult approach when the original volume was published in 1977 and revised in 1985. Modern computers have made this a viable approach and a tool for the current elevator engineers.

All in all Dr Barney has added a substantial volume to the growing library of literature dealing with vertical transportation. The book contributes to the final product of our vertical transportation efforts. We are not dealing with the hardware of elevators and escalators but rather with their use and purpose − that is − moving people safely and swiftly from place to place. The information found in this book will help in accomplishing that goal.

George Strakosch
Elevator Consultant *Emeritus*
c/o Elevator World, Mobile, USA
21 June 2002

Preface

My work in the lift industry started in early 1968, when I was an academic at the University of Manchester Institute of Science and Technology (UMIST). At that time the Encyclopedia Britannica listed four books on lift technology. Of these, the most important to me was "Elevators and Escalators" by George Strakosch (1967). This book, now in its third edition (1998), provided me with an excellent start to what has become my life's work. Strakosch brought together, in one place, a wide wealth of knowledge, much of which still stands today and readers who do not possess a copy would be advised to acquire one.

By 1977 my student, friend and colleague, Sergio dos Santos, now Rector of Minho University, Portugal, and I decided to share the results of our research by writing "Lift traffic analysis, design and control", republished 1985. That book underscored the practice described in Strakosch with mathematical and computer analysis. Although the content of the book remains valid today, other authors have entered the field and have contributed to the theory and practice of lift traffic design and control.

In this book, I have attempted to bring together in one place a complete treatment of elevator traffic design and control. In addition to my work in the subject, the book draws on the work of other authors in the field, which I have included, commented on, extended, agreed with, or disagreed with. The interpretations are mine, as are the mistakes.

The early chapters concern well-established practice that is generally accepted. The latter chapters contain a more philosophical discussion and present controversial aspects that may require further refinement. I would like any comments and challenges to come to me in this continuing debate. Contact details are at www.liftconsulting.org.

My career has lead me through industry, research, teaching, consulting, code making and training. I have written this book for everyone involved in traffic design and traffic control of lift (US: elevator) installations. I have based it on theory arising out of practice. Lift traffic design is a practical science.

Dr Gina Barney
Cumbria, England
11 July 2002

ACKNOWLEDGEMENTS
I wish to thank Michael Godwin, formerly of Wm. Wadsworth & Sons, Bolton, who introduced me to lifts; Sergio dos Santos, who carried out much of the early research and provided a powerful computer analysis program; my countless former students and academic colleagues at UMIST; and the lift industry in general, who have tolerated me and my ideas for nearly 35 years.

This book has greatly benefitted from the support of my numerous associates and colleagues in the lift industry and particularly from many stimulating discussions, corrections and comments from Dr Richard Peters (Peters Research, UK), Craig Pearce (Arup, Australia), Dr Lutfi al-Sharif (WSP Buildings Transportation) and Dr Marja-Liisa Siikonen (Kone, Finland).

CHAPTER ONE

Principles of Interior Circulation

1.1 GENERAL

It could be wondered why the circulation of people in a building is discussed in a handbook principally concerned with vertical transportation. The answer is that the devices which provide mechanical movement of people must be able to receive and despatch their passengers efficiently and effectively. Thus the route taken by people to these circulation elements is important. A further and important consideration is that efficient and effective circulation is safe circulation.

Little is written about the circulation of people in buildings. Tregenza (1971, 1972, 1976) has provided some insight into the UK experience. The classic book on this subject is that of Fruin (1971),[1] and he offers a great deal of information, copiously referenced. A recent paper of interest is that of Kavounas (1993d), which presents an analysis of linearly connected vertical modes. Much of the information in this chapter is due to Fruin, extended by the Author's own work.

The circulation of people in the interior of buildings is a complicated activity. It is affected by a number of factors:

■ Mode: horizontal and vertical movement.

People will generally be walking horizontally, except where they are using passenger conveyors. They will then change mode from horizontal to vertical movement, in order to reach a higher or lower level. To do this they will use stairs, moving walks and ramps (passenger conveyors), escalators or lifts.

■ Movement type: natural or mechanically assisted.

People will move naturally, when walking, and be mechanically assisted, when using moving walks and ramps (passenger conveyors), escalators or lifts.

■ Complications: human behaviour.

The movement of people around a building is complex - because humankind is complex. Individuals have: their own concepts of route; their own purpose for travel; their own level of urgency; their own characteristics of age, gender, culture, handicaps; etc. There is unpredictability in human behaviour.

---

[1] Reprinted by Elevator World in 1987.

The interior circulation in a building must be designed:

■ To consider all circulation routes.

   These include principal and secondary circulation areas, escape routes, service routes and waiting areas.

■ To provide clear and obvious routes.

   Pedestrians should be able to see the route to take, perhaps assisted by good colour coded signs and open vistas.

■ To ensure that the circulation patterns are rational.

   An example is the avoidance of pedestrians passing through a lift lobby, where other persons are waiting.

■ To ensure that incompatible types of circulation do not coincide.

   This would apply to pushing goods trolleys across a pedestrian mall in a shopping centre or sterile/non-sterile movements in a hospital.

■ To minimise the movement of people and goods.

   This would bring related or associated activities together, eg: sales and marketing; personnel and training; etc. in an office building.

The design and location of portals (defined here as: entrances, doorways, gates, etc.), corridors, stairs and mechanical handling equipment (moving walkways, moving ramps, escalators, lifts) must be coordinated to ensure:

■ the free flow of people goods and vehicles

■ they occupy the minimum allocation of space

■ bottlenecks are prevented.

It is important to size each facility. Thus the handling capacities of corridors, which lead to stairs, which in turn lead to a lift, should be adequately sized for their anticipated load. The term "handling capacity" is used here for passive (non-mechanical) building elements in the same way as it is applied to the mechanical elements. The term "demand" is used to indicate the level of usage.

The efficiency of interior circulation is dependant on building shape. Both tall/slender and low/squat buildings are inefficient. The ideal shape is "compact". Factors that influence circulation efficiency are:

■ the relative location of rooms

■ the relationship of major spaces with entrances and mechanical elements

■ the importance of the journey undertaken

- the separation of different traffic types

- the need to group some spaces together

- the conflict of vertical and horizontal circulation modes.

These and other factors lead to an imprecise knowledge of how circulation occurs in a building and permit only empirical methods of design. Much of what this chapter contains should be taken as general guidance only, much of what follows cannot be proved theoretically and many of the conclusions have been empirically derived. Reasons are given for the conclusions so that if new evidence comes to hand (or opinions change) the results can be modified. Regulations may also affect the circulation design, such as fire and safety codes; and these must be taken into account.

Please note that throughout this chapter people are often called pedestrians, when "on foot", and passengers, when they are being mechanically transported.

1.2 HUMAN FACTORS

1.2.1 Interpersonal Distances

The evolution of humankind to its present form has taken a long time ! Along the way various instincts have been implanted within the being, in order for humankind to evolve and survive. Instinct is still a powerful characteristic of the human being and fear is its biggest driving force. The fear of harm conditions all of us. The association of one human to another is predicated on trust, or doubt, and this controls how humans behave one to the other. The spatial separation of humans is thus important.

Consider an individual human subject in an isolated place. What does this subject do if another human approaches ? Fight or flight ? It is a question of interpersonal separation. Hall (1966) classified interpersonal distances with characteristics based on the sensory shifts of sight, smell, hearing, touch and thermal receptivity into seven categories as shown in Table 1.1.

Table 1.1 Interpersonal distances

Separation		Comments
Public distance (far)	> 7.5 m	Little sensory involvement; oral communication loud, exaggerated and stylised (theatrical).
Public distance (near)	3.6−7.5 m	Oral communication less loud, less exaggerated, still stylised; general facial expressions detectable (frown, smile).
Social distance (far)	2.1−3.6 m	Aspects of personal grooming visible; possible to pass objects.
Social distance (near)	1.2−2.1 m	Considerable facial details visible; ease of passing objects; not possible to seize an individual.
Personal distance (far)	0.75−1.2 m	Fine details of complexion, teeth, eyes, etc. visible; occasional detection of body odours; possible to seize a person.
Personal distance (near)	0.45−0.75 m	Details of cleanliness discernable; occasional detection of body odours & perfumes; bodily contact avoidable but easily possible.
Intimate distance	< 0.45 m	Body sounds, smell, heat all perceivable; sight distorted; very difficult to avoid contact.

The public distance classification is sometimes called the flight zone to indicate that an individual can take evasive or defensive action. The social distance (far) represents a zone of potential vulnerability (the *en garde* of sword fighting) and is the distance used for formal meetings in contrast to the social distance (near) used for casual meetings. The personal distance (far) is defined as an individual's circle of trust and can be considered the interpersonal spacing found in a spacious waiting area. The personal distance (near) is commonly encountered in denser waiting areas and queuing situations. Clearly, crowding occurs with the intimate distance classification leading to the "touching" situation found when travelling in a lift car. The seven zones are illustrated in Figure 1.1.

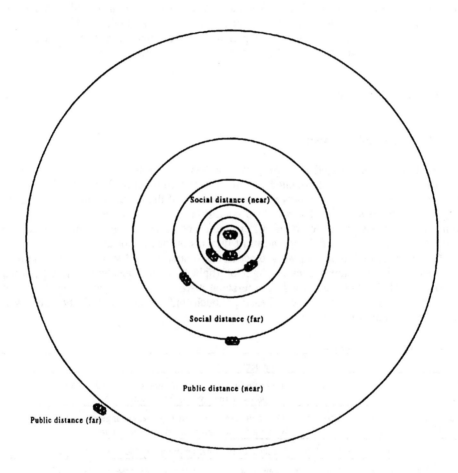

The three inner (unlabelled) rings are:
The innermost ring represents the intimate distance.
    It shows an individual in a 600 mm circle, but only 450 mm from the nearest individual.
The next innermost ring represents the personal distance (near).
The third innermost ring represents the personal distance (far).

Figure 1.1 Interpersonal distances

1.2.2 Human Personal Space and Human Dimensions

Human beings value personal space. This is measured by a personal buffer zone around each individual. The actual size of the buffer zone varies according to an individual's culture, age, status, gender, handicaps, etc. and even their geographical origin.

The physical dimensions of the human body vary widely. It is not politically incorrect to observe that females are generally smaller than males, or that people from the Asia/Pacific region of the world are smaller than Europeans. The space an individual occupies also depends on how the person is clothed and what they might be carrying, or if they are children, what they are doing.

Fruin (1971) indicates four classes of personal space, when persons are queuing or are occupying a waiting area:

(a) circulation zone
(b) personal comfort zone
(c) no touch zone
(d) touch zone.

Class (a) is where individuals can easily pass between other individuals without disturbing other individuals. The area occupied by each person is some 1.2 m^2, ie: 0.8 persons/m^2.

Class (b) is where individuals can still pass between other individuals, but may disturb them. The area occupied by each person is some 0.9 m^2, ie: 1.1 persons/m^2.

Class (c) is where individuals cannot pass between other individuals without disturbing them. The area occupied by each person is some 0.7 m^2, ie: 1.4 persons/m^2.

Class (d) is where individuals cannot pass other individuals. The area occupied by each person is some 0.33 m^2, ie: 3.0 persons/m^2.

Fruin does not mention a crowded situation with individuals occupying 0.25 m^2, ie: 4.0 persons/m^2 or the situation in a lift or crowded train or bus where individuals occupy some 0.2 m^2, ie: 5.0 persons/m^2. This is discussed in more detail in Section 1.3.6.

To determine these zones, Fruin uses an occupancy template. This template was developed by the New York City subway to determine practical standing capacity and also by the US Army. It might, in 1971, have represented a template for 95% of US males. Since then US and UK populations have become larger (ie: overweight) and the template might now be considered to be a ninety percentile (90%). A value for a ninety percentile is statistically a more suitable value, as it is less influenced by extreme maximum values, present when the distribution has a long tail.

Can this template be reasonably used for females and children ? Probably. This is because females usually try to acquire more space in crowded circumstances by folding their arms or enlarging themselves with the objects they are carrying. Children on the other hand are rarely still and therefore demand more space than their weight would imply. It has been observed that, in uncrowded conditions, individual female subjects are comfortable with a personal buffer zone of 0.5 m^2 (0.8 m diameter circle) and individual male subjects with a personal buffer zone of 0.8 m^2 (1.0 m diameter circle). To visualise these sizes, a woman's umbrella occupies an area of approximately 0.5 m^2 and a man's umbrella occupies approximately 0.9 m^2.

To allow for all these circumstances and other factors, such as body sway, it is recommended that the typical occupancy template be considered to be an ellipse of dimensions 600 mm wide by 450 mm deep as shown in Figure 1.2. The area of occupancy is thus 0.21 m^2. This template shape will be assumed to accommodate 90% of all subjects. Note that the actual body template of the individual is inside the ellipse.

If this typical occupancy template is used to represent the ninety percentile individual, then the larger individuals will be compensated by smaller individuals.

These factors must be borne in mind when designing pedestrian waiting areas. For bulk queues, such as when people are waiting for an event, the densities shown in Table 1.2 are recommended. These densities are illustrated in Figure 1.3.

When considering linear queues, ie: people "waiting in line" for a service, it is reasonable to assume two persons per metre length of space. A control barrier can be used to restrain the queue width. The barrier should be at least 600 mm wide. For unrestrained queues, it is necessary to assume they occupy a width of at least 1.5 m.

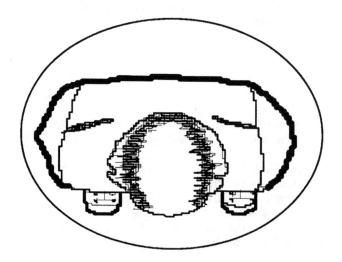

[The ellipse is: 600 mm wide by 450 mm high]

Figure 1.2 Typical occupancy ellipse (showing male subject)

The sizes shown on the template of Figure 1.2 are suitable for the USA, the UK and some European countries, but would need adjustment if used in other parts of the world.

Table 1.2 Density of occupation in waiting areas

Level of density	Comment
Desirable 0.4 persons/m²	Allows individuals to walk more or less where they want to go, or to stand still in one place, without any interference from other individuals.
Comfortable 1.0 persons/m²	Allows individuals to walk, with some deviations necessary, where they want to go and for individuals to stand still, without any interference from other individuals.
Dense 2.0 persons/m²	Individuals, who are walking, must now take care not to collide with other persons; and persons waiting are aware that other individuals are present.
"Crowding" 3.0 persons/m²	It is only possible to walk at a shuffle and with care at the average rate of the "crowd". There is no or little chance of a contraflow. Individuals waiting are very aware of other individuals.
Crowded 4.0 persons/m²	Walking is almost impossible. Individuals waiting are unhappy to be so close to other individuals. This density is only possible where persons are placed in a confined space, such as a lift car, or in a rapid transit train.

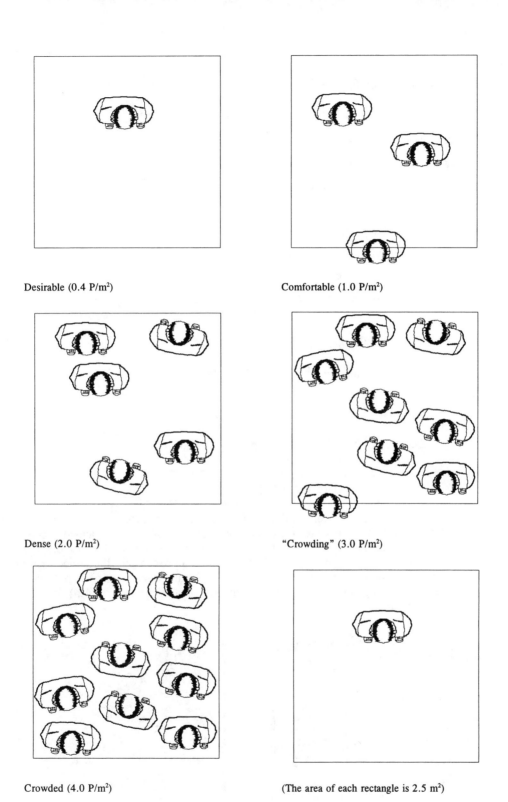

Desirable (0.4 P/m²)

Comfortable (1.0 P/m²)

Dense (2.0 P/m²)

"Crowding" (3.0 P/m²)

Crowded (4.0 P/m²)

(The area of each rectangle is 2.5 m²)

Figure 1.3 Illustration of density of occupation in waiting areas

1.3 CIRCULATION FACTORS

There are a number of factors which affect pedestrian movement. They include:

- pedestrian dimensions
- pedestrian velocities
- unidirectional/bidirectional flow
- cross flows
- patterns of waiting
- site and environmental conditions
- statutory requirements.

These factors will be dealt with in the following sections.

1.3.1 Corridor (handling) Capacity

The capacity of a straight corridor can be given as:

$$C_c = 60\ v\ D\ W \quad \text{persons/minute} \tag{1.1}$$

where:
C_c is the corridor handling capacity (persons/minute)
v is average pedestrian speed (m/s),
D is the average pedestrian density (persons/m²),
W is the effective corridor width (m).

   Equation (1.1) is an empirical relationship with a number of qualifications. Pedestrian speed and density are not independent of each other. For densities below 0.3 P/min, pedestrians can walk freely, which is called free flow design (Figure 1.4). When densities increase above 0.5 P/min there is an approximately linear decrease of average walking speed up to a density of about 3.0 P/min, when walking is reduced to a shuffle. The throughput peaks at densities of about 1.4 P/min (Figure 1.5), which is called full flow design.

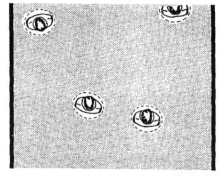

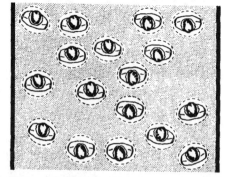

Figure 1.4 Free Flow Design — 0.3 persons/m²
$v = 1.0 - 1.3$ m/s, $C_c = 1080 - 1404$ persons/hour

Figure 1.5 Full Flow Design — 1.4 persons/m²
$v = 0.6 - 0.8$ m/s, $C_c = 3024 - 4032$ persons/hour

Walking speeds vary systematically, ie: statistically, with respect to:

■ type of population (age, gender, grouping, purpose)

■ ability (fitness, handicap)

■ flow direction

■ gradient

■ air temperature and humidity

■ floor finish.

Within each group there will be variations in average speed. Table 1.3 indicates empirically derived average values, as guidance. The table shows the typical pedestrian horizontal speeds (in m/s) and pedestrian flows in persons per minute (persons/minute). The flows assume a corridor width of 1.0 m. The width of corridor is not specific, but must be at least 900 mm and is assumed to be 1.0 m. Equation (1.1) allows for the flow rate to increase/decrease as the corridor width increases/decreases. This factor must be used with care, as small changes in corridor width will have little or no effect. Table 1.4 presents the minimum straight widths of corridors, which have been found to be suitable for different purposes.

Table 1.3 Possible pedestrian flows with grouping

Traffic type	Free flow design (0.3 persons/m²)			Full flow design (1.4 persons/m²)		
	Speed (m/s)	Persons/ minute	Persons/ hour	Speed (m/s)	Persons/ minute	Persons/ hour
Commuters, working persons	1.5	27	1620	1.0	84	5040
Individual shoppers	1.3	23	1380	0.8	67	4020
Family groups, tourists	1.0	18	1080	0.6	50	3000
School children	1.1–1.8	18–32	1080–1920	0.7–1.1	59–92	3540–5520

Table 1.4 Minimum corridor widths

Usage	Minimum width (m)
One-way traffic flow	1.0
Two-way traffic flow	2.0
Two men abreast	1.2
Man with bag	1.0
Porter with trolley	1.0
Woman with pram	0.8
with child alongside	1.2
Man on crutches	0.9
Wheelchair	0.8*

* Very long wheeled vehicles, such as hospital trolleys, require extra width in order to turn at junctions.

Table 1.4 is gender specific owing to the average dimensional differences between the genders. Other minimum widths other combinations, eg: man with twin push chair, etc. can be estimated.

Traffic can only flow freely along unrestricted routes. Corridors are rarely free of obstructions. Table 1.5 provides a number of examples.

Table 1.5 Reductions in corridor widths

Obstruction	Reduction (m)
Ordered queue	0.6
Un-ordered single queue	1.2 − 1.5
Row of seated persons	1.0
Coin operated machine:	
one person	0.6
queue	1.0
Person waiting with bag	0.6
Window shoppers	0.5 − 0.8
Small fire appliance	0.2 − 0.4
Wall-mounted radiator	0.2
Rough/dirty surface	0.2

Example 1.1

In a hospital corridor it is necessary for two trolleys to pass each other. Each trolley is pushed by one porter and another person with a bag of equipment walks alongside. What width should the corridor be ? If a row of seated persons is encountered what effect would this have ? Indicate the probable flow rates at free flow design levels.

From Table 1.4: a trolley and porter occupies 1.0 m width, a man with a bag occupies 1.0 m width. If the traffic is two way, the minimum clear corridor width will need to be at least 4.0 m.

If a large obstruction, such as a row of seated persons, is encountered, the corridor width would need to be increased by as much as 1.0 m. Persons waiting in these circumstances should not be located in a corridor. Unless small obstructions, such as fire appliances, radiators, etc., (see Table 1.5) can be recessed, the width the corridor would need to be slightly increased.

The circulation mix would comprise most people moving slowly and a few others on urgent tasks moving very fast (comparatively). The former would probably have a low speed, perhaps 0.6 m/s and the latter would probably have a higher speed, perhaps 1.5 m/s. A reasonable average would be 1.1 m/s. Using Equation (1.1) the full flow design rate would be:

$$1.1 \times 0.3 \times 5 \times 60 = 99 \text{ persons/min}$$

1.3.2 Portal (handling) Capacities

Portals, which are called by various names, ie: gate, door, entrance, turnstile, etc., form a division between two areas, for reasons of privacy, security, access control, etc. They represent a special restriction in corridor width. Their main effect is to reduce pedestrian flow rates. Table 1.6 indicates the probable range of pedestrian flow rates in persons per minute and persons/hour through an opening of 1.0 m.

Table 1.6 Portal (handling) capacities

Portal type	Flow (persons/minute)	Flow (persons/hour)
Gateway	60 – 110	3600 – 6600
Clear opening	60 – 110	3600 – 6600
Swing door	40 – 60	2400 – 3600
Swing door (fastened back)	60 – 90	3600 – 5400
Revolving door	25 – 35	1500 – 2100
Waist high turnstile:		
free admission	40 – 60	2400 – 3600
with cashier	12 – 18	720 – 1080
single coin operation	25 – 50	1200 – 1800

Note that Table 1.6 indicates flows through a portal of 1.0 m. Most domestic doors are less than this width (approximately 750 mm) and the flow rates would be likely to be the lower values in the range. Doors in non-domestic buildings may be slightly wider than 1.0 m and would permit the higher values in the range to be possible.

1.3.3 Stairway (handling) Capacity

Stairways impose a more stylised and disciplined form of movement on pedestrians. The movement is more regular, it is as disciplined by the steps and permits higher densities than are possible on the flat. Whereas for free movement during walking on the flat a pedestrian requires an area of some 2.3 m² (to account for body sway, etc.), a stair walker only needs to perceive two vacant treads ahead (and room for body sway) and occupies an area of some 0.7 m². Thus free flow design is possible at a density of 0.6 P/m² (Figure 1.6) and full flow design is possible at a density of 2.0 P/m² (Figure 1.7).

The speed along the slope is about half that on the flat, but increased densities are possible. Speed, however, is very much dependent on the slowest stair walker, owing to the difficulty in overtaking under crowded conditions. Higher speeds in the down direction are very often reduced by the need for greater care resulting in similar speeds in both directions. Speed is also affected by the angle of inclination and step riser height. To enable comfortable walking on a stair, a rule of thumb has been established that the sum[1] of the going (g) plus twice the rise (r), ie: $g + 2r$ should lie in the range 550 mm to 700 mm. This approximately matches the average adult stride on a stairway. This results in a range of riser heights of 100 mm to 180 mm and treads of 360 mm to 280 mm, and a range of possible inclinations from 15° to 33°. A private stair often has a rise of 180 mm and a going of 240 mm. In Britain, these dimensions are historical, as two bricks of 3 inch height together with two mortar joints of ½ inch height, used to form a stair, produced a step rise of 7 inches, which is almost 180 mm. An efficient inclination has been found to be 27°.

An empirical formula is given in Equation (1.2) for stairway handling capacity (C_s) and Table 1.7 indicates typical values for pedestrian stairway speeds, along the slope in

---

[1] "Rise" is the height of the step and "going" is the depth of the tread.

metres per second and pedestrian flow rates in persons per minute and persons per hour (bracketed) for each 1.0 m width of stairway.

$$C_s = 0.83\ (60\ v\ D\ W) \qquad \text{persons/minute} \tag{1.2}$$

Other symbols as Equation (1.1). Note that a stair has 83% of the handling capacity of a corridor.

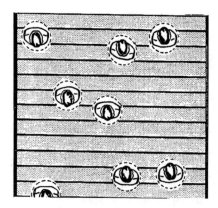

Figure 1.6 Free Flow Design at 0.6 persons/m²
$v = 0.6-0.8$ m/s; $C_s = 3024-4032$ persons/hour

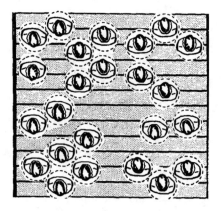

Figure 1.7 Full Flow Design at 2.0 persons/m²
$v = 0.6-0.8$ m/s; $C_s = 3024-4032$ persons/hour

Table 1.7 Stairway (handling) capacity

Traffic type	Free design flow (0.6 P/m²)		Full design flow (2.0 P/m²)	
	Speed	Flow	Speed	Flow
Young/middle-aged men	0.9	27 (1620)	0.6	60 (3600)
Young/middle-aged women	0.7	21 (1260)	0.6	60 (3600)
Elderly people, family groups	0.5	15 (900)	0.4	40 (2400)

1.3.4 Escalator (handling) Capacity

Escalators provide a mechanical means of continuously moving pedestrians from one level to another. Four factors affect their handling capacity.

- *Speed*. This is measured in the direction of the movement of the steps. Commonly available speeds are 0.5 m/s and 0.65 m/s. Most escalators run at one speed only, although some heavy duty escalators can switch-over to the higher speed during heavy traffic. Other speeds are available. A speed of 0.75 m/s is used on the London Underground; and speeds of 0.9−1.0 m/s are used on deep systems in Russia and the Ukraine.
- *Step widths*. Widths of 600 mm, 800 mm and 1000 mm are available, the latter allowing two columns of passengers to be carried. The hip widths, which are measured between the skirting panels are typically 200 mm wider than the step.

Hence the actual width a person can occupy on a 1000 mm step width escalator is some 1200 mm (enough for two people to pass each other).

■ *Inclination*. This is usually 30°, but can range from 27° to, in some cases, 35°. The latter is only available at a maximum speed of 0.5 m/s and a maximum rise of 6 m. The comfortable walking rule for inclination is broken for escalators as the step tread (going) is generally 400 mm, producing a step rise up to 240 mm, in order to achieve the necessary inclination. However, where an escalator can be used for an emergency exit the rise may not exceed 210 mm. Typically an escalator has a rise of 210 mm and a going of 400 mm, which when applied to the stair rule gives a value of 820 mm. This is outside the range quoted in Section 1.3.3 and explains why it is much harder to walk on an escalator.

■ *Boarding and alighting areas*. These areas must encourage pedestrian confidence and assist the efficient and safe boarding of escalators (see Chapter 3). It is recommended that at least one and one, third flat steps (light duty) to two and one, third flat steps (heavy duty) be provided for passengers when boarding/alighting an escalator. The average pedestrian boarding/alighting stride can be assumed to be 750 mm.

The theoretical handling capacity of an escalator (C_e) is given by:

$$C_e = 60 \ V \ k \ s \quad \text{persons/minute} \tag{1.3}$$

where:
V is speed along the incline (m/s)
k is average density of people (people/escalator step)
s is number of escalator steps/m.

For the case where the step depth is 400 mm, k becomes 2.5 and Equation (1.3) is:

$$C_e = 150 \ V \ k \quad \text{persons/minute} \tag{1.4}$$

The factor k allows for theoretical occupation densities of:

$k = 1.0$: 1 person per step for escalators of width 600 mm
$k = 1.5$: 1½ persons per step for escalators of width 800 mm
$k = 2.0$: 2 persons per step for escalators of width 1000 mm.

Table 1.8 gives the theoretical escalator handling capacity values in persons per minute and persons per hour (bracketed), for these values for k.

Table 1.8 Escalator theoretical handling capacity

Speed	Step width 1000 mm		Step width 800 mm		Step width 600 mm	
0.50	150	(9000)	113	(6750)	75	(4500)
0.65	195	(11700)	146	(8775)	98	(5850)
0.75	225	(13500)	169	(10125)	113	(6750)

Observations on the London Underground (Mayo, 1966; Al-Sharif, 1996) have shown that the theoretical occupation densities do not occur in practice. In general, only half the available space on an escalator is occupied, ie: every other step. At this density, on a 1000 mm escalator, this would give a standing person a space of some 400 mm by 1000 mm on which to stand, an area of 0.4 m². This represents the dense level of occupancy of 2.5 persons/m², given in Section 1.2.2. For escalators where all the passengers stand the theoretical handling capacities should be halved. The absurdity of assuming a value for k of 1.5 for an 800 mm escalator is illustrated in Figure 1.8.

Figure 1.8 Theoretical density pattern (step 800 mm, k = 1.5)

v = 0.5 m/s, C_e = 6750 persons/hour

The practice in Britain, Japan and elsewhere of one stationary column and one walking column will not increase an escalator's handling capacity, but will increase the passenger flow rate off the escalator and will decrease an individual's travelling time. Al-Sharif (1996) observed flows of 60% of theoretical.

Example 1.2

On the London Underground it was observed, during peak periods, that passengers stood stationary on the right hand side of the 1000 mm escalator at a density of one passenger on every other step. The left hand side was occupied by a walking column of passengers at a density of one person every third step. Assuming the escalator was running at 0.75 m/s and the speed of the walking passengers was 0.65 m/s, what is the passenger flow rate off the escalator ?

The theoretical flow rate of two stationary columns is given in Table 1.8 as 13500 persons/hour. As there is only one stationary column, the theoretical flow rate for this column will be 6750 persons/hour. However, the likely occupation density (k) will be 1.0 not 2.0, which means the actual flow rate of the stationary column of passengers will be 3375 persons/hour. This can be illustrated using Equation (1.4):

$$60 \times 150 \times 0.75 \times 0.5 = 3375 \text{ persons/hour}$$

The occupancy of the walking passengers is one person for every three steps. Therefore k is 0.33 (one third). But the effective (relative) speed of the passengers is 1.4 m/s (0.75 + 0.65). Then the flow rate using Equation (1.4) will be:

$$60 \times 150 \times 1.4 \times 0.33 = 4200 \text{ persons/hour}$$

The total passenger flow rate is 7575 persons per hour.

At any one time there will be five passengers on six steps giving a value for k of 0.83 and the actual handling capacity of the escalator will be:

$$60 \times 150 \times 0.75 \times 0.83 = 5603 \text{ persons/hour}$$

1.3.5 Passenger Conveyors (Moving Walkways and Ramps)

Walkways have an inclination of $0°$ and ramps have inclinations in the range $3°$ to $12°$. The running speed is determined by the angle of inclination. The speed is again measured in the direction of movement of the steps or pallets, ie: along the horizontal. Commonly available widths are 1000 and 1400 mm. The latter easily allows two stationary files of passengers, or the possibility a stationary file and a walking file of passengers on the moving passenger conveyor.

The theoretical density of passengers assumed for an escalator with two passengers per 1000 mm step, ie: $k = 2$, is 5.0 persons/m^2. In practice this is never achieved on an escalator and half this density is more likely, ie: 2.5 persons/m^2. A passenger conveyor theoretically should permit denser congregations of passengers than an escalator, as the space is not rigidly defined by steps. In practice, the probable density will be the "dense" value given in Table 1.1, at about 2.0 P/m^2. Table 1.9 indicates practical handling capacities in persons per minute and persons per hour (bracketed) assuming a density of 2.0 persons/m^2, using Equation (1.1).

Table 1.9 Handling capacities of passenger conveyors

Incline (degrees)	Speed (m/s)	Width of passenger conveyor (mm)			
		1000		1400	
0	0.50	60	(3600)	84	(5040)
0	0.63	76	(4560)	106	(6350)
0	0.75	90	(5400)	126	(7560)
5	0.70	84	(5040)	—	—
10	0.65	78	(4680)	—	—
12	0.50	60	(3600)	—	—

1.3.6 Handling Capacity of Lifts

The density of occupancy when sizing a lift car can be larger as passengers are constrained (by the car walls) and a greater allowance can be made for averaging. The standard BS EN81: 1998 has a table (1.1) which provides 0.1 m^2 plus 0.2 m^2 per person up to 6 persons, then 0.15 m^2 per person up to 20 persons, and then 0.12 m^2 per person thereafter. These values would require cars to be very crowded and it has been observed that lift cars do not fill to their rated (person) loads. It is recommended that a uniform figure of 0.21 m^2 be assumed, when sizing a lift car, in order to carry out a traffic design. This figure is almost 5 persons/m².

Figure 1.9 illustrates the density pattern for a lift with a rated capacity of 16 persons (rated load 1275 kg) with 16 persons present. It can be seen that the lift is not able accommodate this number of passengers, as the platform area is 2.9 m^2, which would allow some 14 persons to be accommodated. Even with 14 persons present, the passengers would be in the intimate zone discussed in Section 1.2.1. Table 7.2 indicates that the actual car capacity figures are smaller than the rated car capacity suggested by dividing the rated load by 75.

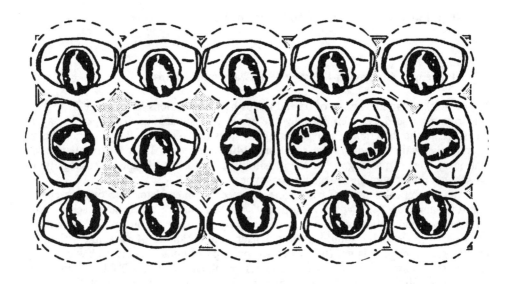

Figure 1.9 16 person lift car occupied by 16 persons

The method of sizing a lift installation to serve a traffic demand is given in Chapters 4−8. However, for completeness some discussion is necessary here. Lifts cannot handle the traffic volumes handled by other facilities; and have a considerable throttling effect on pedestrian movement. For example, the most efficient 8 lift group comprising 21 person rated car capacity cars serving 14 floors can only provide a handling capacity of 50 persons/minute. This is less than a flight of stairs can provide. And a 3 lift group comprising 10 person rated car capacity lifts serving 8 floors can only manage 16 persons/minute. Thus the use of escalators for short travel systems in buildings is recommended. Fortunately, the very high volumes of passenger demand found in bulk transit systems do not occur, when populating or emptying a building. As will be seen later, care must be taken in sizing a lift system for the worst case scenarios.

1.3.7 Comparison of "Handling" Capacities

It might now be interesting to compare the "handling" capacities of the various building elements in persons/minute. Considering all elements to be 1.0 m wide, then Table 1.10 can be obtained. All elements are assumed to be operating at their full flow design density levels, which is the optimal flow level. The values given are rounded for convenience, and apply to average groups of people and facilities.

Table 1.10 Comparison of handling capacities

Circulatory mode	Element	Handling capacity	Density (persons/m²)
Horizontal	Corridor	84	1.4
Horizontal	Portal	60	1.4
Horizontal	Walkway	45	1.5
Incline	Stairs	60	2.0
Incline	Ramp	45	1.5
Incline	Escalator	75	2.5
Vertical	Lift	50 *	5.0

* Passengers constrained by car walls.

1.4 LOCATION OF FACILITIES

1.4.1 General

Having discussed the various circulation elements and their characteristics, it is now necessary to consider their location. The main principles to bear in mind are to minimise the movements of people and goods; to prevent clashes between people and goods and to prevent bottlenecks. Thus the location and arrangement of the passive circulation elements (corridors, portals, etc.) and the active circulation elements (passenger conveyors, escalators, lifts) should take account of:

- the location of entrances and stairs

- the location of lifts and escalators

- the distribution of the occupants in the building.

Case Study CS1 gives an example of conflict between various circulation elements.
Ideally, all circulation activities should be centralised in a main core of a building. This is not always possible. Sometimes the main lobby is close to the main entrance, sometimes the building design places the main lobby some distance into the building. This latter case involves occupants and visitors in a long walk to reach the transportation facilities. However, it may be better for occupants to walk to the centre of a building to access stairs and lifts, since their more frequent usage during the day may outweigh the comparative inconvenience during arrival and departure. Generally the maximum distance to a lift or stair from an occupant's work place should not exceed 60 m with a distance of less than 45 m being preferred. Emergency escape routes are usually closer, but do not necessarily form part of the normally used circulatory routes.

1.4.2 Stairs and Escalators

Where possible, stairs and escalators should not lead directly off corridors, but should be accessed from landing and lobby areas, where people may wait without obstructing a circulation route. Thus the vertical and horizontal modes of circulation can be allowed to merge smoothly.

If it is the intention to encourage the use of stairs for short journeys to/from adjacent floors (interfloor movement), then the stairs should be clearly visible, adequately signed and reached before entering the lift lobby.

The location of escalators should observe the same recommendations as those for the location of stairways. However, it should be noted that escalators occupy a larger footprint than stairs in order to accommodate their inclination, structure and equipment spaces. It is particularly important that the boarding and alighting areas adjacent to an escalator are not part of another circulation route. This will provide a safe area for passengers to board and alight. This topic is discussed further in Chapter 3.

Escalators are typically used for short range movement between adjacent floors (the deep underground railway systems excepted). They are found in offices between principal levels, in shops between trading floors, in shopping centres between malls and elsewhere, such as railway stations, hospitals, museums, etc. They are usually sited in an obvious circulation path making it easy for pedestrians to board them.

There are several standard escalator arrangements, as shown in Figure 1.10. Type (c) is typical of a shop as it allows the shop to deliberately lengthen the circulation route to pass goods for sale. This configuration also takes up less space.

1.4.3 Lifts

Lifts should always be placed together whenever possible, rather than distributed around a building. This arrangement will help to provide a better service (shorter intervals), mitigate the failure of one car (availability of an adjacent car or cars) and lead to improved traffic control (group systems).

Lift lobbies should preferably not be part of a through circulation route, either to other lifts, or other areas in the building. Lobbies should be provided that are dedicated to passengers waiting for the lifts.

Eight lifts are the maximum number which it is considered possible to present to waiting passengers, especially if the lifts are large (<16 person). This constraint allows passengers to ascertain the arrival of a lift easily (from the landing lantern and gong signals), walk across the lobby and enter the lift before the doors start to close.

The distance across a lobby is important. If the lobby is too large, passengers have too far to walk and the closure of the car doors has to be delayed (increasing the lobby door dwell time) to accommodate the increased walking time. BS ISO4190-1: 1999 gives some guidance. For residential buildings, the landing depth (measured in the same direction as the depth of the car) should be at least equal to the depth (d_l) of the deepest car. For office buildings, where lifts are located side by side, the landing should be at least equal to $1.5 \times d_l$ and not less than 2.4 m. In office buildings, where lifts are arranged face to face, the distance between facing walls should be at least equal to the sum of the depths of two facing cars, but not more than 4.5 m.

The ideal lobby size would be one which could accommodate one full car load of passengers waiting and permit the simultaneous disembarkation of one full car load of arriving passengers. This area can be calculated from the information given in Section 1.2.2.

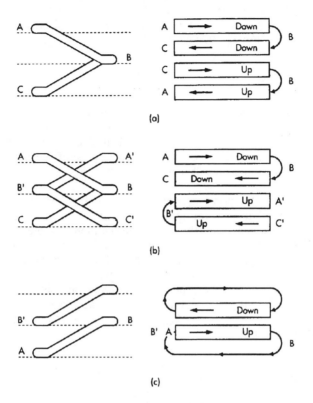

Figure 1.10 Escalator configurations (a) parallel (b) cross-over (c) walk round

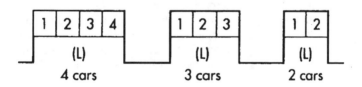

Figure 1.11 Preferred arrangement for lift cars: side by side (or in line)

8		1												
7	(L)	2		6		1								
6		3		5	(L)	2		4	(L)	1				
5		4		4		3		3		2		2	(L)	1

For 7 cars omit #1 or #8. For 5 cars omit #1 or #6. For 3 cars omit #1 or #4.

Figure 1.12 Preferred arrangement for lift cars: facing

The preferred arrangements from BS5655: Part 6: 2002 of two to four lifts arranged side by side are given in Figure 1.11 and for two to eight lifts arranged opposite each other are shown in Figure 1.12. Note all the lobbies, indicated (L), have separate waiting areas with no through circulation.

Example 1.3

Suppose there are four 21 person lifts arranged in a facing (2 × 2) configuration. What distance across the lobby would be appropriate given the car width is 2100 mm and the car depth is 1600 mm ? The cars occupy a 6.0 m length of lobby (along the front of the cars). Assume the persons waiting do so at the dense level of occupancy. Assume that a 21 person car can accommodate 21 persons.

When one of the cars in the group of four arrives, it will disembark 21 persons. Thus for these passengers to be accommodated at the dense level of occupancy of 2.0 persons/m^2 (see Table 1.1 and Figure 1.2) in the lobby, an area of 10.5 m^2 would be required. If it is assumed that there are 21 persons waiting to board, then they will require a further area of 10.5 m^2. Hence a total of 21 m^2 is required. As the lobby length is 6.0 m, then a width of 3.5 m is indicated, which is slightly larger than the ISO recommendation of the sum of the depths of two facing cars, ie: 3.2 m.

As a 21 person car will generally only be filled to an average occupancy of 80% of its rated load ie: 17 persons, and only 17 persons might be waiting, then a 3.2 m width lobby would probably be sufficient.

1.4.4 Stairs, Escalators and Lifts

In general, lifts are used for "long distance" travel over a large number of floors and stairs or escalators for travel over a small number of floors. A judgement is made by the passenger with respect to the waiting time for a lift versus the walking time (and walking effort) with respect to stairs and escalators. Low rise structures, such as shopping centres, sports complexes, conference and exhibition centres, railway stations, airports, hospitals, etc. are good examples of buildings where the provision of stairs and escalators considerably aids circulation. Peters *et al.* (1996) provides guidance on stair usage, which can be summarised (in round numbers) as shown in Table 1.11.

Table 1.11 Stair usage

Floors travelled	Usage up	Usage down
1	80%	90%
2	50%	80%
3	20%	50%
4	10%	20%
5	5%	5%
6	0%	0%

In office environments the usefulness of stairs and escalators is lessened, although they can be a advantageous where (say) a heavily populated trading floor(s) must be accessed. They are also very useful for access to car parks, relieving the lifts of the duty to travel below the main terminal. Escalators may be used for access to the two lobby levels where double decker lifts are installed.

Table 1.12 offers some guidance to the division of passengers between lifts and escalators in offices. The use of escalators is mainly inhibited by the length of time travelling.

Table 1.12 Lifts and escalators: division of traffic

Floors travelled	Escalator	Lift
1	90%	10%
2	75%	25%
3	50%	50%
4	25%	75%
5	10%	90%

The provision of well signed and positioned stairs and escalators can considerably lessen the demands made on the lifts. Designers must take these factors into account.

1.5 FACILITIES FOR PERSONS WITH DISABILITIES

The discussion so far has assumed that all persons circulating in a building are fully able-bodied. However, a large proportion of the population are disadvantaged in some way. The prEN81-70: 1999 standard "Accessibility to lifts for persons including persons with disability" categorises disabilities into physical, sensory and intellectual, as summarised in Table 1.13.

Table 1.13 Categories of disability

Disability category	Sub-category
Physical	Impaired mobility
	Impaired endurance
	Impaired dexterity
Sensory	Impaired vision
	Impaired hearing
	Impaired speech
Intellectual	Learning difficulty

All of the disabilities in Table 1.13 can be temporary or permanent. Impaired mobility could result from a broken leg (temporary), or a missing limb (permanent). Impaired vision could be as the result of illness (temporary), or due to age (permanent). Intellectual ability may be impaired as the result of being intoxicated (temporary), or by being dyslexic (permanent).

Some people are self-handicapped. For example, a person with a bulky package or long clothing will be handicapped, when using an escalator. Another example is a person who refuses to wear spectacles and cannot clearly see the call buttons for a lift. Sometimes a solution for one disability reacts with that for another. Thus a tall, older person finds it difficult to read the legends on car operating panels set at a suitable height for wheelchair users. The BS8300: 2001 standard "Design of buildings and their approaches to meet the needs of disabled people" gives some guidance for facilities within buildings and examples are given in Table 1.14.

Table 1.14 Examples to assist the disadvantaged

Element	Example
Steps and stairs	Colour contrast at interface
Ramps	Maximum gradient and minimum width
Handrails	Suitable dimensions
Lifts	Manoeuvring space

The Disability Discrimination Act: 1995 in the UK and the Americans with Disabilities Act legislation in the USA lay down various provisions and regulations. Approved Document M to the UK Building Regulations and BS8300: 2001 provide actual guidance in the UK. If the main concern is the use of lifts, Figure 1.13 gives a summary of some of the most relevant guidance.

Passenger lifts

■ Provide unobstructed space of 1500 x 1500 mm on landings for wheelchairs.
■ Provide wheelchair space in car of (w x d):
  1000 x 1250 mm (unmanoeuvrable & unaccompanied)
  1100 x 1400 mm (unmanoeuvrable & accompanied)
  2000 x 1400 mm (manoeuvrable & accompanied).
■ Car doors to be at least 800 mm (900 mm is better).
■ Car and landing call dwell times to be lengthened to 5.0 s.
■ Car operating panels placed on a side wall (preferably both sides).
■ Car operating panels placed 900 mm to 1100 mm from floor.
■ Audible announcements and visual displays.
■ Buttons must be easily operable and with tactile indication.
■ Raised numbers on buttons.
■ Good colour contrast between numbers and background on all indications.
■ Emergency alarm button should include a visual acknowledgement of operation.
■ Emergency telephone and inductive coupler for hearing aid users.
■ Lift car interior should be well lit (>100 lux).
■ Mirror on rear wall for small cars to enable wheelchair users to see behind them.
■ Support rail in car at 900 mm above floor and 35 mm in diameter.
■ Controls on landing 900 mm to 1100 mm from floor.
■ Landing controls to include lift coming signal.
■ Area outside car doors to be well illuminated, at least 50 lux at the door threshold.
■ Floor surface to be in colour contrast to adjacent floor outside car doors.

Platform and stair lifts

■ Should be clearly signposted.
■ Should be key controlled operation.
■ Should be by adjacent stairs.
■ Should not compromise means of escape.

Figure 1.13 Facilities for persons with disabilities using lifts

Arrangements made to allow persons with disabilities to make use of circulation elements assist the able-bodied, and should be implemented, wherever possible.

CHAPTER TWO

Circulation
in Shopping Centres

Dober (1969) said "Circulation is the act of passing from place to place" and Beddington (1982) stated that "People flow like liquid, following the line of least resistance and greatest attraction". In a shopping centre, or mall, the former must be enabled and the latter encouraged.

A shopping centre is unlike the conventional high street, where shops line each side of the road, with shoppers on pavements at the sides and vehicular traffic passing along the middle of the street. A shopping centre is usually a purpose built building, where all shoppers are protected from the weather in a climatically controlled environment and are segregated from vehicular traffic. The shops line each side of the malls with several levels of malls above and below. Generally, shopping centres are on one, two or three levels to avoid installing much mechanical people moving equipment. There are places set aside for rest, sustenance and amusement.

2.1 INTRODUCTION

As was indicated in Chapter 1, the interior circulation of people in buildings is a complicated activity. It was shown that the interior movement of people in buildings must be designed to consider all circulation routes to allow the free flow of people, goods and vehicles with the minimal wastage of space and the prevention of bottlenecks.

In a shopping centre, however, some of the good design criteria set out above may be intentionally violated, as they are not necessarily conducive to the selling of goods. For instance, having attracted shoppers into a store, all routes, except the exit from the store, may be clearly marked (emergency exits excepted). The free flow of people may be deliberately reduced by the introduction of display stands along the route offering goods for sale to encourage impulse buying. Circulation may be designed to be irrational, but not obviously so, for instance with regard to escalator layouts to cause shoppers to walk around part of a floor, in order to reach the next facility, thus presenting merchandise to prospective shoppers.

No two shopping centres have the same structure, population or circulation patterns. Most shopping centres are designed to occupy two levels and sometimes three. Two levels are generally considered as much as the average shopper is prepared to contemplate, when in a centre. Centres with three levels often have food courts at the upper or lower levels to form an attraction and a contrast to the main sales areas.

The general intention behind the design of a shopping centre or "mall" is to encourage shoppers to enter the centre, then to stop and browse and hopefully to purchase goods on "impulse". The malls should provide a modulated sequence of conditions through side malls, a range of linking corridors to central squares and features. The purpose is to create a feeling of bustle, excitement, sparkle, competition and a variety of experiences within an organised framework, whereby the shopper has a retreat from the effects of the weather and the motor car, and is cocooned within a relatively safe and comfortable environment.

This chapter looks at the theoretical aspects of circulation, in general terms, relating it to observations made by the Author's researchers and then discusses how the knowledge gained may be applied to shopping centres.

This chapter is not concerned with:

■ the estimation of external traffic flows into a shopping centre
■ the estimation of peak flows down malls
■ shopping centre design, except where it impinges on circulation
■ in-store circulation.

Readers will find Fruin (1971) knowledgeable on some of these matters. The first bullet in the list above concerns road traffic engineers, who have generated model approaches to the problem. Work is being carried out with regard to the other aspects and will be reported later.

This chapter is concerned with circulation and movement within the shopping centre. It is concerned with the two main circulatory aspects: horizontal traffic flows along malls and through entrances, and vertical movement between the different levels in the shopping centre. Two traffic conditions are identified and examined: a low level of shopper occupancy (uncrowded free flow), and a peak value of shopper occupancy (crowded). Contrast these with those given in Chapter 1.

Much of what follows in the theoretical sections cannot be proved and many of the recommendations have been based on observation and experience. In addition, circulation is a human activity, which is subject to unpredictable behaviour patterns. Interior design is also significantly affected by regulations such as Fire and Safety Codes and these must be taken into account.

There are many factors that will affect circulation movements in a shopping centre, some of which are indicated in Table 2.1.

2.2 THEORETICAL ASPECTS OF HORIZONTAL MOVEMENTS

The most likely people in a shopping centre will be shoppers and tourists. The more practical unit of time to be used for this environment is one hour instead of one minute.

2.2.1 Mall (handling) Capacity (reprise)

Equation (1.1) indicates the capacity C_c in persons/minute per metre width of a straight corridor, or mall as:

$$C_c = 60 \, v \, D \, W \tag{1.1}$$

Please refer to Section 1.3.1 of Chapter 1. Tables 1.3 and 1.4 should also be noted. Figures 1.4 and 1.5 illustrate the densities. So as malls are rarely free of obstructions the effective width of a 5.0 m wide mall reduces to 4.0 m, if a row of people are seated along one side.

2.2.2 Entrance Capacity

As Chapter 1 explains, an entrance (gate, door, portal, turnstile, etc.) forms a division between two areas and introduces a constriction in corridor/mall width. Table 1.6 indicates probable flow rates in persons/minute per metre width of entrance.

Table 2.1 Factors affecting movement

Factor	Comment
Simple layouts are best. A good design will enable a shopper to find their way in, through and out of a centre. Simple floor plans overcome the problems of shoppers' unfamiliarity with a centre.	Assists circulation
Provide visual stimulation and variety. This can be provided by the shop fronts themselves. The size of every shop front affects its trading potential (and hence its revenue/m² and its rental).	Inhibits circulation
Design should centre on a series of primary nodes. Include landmarks such as intersecting malls and transfer points such as parking, entrances and exits. Nodes are activity areas where pathways (malls) meet and people relax.	Inhibits circulation
Magnets and anchors. Standard mall designs rely on "magnet" or "anchor" stores to draw shoppers past secondary stores, which provide convenience goods and encourages impulse buying opportunities.	Assists circulation
Points of conflict should be minimised. This will allow shoppers to concentrate on shop displays. For example, cross flows, counter flows and right angle bends all cause conflicts. People in a minor flow will alter pace and timing to fit the gaps in the major flow. Ideally shoppers should be able to pick their own speed and direction.	Inhibits circulation
The length of malls is important. About 200 m is the maximum distance a shopper is likely to walk. The introduction of bends makes a mall appear longer than it really is. The use of magnet stores at each end of a long mall increases the attraction and reduces the apparent length.	Assists circulation
Exploration. A shopping centre should be able to be explored in one trip, so pause points need to be cleverly placed. Additional breaking up of the mall by the use of courts and squares for public space, rest and recreation areas help to reduce the apparent mall length.	Assists circulation
Mall widths. These should be narrow enough not to discourage shoppers from crossing over to shop on the other side.	Assists circulation
Street furniture. Malls are often "landscaped" by the introduction of street furniture (seats, bins, etc.), planters and displays to break up and reduce the perception of space in the mall.	Inhibits circulation
Escalators, moving walkways and ramps and lifts. These need to be carefully sited to invite shoppers onto other levels.	Assists circulation
Access. Shoppers have to enter and leave a shopping centre by means of entrances. These entrances interfere with the flow as they often have either a swing door or an automatic sliding door, and because the shopper may be adjusting to the new environment and even looking at a store directory.	Inhibits circulation
Location. The positioning of vertical circulation elements requires great care to avoid "dead-ends" and "double-back" circulation. The elements should be provided in the natural circulation path of shoppers.	Assists circulation

2.3 THEORETICAL ASPECTS OF VERTICAL MOVEMENTS

2.3.1 Stairway Capacity

Stairways impose a more stylised form of movement on pedestrians (see Section 1.3.3 of Chapter 1). An empirical formula for stair capacity is given in Equation (1.2) as:

$$C_s = 0.83 \ (60 \ v \ D \ W) \ \text{persons/minute} \tag{1.2}$$

Table 1.7 should be consulted and Figure 1.6 and 1.7 illustrate the densities.

2.3.2 Escalator (handling) Capacity

Escalators provide a mechanical means of moving pedestrians from one level to another. The handling capacity of an escalator (in persons/minute) is given by Equation (1.3):

$$C_e = 60 \ V \ k \ s \ \text{persons/minute} \tag{1.3}$$

Table 1.8 gives the theoretical handling capacity (C_e) values. Figure 1.8 illustrates a theoretical density pattern for an 800 mm escalator.

2.3.3 Lift Handling Capacity

In shopping centres lifts are used to transport shoppers to/from car parks and to allow shoppers with prams and pushchairs to access all levels. Quite often observation lifts are used for this latter purpose. The handling capacity of lifts is dealt with in detail in later chapters. Figure 1.9 illustrates a car occupied by 16 persons, which is clearly crowded.

2.4 PRACTICAL LEVELS OF SHOPPER MOVEMENTS

In shopping malls and shopping centres, two levels of occupancy can be observed. These are (1) uncrowded, which is similar to free flow elsewhere and (2) crowded, which is similar to full flow design elsewhere.

2.4.1 Malls and Entrances

- The walking speed of shoppers in uncrowded conditions is generally 1.3 m/s and in crowded conditions is generally 1.0 m/s (see Table 2.2.).

- The density of shoppers in uncrowded conditions is 0.2 persons/m² and 0.45 persons/m² during crowded conditions. The density can increase to 1.0 persons/m² at pinch points (where the mall size is inadequate, eg: at a food court).

- Counterflows reduce mall capacity by 15% compared to unidirectional flows.

- Mall widths should be of the order of 6−8 m wide as a compromise between too wide to cross and too narrow to pass along.

■ The effective mall width reduces (equal to actual mall width minus street furniture and window shoppers) as the condition changes from uncrowded to crowded. This results from more stationary shoppers looking into shop windows.

■ Walking speeds reduce to 0.7 m/s, when shoppers pass through entrances.

Table 2.2 Actual mall pedestrian flows rates
The table shows the likely pedestrian flow rates in persons per hour under uncrowded conditions (0.2 persons/m^2) and crowded conditions (0.45 persons/m^2) per metre width of mall.

Traffic type	Uncrowded (0.2 P/m^2)		Crowded (0.45 P/m^2)	
	Speed	Flow rate	Speed	Flow rate
	(m/s)	(persons/hour)	(m/s)	(persons/hour)
All shoppers	1.3	936	1.0	1620

Figures 2.1 and 2.2 illustrate the uncrowded and crowded shopper density levels.

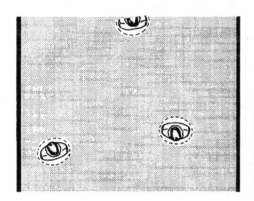 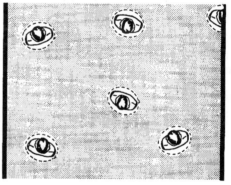

Figure 2.1 Uncrowded mall density - 0.2 persons/m^2
v = 1.3 m/s; C_c = 936 persons/hour

Figure 2.2 Crowded mall density - 0.45 persons/m^2
v = 1.0 m/s; C_c = 1620 persons/hour

2.4.2 Stairs

The dimensions of a stair limit many aspects of locomotion. For instance, pace length is restricted by tread depth (going). More accurate cones of vision are required for step placement and assistance is often required by the use of handrails. The energy consumed is related to the riser height, which should be less than 180 mm, but not too shallow else walking rhythm is affected.

■ Uncrowded density is found to be approximately 0.4 P/m^2 and crowded density reaches 0.8 P/m^2, see Figures 2.3 and 2.4.

■ Shoppers' speeds when using stairs vary according to group with an average of 0.7 m/s, see Table 2.3.

■ The stair capacity under uncrowded conditions is about 900 persons/hour/metre and under crowded conditions is about 1800 persons/hour/metre.

■ There is a tendency for more down traffic than up traffic in the ratio 60:40.

■ A minor contraflow will reduce a major flow by effectively reducing the stairway width by some 750 mm.

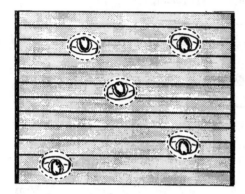 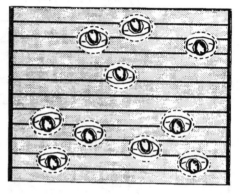

Figure 2.3 Uncrowded stair density at 0.4 persons/m² **Figure 2.4** Crowded stair density at 0.8 persons/m²
$v = 0.6-0.8$ m/s: $C_s = 3024-4032$ persons/hour $v = 0.6-0.8$ m/s: $C_s = 3024-4032$ persons/hour

Table 2.3 Stairway (handling) capacity (persons/hour)
The table shows the likely pedestrian flow rates in persons per hour under uncrowded conditions (0.4 persons/m²) and crowded conditions (0.8 persons/m²) per metre width of stair

Traffic type	Speed (m/s)	Design flow	
		Uncrowded (0.4 persons/m²)	Crowded (0.8 persons/m²)
Men	0.8	960	1920
Women	0.7	840	1680
Elderly men	0.5	600	1200
Elderly women	0.6	720	1440
Children	0.8	960	1920
Push chairs	0.5	600	1200

2.4.3 Escalators

■ About 80% of shoppers will use the escalators to reach other levels in a shopping centre, as there will rarely be a queue to use them.

■ Even under queuing conditions, 100% step utilisation will not be achieved.

■ 800 mm escalators have an assumed step utilisation of $k = 1.5$. However, escalators have been observed to only load to $k = 0.5$ step utilization under uncrowded conditions (Figure 2.5) to $k = 1.0$ step utilization under very crowded conditions

(Figure 2.6). The width of an 800 mm step is not large enough to accommodate two adult people side by side. To achieve this, a 1000 mm is required. Then a higher utilisation can be achieved.

■ Actual handling capacities range from 33% of theoretical for uncrowded conditions to 66% for crowded conditions (Table 2.4).

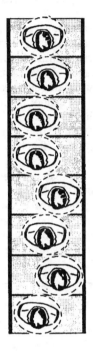

Figure 2.5 Uncrowded density $k = 0.5$
$C_e = 2250$ P/h, step width 800 mm

Figure 2.6 Crowded density $k = 1.0$
$C_e = 4500$ P/h, step width 800 mm

Table 2.4 Actual escalator handling capacity (persons/hour)

Speed	Step width 600 mm		Step width 800 mm		Step width 1000 mm	
	Uncrowded	Crowded	Uncrowded	Crowded	Uncrowded	Crowded
0.50	1500	3000	2250	4500	3000	6000

An interesting point is that it takes just under one second for a step to appear at a boarding point of a 0.5 m/s escalator. This is too fast for most people to get onto each vacant step as it appears. Hence shoppers tend to hesitate and under busy conditions some queuing occurs.

2.4.4 Moving Ramps (passenger conveyors)

Many shopping facilities these days provide trolleys for the shopper to use. If the shopping centre has several levels, shoppers are inconvenienced if they have to leave a trolley at one level to reach another level. Although escalators can be designed to accept a trolley securely they are dangerous in use as often the goods on or in them can dislodge and injure other escalator users. Where space is available, an inclined passenger conveyor (moving ramp) can be installed. This is becoming a commonly applied solution to this problem and greatly improves circulation.

2.4.5 Handling Capacity of Lifts

Observation, pram lifts, car park and other lifts are provided in shopping centres, but not in sufficient quantities to serve more than a fraction of the shoppers. They are mainly used by the elderly, infirm, disabled, mothers with children and push chairs, and people with heavy packages. Observation lifts are sometimes installed as a feature to provide a visual impact in retail complexes. They do contribute to the circulation aspects of a shopping centre, but cannot be considered a major constituent as passengers often ride one simply for the ride.

Lifts cannot handle the traffic volumes handled by other facilities; and have a considerable throttling effect on pedestrian movement. For example, a group of two, 16 person, observation lifts, serving two retail levels and two car park levels, probably has a possible handling capacity of only about 300 persons per hour. This is due to the need to have long door dwell times, slow motion dynamics, the slowness of passenger loading/unloading and lower levels of occupancy near to 50% of rated capacity (see Figure 2.7) owing to the presence of prams, push chairs and baggage. Thus the recommendation to install as many escalators as possible in shopping centres is essential for the traffic handling of the large volumes of traffic.

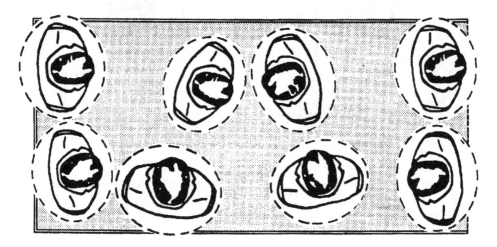

Figure 2.7 16 person lift car occupied by 8 persons

2.5 EXAMPLE 2.1

A 0.5 m/s, 800 mm escalator is installed at the end of a 4.0 m wide side mall in a two level shopping centre. All shoppers reaching the escalator must use it to travel to the other level. What will be the density of shoppers in the mall, when queuing starts at the escalator, assuming the escalator is operating in the crowded condition ?

From Table 2.4 it is possible to handle 4500 passengers each hour on a 800 mm crowded escalator. The mall flow will balance the escalator flow at:

$$4500/4 = 1125 \text{ person/hour/metre of mall width}$$

From Table 2.2 with an uncrowded mall density of 0.2 persons/m^2, the flow rate is 936 persons/hour/metre of mall width and with an crowded mall density of 0.45 persons/m^2 the flow rate is 1620 persons/hour/metre of mall width. If there is a linear relationship between the two known density levels then the shopper density at the balance point will be:

$$0.2 + \frac{(0.45 - 0.2)}{(1632 - 936)} \times (1125 - 936) = 0.27 \text{ P/m}^2$$

2.6 SUMMARY

Shoppers do not populate a shopping centre to the high levels (in density terms) found in other public places, eg: railway stations. Also the walking speeds vary widely and are close to the natural (comfortable) speed of 5 km/h.

Table 2.5 Summary of circulation elements in shopping centres

Element	Characteristic	Range	Criteria
Malls	Density (theoretical)	0.3/1.4	Free/full
	Density (measured)	0.2/0.45	Uncrowded/crowded
	Speed (theoretical)	1.3−0.6	Free/full
	Speed (measured)	1.3−1.0	Uncrowded/crowded
	Speed (measured)	0.7	Through entrances
Stairs	Density (theoretical)	0.6/2.0	Free/full
	Density (measured)	0.4/0.8	Uncrowded/crowded
	Speed (theoretical)	0.9−0.4	Free/full
	Speed (measured)	0.8−0.5	Uncrowded/crowded
Escalators	Handling capacity (theoretical)	6750	
	Handling capacity (measured)	2250/4500	Uncrowded/crowded
Lifts	Not possible to accurately estimate, but ratio theoretical:actual approximately 2:1.		

Densities in persons/m^2: Speeds in m/s: Handling capacities in persons/hour.

2.7 FACTORS AFFECTING CIRCULATION — GOOD PRACTICE

A shopper spends a large percentage of the time in a shopping centre walking and browsing on the level. The levels of density are necessarily lower so that shoppers feel comfortable. The shoppers' primary (and preferred) means of transfer from one level to another is an escalator. The secondary means of transfer from one level to another is a stairway and an additional means of transfer from one level to another is a lift. Persons moving on a stairway or escalator will exhibit more body sway, in order to keep their balance, and therefore require more space. Some guidance as to good practice is given in Table 2.6.

Table 2.6 Good practice for shopping centres

☐ Malls should be designed to avoid "pinch points".

☐ Stairs should have a minimum width of 2.5 m.

☐ Stairs should be "channelised" by use of a separating rail, which aids movement and can divide up and down flows.

☐ Stairs should always be located near to escalators to form a secondary means of vertical circulation. (Important when an escalator is out of service.)

☐ To ease flows, stair risers should be less than 180 mm and the slope less than 30°.

☐ All stairs should have intermediate landings for rest and circulation diversion between flights of no more than 16 steps.

☐ Adequate clear areas should be provided at access points of stairs to allow queuing and safe movement.

☐ Adequate clear areas should be provided at boarding and alighting points of escalators to allow queuing and safe movement, perhaps with barriers to discipline users.

☐ There should be at least two escalators at each location to serve two traffic flows.

☐ Escalators should be located in a parallel arrangement.

☐ The maximum rise for an escalator should be less than 6 m.

☐ Escalator step widths of 1000 mm are to be preferred, not necessarily to permit passengers to stand side by side but to allow shopping to be carried.

☐ The standard escalator step riser is 230 mm which is larger than the recommended maximum height of 180 mm, so walking on the stopped escalator is tiring.

☐ Maintenance of escalators should be carried out when the centre is closed.

☐ The reliability of escalators leaves a lot to be desired, some investigation by manufacturers as to the problem areas should be carried out.

☐ Car operating panels within lifts should be simple in layout and operation.

☐ Maintenance of lifts should be carried out when the centre is closed.

☐ All stairs, escalators and lifts should be adequately illuminated.

☐ All stairs, escalators and lifts should be readily visible.

☐ All stairs, escalators and lifts should be easily identified and well signed.

It is important to realise that a facility which an able-bodied person considers to be a primary means of transportation may not be suitable for a person with disabilities, or a person supervising young children, or a person simply burdened with shopping. Hence the need to provide "pram lifts" and adequate other facilities for persons with disabilities.

2.8 EXAMPLE 2.2

An underground station has a ticket hall one level below the street. At street level and at the ticket hall levels there are extensive shopping facilities (50 shop units at each level). A 1.0 m wide, 0.5 m/s escalator connects the two levels. Calculate the handling capacity of the escalator (a) during the morning rush hour and (b) during a Saturday shopping period. Assume during the rush hour the left hand side has a column of passengers walking at 1.0 m/s, occupying every third step and the right hand side has a stationary column occupying every other step. Assume during the shopping period that passengers stand two abreast, do not walk and occupy every other step.

(a) During the rush hour.

Using Equation (1.4) the flow rate over one hour for the stationary column is:

$$C_e = 150 \times 0.5 \times 0.5 \times 60 = 2250 \text{ P/h}$$

The walking passengers occupy every third step giving k as 0.33, but their effective speed is 1.5 m/s (0.5 + 1.0). Using Equation (1.4) again, the flow rate over one hour for the walking column is:

$$C_e = 150 \times 0.33 \times 1.5 \times 60 = 4500 \text{ P/h}$$

The total handling capacity over the rush hour is 6750 passengers.

(b) During the Saturday shopping period.

The occupancy value k is 1.0.

Using Equation (1.4) the flow rate over one hour for the two stationary column is:

$$C_e = 150 \times 0.5 \times 1.0 \times 60 = 4500 \text{ P/h}$$

The total handling capacity during one shopping hour is 4500 passengers.

CHAPTER THREE

Circulation on Escalator Landings
with Passenger Conveyors

3.1 INTRODUCTION

Escalators are part of the pedestrian flow system and their landings present an interesting problem, as there is an interface between the natural mode of movement and the mechanical mode of movement. There are two interfaces, one at the boarding station and one at the alighting station. At boarding, the natural mode of movement, a walking pedestrian on solid ground changes to the mechanical mode of movement of a stationary passenger on a moving escalator step. At alighting, the standing (passenger) mode alters to the moving (pedestrian) walking mode. Some people walk onto, along and off the escalator and their effect on circulation will be considered later.

It is well documented that most persons boarding an escalator will hesitate as they board. Although they may be walking at a comfortable speed of 1.0 m/s (Table 1.3) at the interface, they may well take about one second, the equivalent of one metre of walking distance to board (Fruin, 1971[1]). The hesitation time can vary according to how many flat escalator steps are presented before the escalator steps are formed. Human factors such as age, gender, agility, size, handicap, purpose, clothing, bags carried, etc. also effect boarding efficiency.

Having boarded, the passenger may then stand and be transported at the escalator rated speed of (say) 0.5 m/s. At the alighting end of the escalator the standing passenger must alight by walking off the escalator. Again some hesitation occurs with most persons, which again varies according to how many flat escalator steps are presented and the human factors.

The initial discussion will consider a commonly installed escalator with a 1000 mm nominal step width running at a rated speed of 0.5 m/s. Such an escalator has a theoretical handling capacity of 150 persons/minute (see Table 1.8). This assumes that two persons are standing on each step. This is not practical as it is generally assumed that the nominal body ellipse (profile) is 600 mm wide by 450 mm deep (Figure 1.1). Thus to accommodate two passengers on each step, the side of each passenger's body must overhang the step edges and intrude towards the balustrade interior panels. There is usually an extra width here, sometimes called the hip width, of some 200 mm, giving an overall hip width of some 1200 mm. This allows space for side by side occupation of a step by two passengers and, if necessary, a moving file of passengers alongside the stationary file. The back/front of each passenger's body will also certainly touch the passengers standing behind and in front, as an escalator step depth is usually 400 mm.

The more realistic and observed occupancy is one person per step or two persons every other step (see Figure 1.8). This level of occupancy could, to some extent, have been naturally imposed by the problems of boarding. The practical handling capacity is thus about half of the theoretical, ie: 75 persons/minute. It should be noted that the

---

[1] Page 102.

hesitations at boarding often result in an escalator not delivering its potential practical handling capacity, as boarding can only proceed at the speed of the slowest person. This may result in queues developing at the boarding station and the need to provide space to accommodate these waiting persons.

At the level of occupancy of one passenger a step, and assuming a step is 400 mm deep, two passengers occupy an area of 800 mm deep by 1000 mm wide, ie: the density of occupancy of the passengers is 2.5 persons/m². Table 3.1, which summarises Table 1.2, indicates that this density, for queuing persons, would be at a density level between "dense" and "crowding". The escalator passengers are queuing on the escalator in what is generally accepted as crowded conditions.

Table 3.1 Density of occupation by persons waiting

Level	Density
Desirable	0.4 P/m²
Comfortable	1.0 P/m²
Dense	2.0 P/m²
"Crowding"	3.0 P/m²
Crowded	4.0 P/m²

When the passengers reach the end of the escalator they are mechanically fed off the escalator. There is no option — they must leave regardless of the space available to accommodate them ! At the boarding station a passenger can choose not to board, but this option is not available at the destination landing. The potential for conflicts with other passengers (caused by collisions between moving and stationary persons; obstruction caused by persons hesitating or standing on the landing; etc.) becomes significant.

Thus at the alighting interface the passenger must start to walk off the moving escalator. They become pedestrians and begin to move naturally, and after some hesitation owing to the change of mode from mechanical movement to natural movement, they reach their comfortable walking speed. This is likely to be 1.0 m/s (Table 1.3).

There is a similar interface between the natural and mechanical modes at the boarding and alighting points of passenger conveyors (moving walkways or ramps). The problem of boarding onto and alighting from moving walkways is less severe. The alighting/boarding stations are level, or slightly ramped, onto a set of flat horizontal pallets, which are often wider than escalator steps. Ramps are similar except the pallets move at an incline of 5°, 10° or 12°. There is less passenger hesitation and uncertainty and passenger conveyors generally deliver their practical potential handling capacity (Table 1.9).

It is common in Britain for one file of passengers to stand and one file to walk on an escalator. In this case the inter-passenger distance increases to one every three steps. The escalator handling capacity does not increase, but the transit time for the walking passengers is decreased. The same number of passengers will alight. One advantage, with these passengers, is that they are less likely to hesitate, on boarding and alighting, as they do not stop walking, but they may come into conflict with the stationary passengers. Passenger conveyors exhibit a similar scenario.

3.2 SIZING OF ESCALATOR LANDINGS

It is essential to present alighting passengers from an escalator with an opportunity to move freely. They should therefore enter a larger space at the landing. But what size should this space be in order to achieve a free movement ? Where may guidance be found ?

It is worth recalling at this point the space required for moving about as indicated in Chapter 1. Pedestrian speed and density are not independent of each other. For densities below 0.3 P/min pedestrians can walk freely and this is called free flow design. When densities increase above 0.5 P/min there is an approximately linear decrease of average walking speed up to a density of about 3.0 P/min, when walking is reduced to a shuffle. The throughput peaks at densities of about 1.4 P/min and this is called full flow design.

The European standard for escalators "Safety rules for the construction and installation of escalators and passenger conveyors", EN115: 1995 (A1: 1998) offers the following guidance:

EN115: 1995, Clause 5.2.1

5.2.1 At the landings of the escalator and passenger conveyor, a sufficient unrestricted area shall be available to accommodate passengers. The width of the unrestricted area shall at least correspond to the distance between the handrail centrelines (see b_1 in figure 2). The depth shall be at least 2.50 m, measured from the end of the balustrade. It is permissible to reduce it [the depth] to 2.00 m if the width of the unrestricted area is increased to at least double the distance between the handrail centrelines. Attention is drawn to the fact that this free area has to be considered as part of the whole traffic function and, thus, needs sometimes to be increased.

In the case of successive escalators and passenger conveyors without intermediate exits, they shall have the same theoretical capacity (see **14.2.2.4.1j**).

The wise intention of this clause is to ensure the safety of passengers alighting from a moving escalator (or passenger conveyor). This is achieved by providing each escalator with an unrestricted area or reserved space. Should the reserved space be too small and it becomes crowded: with the alighted passengers; or other persons attempting to join a successive escalator; or other persons crossing in front of the escalator to an adjoining escalator; then passengers currently on the moving escalator will be unable to leave the escalator, with significant safety hazards resulting. If circulation is restricted, conflicts occur. These conflicts can be aggravated by passengers who have walked the escalator rather than standing. If these conflicts are likely to cause people to be brought into contact with moving machinery, as is the case here, with the risk of entrapment, then there are considerable safety implications. If efficient and effective circulation resolves such conflicts then it follows that safety is also improved.

How will Clause 5.2.1 help in the understanding of the circulation of persons ? Consider Option 1 of EN115, shown diagrammatically in Figure 3.1. The distance between the handrail centrelines is usually 1.5 m and, as the depth is to be 2.5 m, gives an available area of 3.75 m². If a pedestrian were to walk straight across the reserved space at 1.0 m/s then the transit time would take 2.5 s. In this time 3.125 passengers could alight (escalator handling capacity 75 persons/minute). This gives a density of 0.83 persons/m². Free flow movement in a circulation space without the potential for conflicts occurs at about 0.3 persons/m², as indicated above.

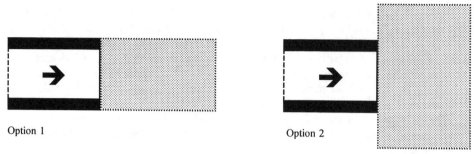

Option 1   Option 2

Figure 3.1 EN115 illustration of unrestricted space

The density of 0.83 persons/m² is higher, but it would be satisfactory for four main reasons:

(1) The movement of persons off the escalator is more disciplined than free movement in an unbounded circulation space.

(2) There is likely to be more space available outside the handrail centrelines and hence the reserved space, such as that used for the escalator enclosure and clearance to the building fabric, which increases the width for this option.

(3) Persons may leave the reserved space to the side rather than straight ahead and enter other circulation space more quickly.

(4) The escalator passenger perceives a three fold increase in circulation space reducing the density of occupation from 2.5 persons/m² to 0.8 persons/m².

If Option 2 of EN115 is taken, the transit time to walk 2.0 m would be 2.0 s. During this time 2.5 passengers would alight. The reserved space they would enter has a width equal to twice the distance between the handrails centrelines, ie: 3.0 m and a depth of 2.0 m giving an area of 6.0 m², which is 60% larger than Option 1. This results in a lower density of occupation of 0.42 persons/m².

Option 2 offers more reserved space to the sides than Option 1. This extra space assists the pedestrians to turn to the side (their turning circle), if the forward direction should be bounded by, for example, a wall.

In summary, a 1000 mm escalator with a rated speed of 0.5 m/s transporting 75 persons/minute will offer occupation densities of between 0.4 and 0.8 persons/m², dependent on the option selected.

3.3 MATHEMATICAL ANALYSIS

What happens if escalators of different rated speeds and nominal step widths are considered ? The procedure in Section 3.2 can be expressed mathematically.

The transit time (t_t) for a pedestrian (previously a passenger) to walk at a speed of v_p, in metres per second, across the depth d_t, in millimetres, of the escalator landing is:

$$t_t = \frac{d_t}{v_p} \quad \text{s}$$

(3.1)

The practical handling capacity (C_e) of an escalator (see Chapter 1) is about 50% of the theoretical:

$$C_e = 0.5 \times 2.5vk \quad \text{persons/s} \tag{3.2}$$

where v is the rated speed of the escalator
and k is number of passengers per step.

Assume the flow rate of passengers leaving the escalator is C_e, then the number of pedestrians (Q) on the escalator landing is:

$$Q = C_e \times t_l = 0.5 \times 2.5vk \times \frac{d_l}{v_p} \quad \text{persons/s} \tag{3.3}$$

For Option 1, the area of the landing is given by the distance between the handrail centres X by the depth d_l (taken as 2.5 m):

$$A_l = d_l \times X \quad \text{m}^2 \tag{3.4}$$

Then the density of occupation (D_l) on the landing for Option 1 is:

$$D_l = \frac{Q}{A_l} = 0.5 \times 2.5vk \times \frac{d_l}{v_p} \times \frac{1}{Xd_l} = 1.25 \times \frac{vk}{v_pX} \quad \text{persons/m}^2 \tag{3.5}$$

For Option 2, the area of the landing is given by twice the distance between the handrail centres and the depth d_l (taken as 2.0 m):

$$A_l = d_l \times 2X \quad \text{m}^2 \tag{3.6}$$

Then the density of occupation (D_l) on the landing for Option 2 is:

$$D_l = \frac{Q}{A_l} = 0.5 \times 2.5vk \times \frac{d_l}{v_p} \times \frac{1}{2Xd_l} = 0.625 \times \frac{vk}{v_pX} \quad \text{persons/m}^2 \tag{3.7}$$

Notice that Option 2 always has half the density of Option 1. Thus Option 2 will always provide a more efficient and effective, and safer, circulation than Option 1.

Using this mathematics, Table 3.2 can be formed. This shows in its final column the density on the escalator landings. It can be seen that they range from 0.42 persons/m^2 for Option 2 with a 1000 mm escalator at a rated speed of 0.5 m/s to 1.71 persons/m^2 for Option 1 with a 600 mm escalator at a rated speed of 0.75 m/s.

3.4 CONSIDERATIONS FOR A DESIGN DENSITY

What would be a recommended value for the occupation density on an escalator landing, where passengers alight, that allows efficient, effective and safe circulation ?

At boarding, a hesitating person or sudden surges of intending passengers can hold up efficient boarding. The intending passengers are inconvenienced by not being able to board and the escalator may fail to deliver even its practical handling capacity, but there is no danger as they are on "solid ground".

Table 3.2 Density of landing occupation values for escalators of different rated speeds and step widths

Nominal escalator step width (mm)	Passenger space on escalator (m²)	Passenger density on escalator (P/m²)	Option from EN115	Available area on landing (m²)	Pedestrian transit time on landing (s)	Escalator rated speed (m/s)	Passenger flow off escalator (P/s)	Number of pedestrians on landing	Density of landing occupation (P/m²)
1000	0.40 One person per step	2.5	1	3.75 (1.5 × 2.5)	2.5	0.50	1.25	3.13	0.83
						0.65	1.63	4.08	1.09
						0.75	1.88	4.70	1.25
			2	6.00 (2.0 × 3.0)	2.0	0.50	1.25	2.50	0.42
						0.65	1.63	3.06	0.55
						0.75	1.88	3.76	0.63
800	0.43 Three persons per four steps	2.3	1	3.25 (1.3 × 2.5)	2.5	0.50	1.25	3.13	0.96
						0.65	1.63	4.08	1.25
						0.75	1.88	4.70	1.45
			2	5.20 (2.0 × 2.6)	2.0	0.50	1.25	2.50	0.48
						0.65	1.63	3.06	0.63
						0.75	1.88	3.76	0.72
600	0.48 One person every two steps	2.1	1	2.65 (1.1 × 2.5)	2.5	0.50	1.25	3.13	1.14
						0.65	1.63	4.08	1.48
						0.75	1.88	4.70	1.71
			2	4.40 (2.0 × 2.2)	2.0	0.50	1.25	2.50	0.57
						0.65	1.63	3.06	0.74
						0.75	1.88	3.76	0.86

All numbers are rounded. P is persons.

The problem is similar at the alighting landing, where a hesitating person can hold up efficient circulation. Here, however, hesitations or sudden surges of persons into the reserved space will bring movement down to a shuffle, with the dire consequence that a jam will occur and passengers on the escalator cannot leave.

It has previously been indicated in Chapter 1 that free flow circulation occurs at a density of 0.3 persons/m^2, a peak flow occurs at a density of 1.4 persons/m^2, and a shuffle occurs at 3.0 persons/m^2. The nearer the occupation density is to 0.3 persons/m^2 the better the circulation and the safer the movements. These flows are particular to open circulation spaces.

There are four mitigating factors at escalator landings indicated in Section 3.2. The most important is that the passengers are "more disciplined" as they are on a stair. It was indicated in Chapter 1 that stairs allowed a free flow density of 0.6 P/m^2 and a full design flow density of 2.0 P/m^2. As no rule of thumb is available it is suggested that half the stairway full design density of 1.0 P/m^2 be adopted. This would invoke the use of the penultimate sentence of Clause 5.2.1 of EN115, which says that "... the whole traffic function ..." has to be considered.

Option 1 presents the worse situation in all cases and Option 2, with its bigger reserved area, is much to be preferred. This might involve increasing Option 1 areas to deal with the whole traffic function.

3.5 ARRANGEMENT OF ESCALATORS: SPECIFIC CASE

A number of arrangements will be considered for a 1000 mm step escalator running at a rated speed of 0.5 m/s. Option 1 will be considered first then Option 2. Remember, for this situation Table 3.1 shows that the passengers are standing on the escalator at a density of 2.5 persons/m^2.

3.5.1 Arrangement 1: Single Escalator

Figure 3.1 illustrates this case and the discussion in the section above indicates that no significant problems would be expected.

3.5.2 Arrangement 2: Pair of Escalators Side by Side

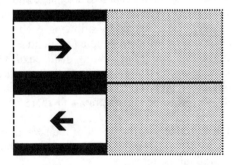

Figure 3.2 Arrangement 2: Pair of escalators side by side

Consider Figure 3.2, where one escalator serves in the up direction and the other in the down direction. In this case it is possible for persons to cross over from one side of the escalator landing to the other. They may do this through the reserved space belonging to the escalator they are not using. This would increase the density of occupation in that space and the likelihood of conflicts. Extra circulation space should be provided on the landing around the periphery of the reserved space to avoid this situation. This space would ideally extend the depth of the reserved space by 1.0 m, or the overall width by 2.0 m. Another alternative is to place a stub wall (as illustrated by single lines) between the escalators to protect their reserved space from intrusion.

Option 2. If the pair of escalators are arranged so that the reserved areas did not overlap then they would comply with Option 2 of EN115 and would be satisfactory. If the two escalators were arranged side by side the depth or the overall width should be increased in order to maintain a reserved area of 12.0 m².

3.5.3 Arrangement 3: Successive Escalators without an Intermediate Exit

Figure 3.3 Arrangement 3: Successive escalators without an intermediate exit

Consider Figure 3.3. Here the first escalator feeds passengers into the second escalator successively. Both escalators must be the same rated speed and size (EN115, Clause 5.2.1). In this case there are no exits from the mid point escalator landing, ie: it acts in a similar way to an intermediate landing on a flight of stairs. The EN115 rule would suggest that at least 4.0 m or 5.0 m would be required. However, the purpose of the mid point landing is simply to allow passengers to change escalators and the passengers are simply alighting from one and boarding the other. There is no possibility of persons leaving or entering.

As the "corridor" between the successive escalators is the same dimension as that discussed in Section 3.3.2, the pedestrian density will be the same. The distance recommended by EN115 is 5.0 m. This would seem excessive, so what should the inter-escalator spacing be ?

Remembering the problems of the mechanical discharge and the natural boarding at alighting, it is important to provide sufficient space between the escalators to accommodate this. This might be achieved by allowing each passenger four walking steps[1] (say) between each escalator. Modern people (Tutt and Adler, 1990[2]) might have a average step of 0.75 m giving a mid point landing distance of 3.0 m. This should be considered the minimum spacing. To a great extent the length of the inter-escalator spacing will be assisted by the number of "flat" steps provided on each escalator. This minimum spacing should therefore be satisfactory and there will be little circulation problems in this case.

Option 2. This option is unlikely as the width of the "corridor" would be too large, but if this option were to be offered, then the 3.0 m spacing would still apply as it is principally dependent on walking patterns.

---

[1] Two steps equal one pace.

[2] Pages 23−28.

It should be noted that the rules of EN115 have been broken here. EN115 gives discretion to increase the recommended reserved space (penultimate sentence), but does not allow a decrease to meet the traffic function (see later).

3.5.4 Arrangement 4: Successive Escalators with One Intermediate Exit

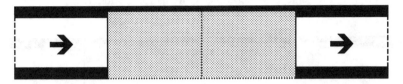

Figure 3.4 Arrangement 4: Successive escalators with one intermediate exit

Consider Figure 3.4. Here the escalators have an exit to one side onto a floor landing allowing passengers to leave the first escalator and others to join the second escalator. The floor landing can be considered as part of the exit area provided it is also unrestricted. If it provides an area of 12.0 m² then a 4.0 m inter-escalator spacing would be suitable, and basically Option 2 would have been adopted.

For Option 2 the same considerations apply as for Option 1.

3.5.5 Arrangement 5: Successive Escalators with Two Intermediate Exits

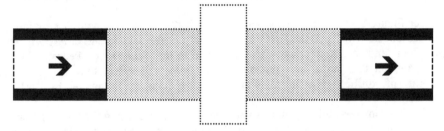

Figure 3.5 Arrangement 5: Successive escalators with two intermediate exits

Consider Figure 3.5. Here the escalators have an exit on each side onto a floor landing allowing passengers to leave the first escalator and others to join the second escalator. The floor landing can be considered part of the exit area, provided it is unrestricted. The discussion made for Arrangement 4 is still valid and a 4.0 m landing depth might be suitable. However, if there is any possibility of persons (and it is most likely) using the escalator landing to cross from one side of the escalators to the other, extra circulation space 1.0 m wide (shown unshaded) should be provided. In this case the escalator spacing should be 5.0 m.

For Option 2, the same considerations apply as for Option 1.

3.5.6 Arrangement 6: Pair of Successive Escalators Side by Side with One Exit

Consider Figure 3.6. Here the escalators serve in both directions and provided all passengers leaving the first escalator joined the second escalator, the situation would be similar to Arrangement 3. However, this is unlikely and persons will want to alight from escalators arriving at the escalator landing and persons from the floor will want to join the escalators leaving the floor. Thus the escalator pair nearest to the exit side of the landing will continually have their reserved space invaded by pedestrians crossing to the other (far pair) escalators. The possibility for conflicts is very great. The EN115 separation of 5.0 m is essential just to ensure satisfactory circulation to the escalators. Also, further unrestricted space should be provided next to the escalators.

For Option 2, the same considerations apply as for Option 1.

Figure 3.6 Arrangement 6: Successive escalators side by side with one intermediate exits

3.5.7 Arrangement 7: Pair of Successive Escalators Side by Side with Two Exits

Consider Figure 3.7 which is Arrangement 6 with two exits. As with Arrangement 6 some passengers will leave the first escalator and join the second and others will leave at the floor landing. New passengers will arrive to leave the floor. Pedestrians from either pair can thus cross into the other pairs' reserved space, thus increasing the possibility of conflict. The 5.0 m inter-escalator spacing recommended by EN115 is essential.

If there is any possibility (and it is very likely) of persons using the escalator landing to cross from one side of the escalators to the other, more circulation space would need to be provided. This additional circulation space is shown unshaded. In this case the escalator spacing would be some 6.0 m.

For Option 2, the same considerations apply as for Option 1.

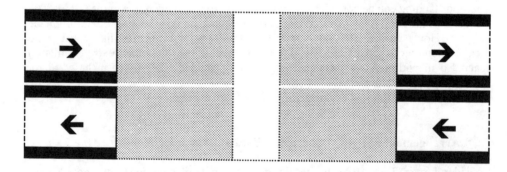

Figure 3.7 Arrangement 7: Pair of successive escalators side by side with two exits

3.6 ARRANGEMENT OF ESCALATORS: ALL TYPES

Section 3.5 considered a number of arrangements for a 1000 mm step escalator running at a rated speed of 0.5 m/s and Options 1 and 2. The method of analysis can easily be applied to other step widths and speeds. The escalator arrangements specified in Section 3.5 produced results at the most "comfortable" end of the range of possibilities as far as occupation densities are concerned. Examination of Table 3.1 shows that the other escalator specifications can produce higher landing densities. Care must therefore be exercised, when designing an escalator arrangement.

3.7 ANOTHER INTERPRETATION OF EN115, CLAUSE 5.2.1

The interpretation of clause 5.2.1 has another school of thought. In the discussion above, each escalator is considered separately and possesses its own unrestricted space. The reasoning is that if a wall is placed at the distant end of the unrestricted space then the first escalator complies with the clause. Another escalator can then be provided on the other side of the wall with its own unrestricted space. This would not be sensible for a smooth flow of passengers, so the wall is removed and the inter-escalator spacing is either 5.0 m or 4.0 m, dependent on the option selected.

The other school of thought considers that the unrestricted space can be shared, ie: an inter-escalator distance of 2.5 m or 2.0 m dependent on the option selected. This might be satisfactory for successive escalators with no intermediate exits, but not a safe solution for exits with high traffic flows and cross traffic.

The density of pedestrians on the escalator landing is given by Equation (3.5) can be simplified to the following formula:

$$\frac{\text{passenger flow}}{\text{unrestricted area}} \quad \text{persons/m}^2 \tag{3.8}$$

ie: it is not dependent on the distance between escalators.

Clearly a zero depth escalator landing would mean a continuous escalator. There has to be somewhere for the alighting passengers to stand ! What is important is that there is enough space to accommodate surges of people and for the pedestrians to move as freely as possible. The idea of separating escalators of any size or rated speed by 2.0 m or 2.5 m cannot be right. The absolute minimum should be 4.0 m and then only where there is a suitable arrangement. The discussions in Section 3.3 should be the guidance considered.

3.8 PASSENGER CONVEYORS

Passenger conveyors do not present as many problems as escalators as they are either horizontal (moving walkways) or ramped (moving ramps). They are also often wider, ie: 1400 mm. Usually the spacing will be considerable and more than that suggested for escalators, as they are often used to move people along rather than upwards. Passengers are not so intimidated at boarding or alighting as it is common for the area in which they are located to be more spacious than escalator "stairways", thus removing any possible claustrophobic feelings.

Occupation densities on passenger conveyors can be much lower than that on escalators and probably the density on the landings could be 50% larger than that recommended for escalators.

3.9 DISADVANTAGED CIRCULATION

The disadvantaged person, as defined in Chapter 1, may be able to use an escalator or passenger conveyor dependent on their particular disability or handicap. Persons with permanent or temporary mobility problems, the partially sighted and the elderly and infirm will be particularly disadvantaged with respect to escalators, but less so for passenger conveyors. The problem for them will be to alight and board safely and this will increase the time and space they need, which may not be available during peak periods.

Some persons may be temporary handicapped, when carrying bulky items or when wearing clothing that reaches to the floor. The latter circumstance is particularly dangerous. Persons with mobility problems (and those without) will find great difficulty in walking on a stationary escalator as the step height is 230 mm compared to the usual 180 mm. Persons with significant disadvantages should seek other means of circulation, wherever possible.

CIRCULATION CASE STUDIES

Contents

The case studies draw on the data and discussion of Chapters 1 – 3 and use the summary data tables following to illustrate practical aspects of interior circulation.

Points of Conflict

CS1.1 BUILDING DATA

Type of building: Speculative Office - design and build.
Occupant: Merchant Bank Headquarters.
Population: 2000 persons (above ground), 140 ground.
Number of floors: Basement, Lower Ground, Ground, Floors 1 − 6.
Lifts: 6no; 2000 kg; 1.6 m/s.
Escalators: 2no; 1000 mm; 0.5 m/s.
Parking: None.

CS1.2 DESCRIPTION OF CIRCULATION AREA

As shown in Figure CS1.1, the main entrance (A) has two 2.0 m wide swing doors and two 1.8 m wide revolving doors. The side entrance (B), used for "drop offs", has one 1.5 m wide swing door and one 1.8 m wide revolving door. A further entrance (H) has a 1.5 m wide swing door, which leads along a corridor (I) beside the escalators to the circulation space (E). This entrance is used as a service route for the ground floor.

The lobby area (C) has some 200 m² clear space.

A set of eight turnstiles (D) leads into a small circulation space (E) of some 15 m² clear space. Leading from this space is a lift lobby (F) of some 30 m² serving a group of six lifts, with doors leading to the ground floor office area at the other end.

From the circulation space (E), a pair of escalators (G) connect to the first floor (one escalator up and one escalator down).

CS1.3 ANALYSIS

Table 1.6 indicates that a swing door will allow a flow of 40 − 60 persons per minute for each 1 m wide door. Taking the mid value of 50 persons per minute, indicates that the two Entrance A swing doors will permit the movement of 200 persons per minute. Table 1.6 also indicates that a revolving door will permit a flow of 25 − 35 persons per minute. Again taking the mid value, the two Entrance A revolving doors will permit the movement of 60 persons per minute. This gives a total Entrance A entry/exit potential of 260 persons per minute. Entrance B will provide entry/exit opportunities for the occupants. The swing door is 1.5 m wide and will not permit the flow of 100 persons per minute possible for a 2 m wide door. A de-rating should be applied to accommodate this, ie: to 60 persons per minute. The Entrance B revolving door is the same size as those at Entrance A and it will permit a movement of 30 persons per minute. This gives a total Entrance B potential of 90 persons per minute.

Entrance H has been ignored.

Calculations would show the lifts have a 5-minute handling capacity of 240 persons at an interval of 20 seconds.

A single escalator has a theoretical handling capacity of 150 persons per minute, see Table 1.8. However, the practical handling capacity is likely to be half of this, ie: 75 persons per minute.

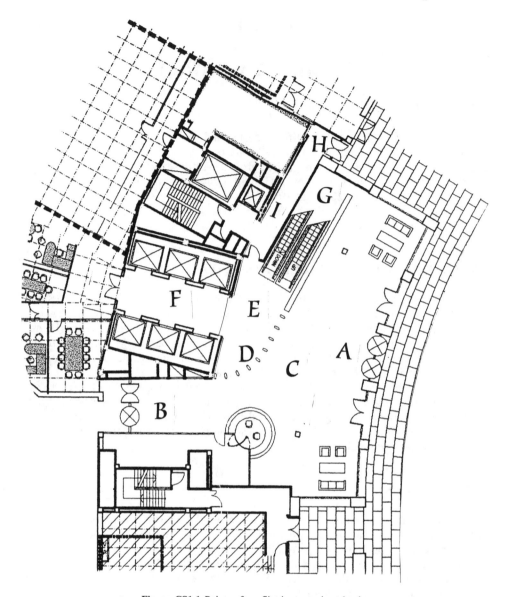

Figure CS1.1 Points of conflict in a merchant bank

Coin operated turnstiles allow a throughput of 25–50 persons per minute (Table 1.6). The installed turnstiles use card readers and will operate very similarly to coin operated turnstiles. However, it would be wise to assume a low end throughput of 30 persons per minute. Assuming six turnstiles set in one direction and two in the counter direction, the potential incoming throughput is 180 persons per minute and the outgoing throughput is 60 persons per minute.

There is no parking provided for the occupants, so they must all arrive through the various entrances. The potential population of the building is 2000 persons and, assuming a maximum of 80% arrive on any day with a peak arrival rate of 15%, then the probable 5-minute arrival rate would be 240 persons. This is 48 persons per minute. Surges of twice this size might be expected.

CS1.4 FIRST CONCLUSIONS

Consider the entry requirement of 240 persons per 5-minutes, ie: 48 persons per minute. The exit requirement will be assumed to be 50% larger at 72 persons per minute.

The main entrance provides 260 persons per minute capability. This is more than adequate and would easily allow for surges. The side entrance will also contribute.

The six incoming turnstiles can handle 180 persons per minute. Again, more than adequate for normal and surge conditions.

The escalators serve to the first level only. They can handle the all likely incoming traffic. This is not likely as only one sixth of the building population occupies the first floor.

The lifts are arriving every 20 seconds and filling with 16 people. The lifts can thus handle 48 persons per minute. This is the likely average arrival rate. They would not handle a surge of arrivals on their own. However, some persons would use the escalators and maybe the stairs.

The entrance lobby (C) has 200 m^2 of clear space and will accommodate 60 persons with free movement (Table 1.3) and 280 persons, if crowded.

The circulation space (E) of 15 m^2 will accommodate 5 to 21 persons, dependent on the occupant density.

The lift lobby (F) has an area of 30 m^2 and will accommodate some 30 people queuing comfortably at a density of one person per square metre (see Table 1.1).

There is more than adequate provision of entrances, turnstiles, lifts and escalators to serve the maximum arrival rate of 240 persons and to reasonably deal with surges. There is also adequate provision for the exit requirements.

There may not be sufficient circulation space.

CS1.5 SECOND CONCLUSIONS

The main problem is that the circulation space (E) after the turnstiles is very small and the space (F) is the lift lobby and is a waiting area and not a circulation space.

There are two escalators (G) connecting into circulation space (E). People from the first floor may be passing down the DOWN escalator to reach the ground floor offices, or to exit from the building. People from the ground will be travelling up the UP escalator to reach the first floor.

Note that the DOWN escalator is on the left hand side and people leaving it will immediately conflict with people moving in and out of corridor (I). People have to leave an escalator, but can chose not to board. The left hand escalator should be the UP escalator, to reduce this conflict.

The escalators are so close to the lift lobby (F) that people wishing to alight from the escalators will inevitably come into conflict with people entering or leaving the lift lobby. Also, people wishing to board the UP escalator will be obstructing the way into the lift lobby, which ever side it is. The escalators should have larger unobstructed boarding and alighting space.

People entering or leaving the circulation space (E) via the corridor (I), will also conflict with both the escalator and lift passengers. This corridor should not access the circulation space at this point.

For the lift lobby consider a worst case scenario where two car loads of passengers (32 persons) are waiting in or entering the lift lobby. As indicated above, the lift lobby has an area of 30 m^2 and so will accommodate some 30 people queuing comfortably.

Some people will be moving across the circulation space (E), which could accommodate a further 21 people. So the lift lobby will be crowded and the vicinity will be busy. Ideally, the lift lobby needs to be a little larger.

It will not be workable, however, because the lift lobby is used as a corridor to the ground floor offices (occupancy 140 persons with a potential 5-minute arrival rate of about 21 persons). So circulation space must be provided to let ground floor staff pass through.

The lift lobby should be blind, to avoid conflicts with through traffic.

CS1.6 POINTS OF CONFLICT — LESSONS TO BE LEARNT

- The turnstiles (C) are too close to the lift lobby (F) and the escalators (G).

- The escalators (G) should have larger unrestricted boarding and alighting spaces in circulation space (E).

- The left hand escalator should be the UP escalator.

- The service corridor (I) should be removed and not access the circulation space (E).

- The escalators are placed too close to the other circulation elements.

- The lift lobby (F) needs to be a little larger.

- The lift lobby (F) should be blind and not have a corridor route through it.

CASE STUDY TWO

Commuter Railway Station Concourse

CS2.1 BUILDING DATA

Type of building: Commuter railway station concourse.
Occupant: The public.
Population: Variable numbers, but the flow in one direction was estimated at
                     160 persons/minute and in the other direction at 80 persons/minute.

CS2.2 DESCRIPTION OF CIRCULATION AREA

The area of interest for this case study is the concourse of a commuter railway station.
Figure CS2.1 indicates the main elements to be considered. On the left hand side there
is a ticket office. On the right hand side there is a vending machine, a row of seats, a
heating radiator and a fire appliance. Persons will queue for tickets at the ticket office.
Although there may be a single ordered queue at the ticket windows, the queue will
probably straggle out into the centre of the concourse into an unordered queue. Standing
persons after the ticket office can also present an obstruction. The obstruction on the
right hand side caused by the legs of people sitting on the seats will have to be avoided.
Obstructions will also be caused by persons using the vending machine.

    Not shown in Figure CS2.1 are the swing doors at the entrance to the concourse from
the street and the stairs leading from the concourse into the platform area.

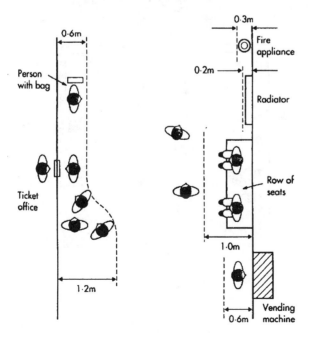

Figure CS2.1 Elements to be considered

CS2.3 ANALYSIS

Figure CS2.1 indicates some dimensional values for some of the obstructions taken from Table 1.5. On the left hand side the unordered queue straggles back and becomes 1.2 m, with the front of the queue reducing to 0.6 m at the ticket window. The single standing person also represents a reduction in concourse width of 0.6 m. On the right hand side the single person at the vending machine causes an obstruction of 0.6 m and the row of seated persons extend for 1.0 m into the main concourse. The radiator and the fire appliance reduce the effective width of the concourse by 0.2 m and 0.3 m respectively.

CS2.4 FIRST CONCLUSIONS

There must be sufficient width of concourse and stairs and an sufficient number of swing doors to allow the free movement of a total of 240 persons (two-way flow) through the concourse. Table 1.7 indicates that a flow of 60 persons per minute is possible for most people for every one metre width of stair. Thus at least 4.0 m of stair will be needed.

Table 1.6 indicates that 1.0 m of swing door can allow a flow 40−60 persons per minute. Assuming that the commuters would achieve the higher value, then at least four, 1.0 m width doors are needed.

Provided there is a clear 4.0 m width of concourse from the entrance doors through to the stairs, then there should be few conflicts between commuters walking across the concourse.

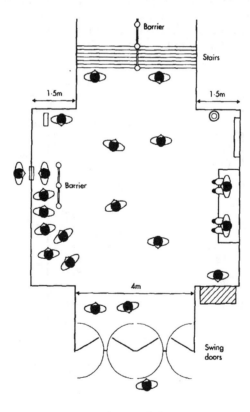

Figure CS2.2 A possible solution

CS2.5 SECOND CONCLUSIONS

Figure CS2.2 shows a possible arrangement. The entrance has been provide with four, 1.0 m wide swing doors, which will allow a flow of up to 240 persons per minute. It will be seen that all the obstructive elements have been moved into 1.5 m deep recesses on each side of a 4.0 m wide unobstructed concourse. A barrier has been positioned by the ticket office window to make the queue more orderly. Standing persons are accommodated in the recessed area. The vending machine has been repositioned on the right hand side and the seats moved back. The radiator and fire appliance now present no obstruction. A central handrail has been placed on the 4.0 m wide stair to impose some discipline in the movement of persons on the stair. It also assists safe passage.

CS2.6 POINTS OF CONFLICT

■ Queues at the ticket office.

■ Standing persons with bags.

■ Persons using vending machines.

■ Seated people.

■ Radiators and fire appliances.

All these elements prevent the free movement of people.

CASE STUDY THREE

Sizing Doorways

CS3.1 BUILDING DATA

Type of building: Concert Hall.
Occupant: The visiting public.
Population: 3500 + staff, etc.
Number of floors: Basement, Ground, Floors 1 − 5.
Lifts: 5no passenger, 2no goods.

CS3.2 DESCRIPTION OF CIRCULATION AREA

The Concert Hall comprises:

(a) a 2500 seat theatre
(b) a rehearsal hall accommodating 450 persons
(c) an assembly hall with a capacity for 550 persons
(d) various entertainment suites at Floor 1
(e) a number of workshop and teaching rooms in the basement.

The three halls are connected by a concourse with entrances at the east and west ends. Each entrance has one 2.0 m diameter revolving door and a pair of outward opening swing doors providing a 2.0 m clear opening. The swing doors cannot be fastened back to provide a continuous clear opening. Most people (80%) will arrive and depart via the east entrance, where the main coach and drop off points are situated. If all halls are occupied simultaneously then some 3500 people may be present plus staff, actors, etc. Figure CS3.1 illustrates the circulation area.

The Concert Hall design team wish to know if the doorways are suitable to deal with the volumes of people expected to attend the events.

CS3.3 ANALYSIS

Consider two circulation scenarios: before performances (arrivals) and after performances (departure). It is likely that all three halls will be occupied simultaneously on occasions. Thus some 3500 persons could arrive and depart at the same time.

Theatre Consultants advise that in a major city, it is likely that the main arrival period will be of some 30 minutes duration, with about 50% (ie: 1750 persons) of the audience arriving in the last 10 minutes before a performance. These figures can be considered as a worse case scenario. In a more relaxed environment the audience arrivals are likely to be over a period of 45 minutes with 50% arriving over the last 15 minutes. The arrival period is affected mainly by the entry arrangements and circulation elements leading to the auditoria.

Everyone will expect to depart simultaneously. The period over which the departure occurs is affected mainly by the number and size of the exit portals. The flows to these circulation elements will be restricted by the internal circulation elements, eg: stairs, lifts, corridors and internal doorways. However, it would be reasonable to expect the dispersal rate for 50% of the audience to be achieved over 10 or 15 minutes, ie: to be no worse than the arrival rate.

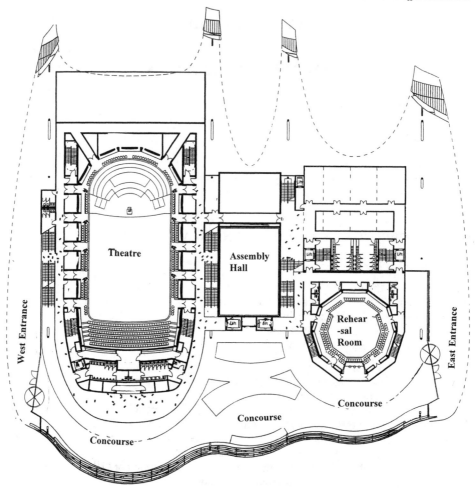

Figure CS3.1 Circulation area in a concert hall complex

CS3.4 FIRST CONCLUSIONS

The worst case scenario for arrivals is 1750 persons over 10 minutes, ie: 875 persons over 5 minutes. The east entrance has to deal with 80% of these arrivals, ie: 700 persons over 5 minutes. The equivalent number for the more relaxed scenario is 466 persons over 5 minutes.

The revolving doors will provide a throughput of approximately 150 persons per 5 minutes (Table 2.5).

The swing doors will provide a throughput of approximately 500 persons per 5 minutes (Table 2.5).

The total east entrance door provision is 650 persons per 5 minutes and will satisfy the relaxed scenario, but not the major city scenario.

Removal of the revolving door and its replacement by a swing door would solve the problem, increasing the throughput to 1000 persons per 5 minutes. However, a revolving door is preferred where there are low traffic levels, such as during the day, as it forms a barrier to the weather. An alternative is to use "lobbying" to provide a weather shield, ie: the provision of two sets of swing doors with a lobby area between them. This alternative will, however, reduce the traffic flows for each storm lobby, by as much as one third, to 333 persons per 5 minutes, ie: a total for two sets of swing doors is 666 persons per 5 minutes. Not quite enough for the worst case scenario.

CS3.5 SECOND CONCLUSIONS

As weather protection was desired and revolving doors were considered to retard easy circulation, it was decided to install three sets of doors with storm lobbies at the east entrance. This gives a total throughput of 1000 persons per 5 minutes, which will cope with both the occupancy scenarios.

The main departure demand will be by 2800 persons, ie: 80% of 3500, leaving via the east entrance. They will be able to leave in a period of some 14 minutes. This dispersal rate meets the departure criterion.

CS3.6 RECOMMENDATIONS

Install three sets of swing doors with a storm lobby.

C3.7 POINTS OF CONFLICT — LESSONS TO BE LEARNT

■ Consider carefully the throughput rates of the different types of door configurations.

■ Take into account weather protection.

■ Make reasonable assumptions for arrival and departure processes.

■ The model split between different facilities may not be equal.

Restaurant Access

CS4.1 BUILDING DATA

Type of building: Multi floor office building.
Occupant: Commercial offices.
Population: 6500.
Number of floors: 33 above Ground.
Restaurant: 500 covers.
Lunch time period: 1½ hour.
Lifts: 12no passenger, 2no goods.
Escalators: 4no.

CS4.2 DESCRIPTION OF CIRCULATION AREA

Please refer to Figure CS4.1. A restaurant facility is situated on the Upper Ground (UG) level of a 33 storey building. The entrance to the restaurant is across a 5.0 m wide bridge linking to two lift lobbies containing four car groups placed off a communicating corridor. Two pairs of escalators serve the restaurant end of the bridge between Ground and UG, and UG and First floor. The restaurant is reached through a restaurant lobby leading off the bridge.

The entrance to the restaurant lobby is via a pair of single swing doors giving a 2.0 m clear opening and which open towards the bridge. The restaurant lobby has an area of 50 m² and contains a menu board and two card validators. It provides for acoustic isolation, kitchen odour control and, importantly, fire protection.

Access from the restaurant lobby into the restaurant itself is via one pair of single swing doors, giving a 2.0 m clear opening to the restaurant seated area. These doors swing into the restaurant lobby. A pair of single swing doors, giving a 2.0 m clear opening, open into a take away section. These doors swing into the restaurant lobby.

A further pair of single swing doors, giving a 2.0 m clear opening, provide an exit route from the restaurant seating area. These doors swing into the restaurant lobby.

Some 15% of staff are absent each day and only some 70% of staff use the restaurant.

CS4.3 ANALYSIS

The 15% absentee factor is often used for lift traffic calculations and the 70% participation is likely. There are 500 "covers" in the seated part of the restaurant allowing 1500 to use this area during a 1½ hour lunch period. The remaining staff, some 2368, use the take away area during the 1½ hour lunch period. The likely demand (persons) is:

Design population	6500
Absentees (15%)	975
Actual population	5525
Restaurant	1500
Take away	2368
Participation (70%)	3868

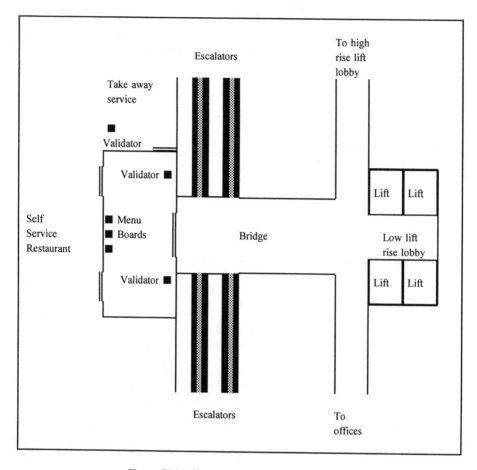

Figure CS4.1 Circulation area for restaurant access

The likely scenario is a fairly steady flow into the take away of 2368 persons and a peaky flow into the restaurant of 1500 persons. The circulation elements to be considered are:

■ low rise lifts to lift lobby

■ high rise lifts to lift lobby

■ one pair escalators to/from bridge and G

■ one pair escalators to/from bridge to First floor

■ corridor from high rise lift lobby to bridge

■ swing doors from bridge to restaurant lobby

■ one pair entrance doors restaurant from restaurant lobby into restaurant

■ one pair exit doors restaurant from restaurant to restaurant lobby

■ one pair entrance/exit doors take away to/from restaurant lobby.

The area at the lobby end of the bridge is very busy and will become very congested as a result of the pedestrian flow, over bridge, to restaurant:

■ bidirectional flow across bridge

■ bidirectional flow off and onto escalators

■ bidirectional flow through bridge to/from restaurant lobby swing doors

■ standing persons in front of card validators

■ standing persons in front of menu board

■ bidirectional flow through take away swing doors

■ unidirectional flow into restaurant

■ unidirectional flow out of restaurant.

CS4.4 FIRST CONCLUSIONS

Assume a fairly steady flow into the take-away and a peaky flow into the restaurant. If there were to be a steady flow 3868 staff over 1½ hours, this gives an average flow of 43 persons per minute. It is conventional to take twice this rate in order to represent peaks, ie: 86 persons per minute. At restaurant sitting changeover times there will be a bidirectional flow, ie: a possibility of 86 persons in both directions. All this flow must pass over the bridge and through the swing doors of the restaurant lobby.

In general terms (Table 1.6) openings of 1.0 m width will allow the passage of:

60 persons per minute   clear opening
40 persons per minute   a swing door
60 persons per minute   a swing door fastened back.

Thus the pair of 1.0 m wide single swing doors (2.0 m total) will only permit 80 persons per minute. This is a classical "pinchpoint" and the access is not adequate.

The presence of the outwards opening swing doors at the restaurant lobby entrance, which are only some 1.0 m away from the boarding/alighting area for the escalators, is dangerous. Pedestrian confusion and congestion will result. Persons arriving by escalator have no option but to alight at the end of the flight. They cannot stand and wait. On this count, the restaurant lobby doors should be removed and a more open access allowed into the restaurant lobby.

CS4.5 SECOND CONCLUSIONS

Please refer to Figure CS4.2. A clear opening should be provided into the restaurant lobby from the bridge of at least 4.0 m clear opening width. This will meet the likely peak bidirectional demand of 172 persons per minute and allow for some congestion. Note also the increase in escalator alighting/boarding area by the use of deflection barriers.

The restaurant lobby is about 50 m^2 in area and at a full design pedestrian density of 1.4 persons/m^2 would accommodate 70 moving persons at any one time. This is too small for the anticipated peak flow. The area should be increased in size to 100 m^2.

This might be achieved by occupying the area behind the present position of the menu board back to the structural elements.

The position of the menu board itself will also result in clashes between standing (queuing) pedestrians and moving (walking) pedestrians and should be relocated in the restaurant lobby. It can be positioned on the new back wall and would then not cause any congestion.

The card validators should be removed from the restaurant lobby and re-located in the restaurant again to reduce any colllisions between standing (queuing) pedestrians and moving (walking) pedestrians.

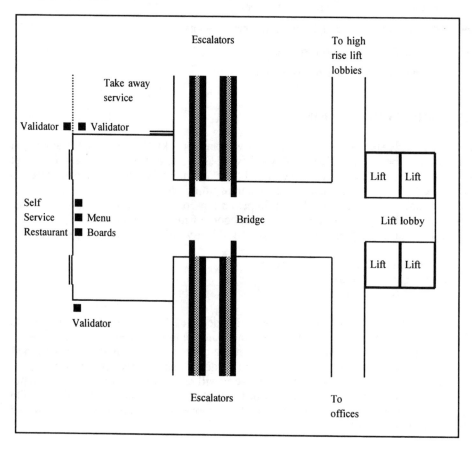

Figure CS4.2 Improved circulation area for restaurant access

The three sets of lobby doors can then continue to open into the restaurant lobby (the direction is selected for fire escape purposes) without causing problems.

CS4.6 POINTS OF CONFLICT — LESSONS TO BE LEARNT

■ The pair of 2.0 m wide single swing doors into the restaurant lobby.

■ The small area of the restaurant lobby.

- The siting of the menu board in the restaurant lobby.

- The siting of the card validators in the restaurant lobby.

- Separate the take away section from the restaurant section.

- Safety at the top of the escalators.

Escalator Landing in an Office Building

CS5.1 BUILDING DATA

Type of building:	International Headquarters.
Occupant:	Banking and trading.
Population:	8500 persons.
Number of floors:	4 Basements, Ground, Floors 1 − 42.
Lifts:	24no passenger, 4no goods.
Escalators:	6no.

CS5.2 DESCRIPTION OF CIRCULATION AREA

The area of interest for this case study is the escalator service to ground, second, third and fourth floors. The second, third and fourth floors accommodate financial traders and each have a likely daily population of 434 persons, with a potential daily population of 536 persons.

A pair of 10 m rise escalators (A-up and B-down) located at the east end of the ground floor serve between ground and second floors, reaching a second floor landing at the west end of the second floor.

A pair of 4 m rise escalators (C-up and D-down) located at the east end of the second floor serve between second and third floors and reach a third floor escalator landing at the mid point of the third floor.

Immediately leading from this third floor escalator landing, a third pair of 4 m rise escalators (E-up and F-down) serve between the third and fourth floors reaching a landing at the west end of the fourth floor.

The two pairs of escalators (C and D, E and F) must be considered as successive sequential escalators as shown in Figure CS5.1. One escalator of the first pair of escalators feeds passengers into an escalator of the second pair in sequence, ie: in the up direction C feeds E and in the down direction F feeds D.

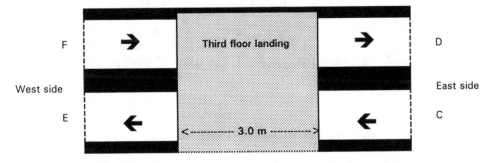

Figure CS5.1 Successive escalators side by side with one intermediate exit

There is unrestricted access for boarding and alighting the escalators at the ground floor, second floor and fourth floor escalator landings. The third floor escalator landing leads into a 15 m² lobby on the southern side of the escalators from which two sets of glazed, 2 m wide, double leaf, double swing, doors lead onto the third floor. There is a wall across the northern end of the third floor escalator landing, so that boarding and alighting must be through the lobby. The doors from the third floor escalator lobby are sometimes locked for security reasons. The distance between the balustrade ends of the two successive escalators (C and D, E and F) at the third floor escalator landing is 3.0 m.

CS5.3 ESCALATOR HANDLING CAPACITY

The rated speed of all escalators is 0.6 m/s[1] and their nominal width is 1.0 m. The theoretical handling capacity from Equation (1.4) is 180 persons per minute and assumes that two persons are standing on each step. This is not practical, as discussed in Section 3.4, and a practical handling capacity will be 90 persons per minute.

As indicated above, there is a likely daily potential for 1302 persons to occupy the second, third and fourth floors. Suppose that an "average" 5-minute peak of 15% of the building population is anticipated (similar to that experienced by lifts), ie: 195 persons per 5 minutes or 39 persons per minute would leave the ground floor. It is unlikely that a smooth arrival process will occur and surges above the average peak might be expected. Common practice is to assume a surge to average peak ratio of two to one. Thus Escalator (A) could be required to carry 78 persons per minute. This is below the handling capacity of the escalators, so there should be little queuing. If, however, the daily occupancy of the three floors was the higher value of 1608 persons, a surge of 97 persons would be possible and there could be some queuing on occasions.

One third of the passengers could stay on the second floor and the remainder (52) could travel up to the third floor escalator landing. Half of these passengers could leave at the third floor and the remainder (26) continue on up to the fourth floor escalator landing. This is a balanced case, and it is quite possible for no one to leave at the second and third floor landings, but that all the passengers travel to the fourth floor.

It would appear that the escalators can handle the average demand and also cope with surges at twice the average demand.

CS5.4 CIRCULATION AT THE THIRD FLOOR LOBBY

The third floor lobby has an area of 15 m² and at a full design flow of 1.4 persons/m² (Table 1.3) could accommodate 21 persons. This should be sufficient to accommodate passengers wishing to board and those who are alighting at this landing. However, if at any time the doors were locked for security reasons, the escalators must stop running as the area could fill in a very short time, which would present an unsafe situation.

Each set of doors leading from the lobby are capable of handling 100 persons per minute. Thus the two sets can handle more people than the escalators can deliver to the landing.

CS5.5 CIRCULATION AT ESCALATOR LANDINGS

As indicated in the description, the ground floor, second floor and fourth floor escalator landings have unrestricted access, and do not present a problem. The third floor escalator landing is shared by four escalators.

---

[1] This is not a standard rated speed, see Table 3.8.

In the worst case scenario it is possible for the up escalator (C) from the second floor to third floor to deliver 90 persons per minute (its maximum handling capacity) onto the third floor escalator landing and the down escalator (F) from the fourth floor to the third floor to deliver the same number of passengers. Thus a total of 180 passengers per minute travelling to the third floor escalator landing is possible. In addition, some persons might also be leaving the third floor to travel up to the fourth floor, or travel down to the second floor. This means that over 180 persons could need to traverse the third floor escalator landing every minute.

As Option 1 has been adopted, then using Equation (3.5) the density of occupation on the third floor landing will be 1.0 persons/m^2. This is the value suggested as the maximum design density in Section 3.4. The passengers therefore move from the escalator at a density of 2.5 persons/m^2 to a landing density of 1.0 persons/m^2. If Option 2 had been possible then the density of occupation would be half the Option 1 value at 0.5 persons/m^2.

There is an exit to one side of the pedestrian pathways. Suppose that half the passengers leave the first escalator at the third floor escalator landing and a number of passengers arrive to board either of the second escalators. The passengers walking from the first to second escalators could come into conflict with the boarding passengers.

It is noted that the third floor escalator landing has a 3.0 m distance between the ends of the balustrades of the successive escalators. If this distance is shared between the successive escalators then each has 1.5 m of unrestricted space instead of 2.5 m. The problem is the opportunity for circulation conflicts, as discussed above, occurring in too small a space. The main concern must be passenger safety. The distance between the successive escalators should be at least 5.0 m as Option 1 has been adopted. Is this sufficient ?

The traffic function is complicated as there is cross traffic from one escalator pair to the other. EN155 states: "consider ... the whole traffic function ...". Thus it would be sensible to increase the unrestricted landing depth by (say) another 1.0 m. This would be a more acceptable situation.

CS5.6 CONCLUSIONS

■ The escalators provide sufficient handling capacity.

■ The third floor escalator landing lobby and its entry/exit doors are sufficient to meet the circulation demands made on them.

■ Where successive escalators are installed it is important to provide unrestricted areas at landings, at least to that suggested by EN115.

■ The traffic function must also be considered.

CS5.7 POINTS OF CONFLICT — LESSONS TO BE LEARNT

■ Insufficient spacing of successive escalators.

■ Ensure that whenever the entry/exit doors to the third floor escalator lobby are locked, that the four escalators are switched off.

■ This case study indicates the need to consider safety as well as circulation.

Escalator Landing in Court Building

CS6.1 BUILDING DATA

Type of building:	Court House.
Occupant:	Court officials, public and prisoners.
Population:	Not quantifiable.
Number of floors:	4 plus Basement.
Lifts:	n/a.
Escalators:	4no.

CS6.2 DESCRIPTION OF CIRCULATION AREA

The area of interest is the escalator provision from the entrance level (Ground) to the court level (Level 2) as shown in Figure CS6.1. These two levels are linked by two pairs of successive escalators, with a landing at Level 1. The nominal step width of the escalators is 600 mm and they run at a rated speed of 0.5 m/s over an 8 m total rise. The maximum proposed depth of the intermediate landing was 3.0 m, in order to meet footprint requirements.

The Client sought advice as to the efficiency and safety of the arrangement.

Figure CS6.1 Successive escalators without an intermediate exit

CS6.3 ANALYSIS

The escalators will provide a practical handling capacity of 37 persons per minute. This is likely to be sufficient as lifts and stairs are provided nearby. The depth of 3.0 m on the landing does not comply with the recommendations in EN115, but does follow the suggestions in Section 3.4. This should be satisfactory.

CS6.4 DESIGN CHANGE

At a late point in the design it was suggested that the "blind" escalator landing at Level 1 be opened up to allow access to the catering facilities located there from both the court and the ground levels. This suggestion was questioned. The client indicated the poor circulation provided to the Level 1.

The suggestion did not meet the safety requirements outlined in Chapter 3. In addition, the likelihood of persons crossing from one side of the escalators to the other was very strong with consequential conflicts.

A compromise was to restore the central dividing walls, but to allow passengers to leave and join in one direction only. This compromise would be reasonable in the circumstances as the step width of the escalator was 600 mm and hence all passengers could only alight and board in single file, thus easing the competition for landing circulation space.

CS6.5 CONCLUSION

The design change could be accommodated.

CS6.6 POINT OF CONFLICT

■ Insufficient spacing of successive escalators for proposed circulation.

Terminal Railway Station

CS7.1 BUILDING DATA

Type of building: Terminal railway station.
Occupant: Passengers.
Population: 500 passengers per train.
Number of floors: 2: Platform and Concourse (below).
Length of train: 130 m.
Lifts: 1no passenger (disabled access).
Escalators: 8no total (see Figure CS7.1).
   2no up, 2no down at the one quarter point of the platform.
   2no up, 2no down at the three quarter point of the platform.

CS7.2 DESCRIPTION OF CIRCULATION AREA

This case study considers a terminal railway station with the platforms above a concourse area. The trains are 130 m in length and arrive with a 180 s (three minute) headway. Each train accommodates 500 passengers.

A set of two up escalators and two down escalators are situated at both the one quarter and the three quarter points along the platform length. The escalators run at a rated speed of 0.5 m/s and are 1000 mm wide.

Can the number of escalators be reduced ? The most important time of the day to consider is a morning arrival peak period, with little or no boarding passengers.

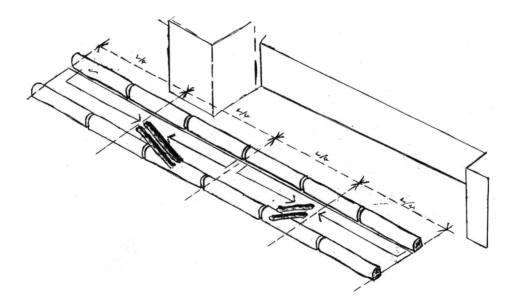

Figure CS7.1 Sketch drawing of original proposal

CS7.3 ANALYSIS

The most important consideration is to clear the platform before the next train arrives and the next important consideration is to avoid passenger queuing. From Table 1.3 assume that the passengers walk at an average speed of 1.3 m/s (3 mph). This is an average between commuters at 1.5 m/s and tourists at 1.0 m/s and is a likely average. Assume 1000 mm, 0.5 m/s escalators can provide half of the theoretical handling capacity (Table 1.8), ie: 75 persons per minute (1.25 persons/second).

Assume half the passengers walk to each escalator set.

Maximum distance a passenger will walk is:	130/4 = 32.5 m
Time for last passenger to reach the escalators is:	32.5/1.3 = 25 s
Passenger arrival rate at the escalators:	250/25 = 10 persons/second

With two up escalators at each position.

Handling capacity of two escalators:	2.5 persons/second
The time for complete clearance will be:	250/2.5 = 100 s
The last passenger will wait:	75 s
The number of passengers queuing will be:	(10 − 2.5) × 25 = 188

Thus the platform clearance time is 100 s and the platform will be clear by the time the next train arrives.

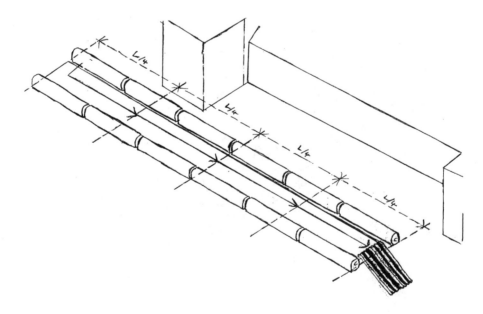

Figure CS7.2 Skectch drawing of final proposal

CS7.4 FIRST CONCLUSIONS

With bulk transit systems large numbers of passengers arrive simultaneously. To handle such a situation it is necessary to spread the load by introducing something for the passengers to do, before they reach the mechanical handling equipment. A good technique is to make them walk a reasonable distance. If possible, some sales points should be introduced along the platform to further delay the arrival of passengers to the escalators, but not so much as to fully occupy them during the train headway period of three minutes.

Consider a set of two up escalators at one end of the platform (see Figure CS7.2).

Maximum distance a passenger will walk is:	130 m
Time for last passenger to reach the escalators is:	130/1.3 = 100 s
Passenger arrival rate at the escalators:	500/100 = 5 persons/second
Handling capacity of the two up escalators:	2.5 persons/second
Thus the clearance time will be:	500/2.5 = 200 s
The last passenger will wait:	100 s
The number of passengers queuing will be:	(5 − 2.5) × 100 = 250

Thus the platform clearance time is 200 s and the platform will not be clear by the time the next train arrives. This is not a solution.

Consider a set of three up escalators at the end of the platform.

Handling capacity of the three up escalators:	3.75 persons/second
Thus the clearance time will be:	500/3.75 = 133 s
The last passenger will wait:	33 s
The number of passengers queuing will be:	(5 − 3.75) × 100 = 125

Thus the platform clearance time is 133 s and the platform will be clear by the time the next train arrives.

Consider a set of four up escalators at the end of the platform.

Handling capacity of the four up escalators:	5.0 persons/second
Thus the clearance time will be:	500/5 = 100 s

There will not be a queue, or passenger waiting time.

The platform clearance time is 100 s and the platform will be clear of passengers by the time the next train arrives.

CS7.5 SECOND CONCLUSIONS

Effective load spreading is achieved where the escalators are placed at the end of the platform.

It is not reasonable to design for a peak load. Queuing would only occur at peak times, when some queuing can be tolerated, provided the platform is cleared before the next train arrives. With four escalators there is no queuing. With three escalators there would be 125 passengers queuing for up to 33 s. With two escalators the platform does not clear in time.

The three up-escalator scenario would be acceptable and reduces the number of escalators that would be required if the distributed proposal had been adopted.

Probably a set of four escalators would be ideal. The outer escalators could be permanently set for up and down directions with the inner pair operating on a tidal flow basis or switched off outside the peak periods. Remember, escalators are taken out of service for maintenance and inspection (and sometimes fail) and this arrangement would allow this.

CS7.6 POINTS OF CONFLICT — LESSONS TO BE LEARNT

■ Platform clearance during a train headway.

■ Queues of containable proportions.

■ Passenger waiting times that are tolerable.

■ Escalator tidal flow possibilities.

Lifts versus Escalators

CS8.1 BUILDING DATA

Type of building: Stock Exchange.
Occupant: Financial Traders.
Arrival rate: 375 persons/5-minutes.
Number of floors: Ground, 1, 2, 3, 4.
Interfloor distance: 6.0 m (rise).
Lifts   Rated speed: 1.6 m/s.
             Rated load: 2500 kg.
             Cycle time: 12 s.
Escalators   Step width: 1000 mm.
             Rated speed: 0.5 m/s.
             Handling capacity: 375 persons/5-minutes (practical).

CS8.2 DESCRIPTION OF CIRCULATION AREA

During uppeak, do lifts or escalators provide the fastest travel time, when serving a high density, low rise building ?

CS8.3 ANALYSIS

The conventional wisdom is that passengers will be influenced to travel either by lifts or escalators mainly by the distance to be travelled, *viz*:

Table CS8.1 Division of traffic between escalators and lifts

Floors travelled	Division of traffic (%)	
	Escalator	Lift
1	90	10
2	75	25
3	50	50
4	25	75
5	10	90

CS8.4 ASSUMPTIONS

For an escalator with a rated speed of 0.5 m/s, the travel time between boarding and alighting stations is 24 s. Assume that the passenger boarding and alighting time is 2 s. If the escalators are sequentially arranged, the walk time from one to the next will be assumed to be 2 s. The practical handling capacity of the escalators is 375 persons per 5 minutes.

The lift system is to be sized to provide the same capacity as the escalators. Calculations can show that three lifts are required to serve a single floor above ground; four lifts to serve two floors above ground; five lifts to serve three floors above ground; and six lifts to serve four floors above ground. Further assumptions are: the practical occupancy is 19 persons, the passenger loading time for 19 passengers is 12 s, the interval will be 15 s and an average waiting time is 12 s regardless of the number of lifts.

CS8.5 CALCULATIONS

It is also assumed that the escalator passengers stand and do not walk on the escalator and that the escalators are arranged in either a parallel or a cross over configuration (see Figure 1.10). Using the assumptions above, calculations can show that the travel time by lift or escalator is as shown in Table CS8.2.

Table CS8.2 Travel times of lifts versus escalators

Travel to:	First floor		Second floor		Third floor		Fourth floor	
Equipment (L = lifts, E = Escalator)	L = 3	E = 1	L = 4	E = 2	L = 5	E = 2	L = 6	E = 3
Average waiting time (s)	13	0	13	0	13	0	13	0
Passenger boarding time (s)	12	2	12	4	12	6	12	8
Average passenger alighting time (s)	6	2	3	4	2	6	1.5	8
Travel time (s)	12	24	24	48	36	72	48	96
Lift delay at stops/walking time (s)	0	0	6	2	8	4	9	6
Total	43	28	58	58	71	88	84	118

Notes to Table CS8.2. The average waiting time is calculated as 85% of interval. Passengers are assumed to take 19 s to board a lift and 2 s to board/alight from an escalator. The travel time for a lift is the cycle time and for an escalator is the time on the escalator. The lift delay time is the time spent at intermediate landings for passengers to exit, assuming the lifts stop at all floors. The average passenger alighting time is the time for the remaining passengers to leave a lift at the final destination.

CS8.6 FIRST CONCLUSIONS

It can be seen that the escalators are quickest when there is only one floor above, equal for two floors above, but slower thereafter. If one line of passengers were to walk on the escalator, their travel time would be reduced by 12 s per rise, ie: reducing the average travel times by 6 s per rise. However, if a walk-around escalator configuration were to be provided, the escalator travel times could increase by another 25 s per rise.

CS8.7 SECOND CONCLUSIONS

Other considerations are space and cost. Probably the lifts will occupy more space for the low rise scenarios and less for the high rise. Probably the escalator option is cheaper for the low rise scenario. It is important to note that the lift and escalator installations considered here were both able to provide equal handling capacities. This may not always be the case.

Circulation Data Tables

The tables in this Data Sheet have been extracted from Chapters 1 and 2. They are presented here to provide a quick reference to all the circulation data discussed. The text should be referred to for specific details of use.

Circulation Data Tables

Table 1.1 Interpersonal distances

The table shows the empirical psychological classifications.

Classification	Separation
Public distance (far)	> 7.5 m
Public distance (near)	3.6–7.5 m
Social distance (far)	2.1–3.6 m
Social distance (near)	1.2–2.1 m
Personal distance (far)	0.75–1.2 m
Personal distance (near)	0.45–0.75 m
Intimate distance	< 0.45 m

Table 1.2 Density of occupation in waiting areas

Table shows densities for persons waiting in circulation spaces. This may change for queues.

Level	Density
Desirable	0.4 P/m^2
Comfortable	1.0 P/m^2
Dense	2.0 P/m^2
"Crowding"	3.0 P/m^2
Crowded	4.0 P/m^2

Table 1.3 Possible horizontal pedestrian flows

Flows in persons per minute (P/min) and persons per hour (P/h) and typical pedestrian speeds in metres per second (m/s) for a free flow design density of 0.3 persons per square metre (P/m^2) and for a full flow density of 1.4 persons per square metre (P/m^2) with grouping.

Traffic type	Free flow (0.3 P/m^2) design density			Full flow (1.4 P/m^2) design density		
	Speed	P/min	P/h	Speed	P/min	P/h
Commuters, workers	1.5	27	1620	1.0	84	5040
Individual shoppers	1.3	23	1380	0.8	67	4020
Family groups, tourists	1.0	18	1080	0.6	50	3000
School children	1.1–1.8	18–32	1080–1920	0.7–1.1	59–92	3540–5520

Table 1.4 Minimum corridor widths

Minimum width of corridors to accommodate various types of traffic; some compensation can be allowed for mixed two way traffic situations. A 3 m wide corridor is often suitable.

Usage	Minimum width (m)
One way traffic flow	1.0
Two way traffic flow	2.0
Two men abreast	1.2
Man with bag	1.0
Porter with trolley	1.0
Woman with pram	0.8
with child alongside	1.2
Man on crutches	0.9
Wheelchair	0.8

Table 1.5 Reductions in corridor widths

Reductions in corridors widths in metres caused by various obstructions.

Obstruction	Reduction (m)
Ordered queue	0.6
Un-ordered single queue	1.2–1.5
Row of seated persons	1.0
Coin operated machine:	
one person	0.6
queue	1.0
Person waiting with bag	0.6
Window shoppers	0.5–0.8
Small fire appliance	0.2–0.4
Wall mounted radiator	0.2
Rough/dirty surface	0.2

Circulation Data Tables

Table 1.6 Portal (handling) capacities

Handling capacities in persons per minute (P/min) and persons per hour (P/h) through an opening of 1 m width.

Portal type	P/min	P/h
Gateway	60–110	3600–6600
Clear opening	60–110	3600–6600
Swing door	40–60	2400–3600
Swing door		
(fastened back)	60–90	3600–5400
Revolving door	25–35	1500–2100
Waist high turnstile:		
free admission	40–60	2400–3600
with cashier	12–18	720–1080
single coin operation	25–50	1200–1800

Table 1.9 Handling capacities of passenger conveyors (moving walkways and ramps)

The table indicates the likely handling capacity of passenger conveyors (walkways and ramps) based on a standing density of 2.0 persons/m^2.

Incline (°)	Speed (m/s)	Width (mm) 1000		Width (mm) 1400	
0	0.50	60	3600	84	5040
0	0.63	76	4560	106	6350
0	0.75	90	5400	126	7560
5	0.70	84	5040	–	–
10	0.65	78	4680	–	–
12	0.50	60	3600	–	–

Table 1.7 Stairway (handling) capacities

Flows in persons per minute (P/min) and persons per hour (P/h) and typical pedestrian speeds in metres per second (m/s) for a free flow design density of 0.6 persons per square metre (P/m^2) and for a full flow density of 2.0 persons per square metre (P/m^2) with grouping.

Traffic type	Free flow (0.6 P/m^2) Speed	P/min	P/h	Full flow (2.0 P/m^2) Speed	P/min	P/h
Young/middle-aged men	0.9	27	1620	0.6	60	3600
Young/middle-aged women	0.7	21	1260	0.6	60	3600
Elderly people, Family groups	0.5	15	900	0.4	40	2400

Table 1.8 Theoretical escalator handling capacity

Theoretical escalator handling capacities for various factors of k in persons per minute (P/min) and persons per hour (P/h).

Speed	Step width 1000 mm		Step width 800 mm		Step width 600 mm	
	P/min	P/h	P/min	P/h	P/min	P/h
0.50	150	9000	113	6750	75	4500
0.65	195	11700	146	8775	98	5850
0.75	225	13500	169	10125	113	6750

Circulation Data Tables

Table 1.11 Stair usage

The table shows the likely attraction of traffic using stairs and using escalators and lifts depending on the distance the person wishes to travel.

Floors travelled	Usage up	Usage down
1	80%	90%
2	50%	80%
3	20%	50%
4	10%	20%
5	5%	5%
6	0%	0%

Table 1.12 Lifts and escalators: division of traffic

The table shows the likely attraction of traffic between escalators and lifts depending on the distance the person wishes to travel.

Floors travelled	Escalator	Lift
1	90%	10%
2	75%	25%
3	50%	50%
4	25%	75%
5	10%	90%

Table 2.2 Actual mall pedestrian speed and flows

The table shows the likely pedestrian flow rates in persons per hour (P/h) and typical pedestrian speeds in metres per second (m/s) for uncrowded conditions of 0.2 persons per square metre (P/m^2) and for crowded conditions of 0.45 persons per square metre (P/m^2)

Traffic type	Uncrowded (0.2 P/m^2)		Crowded (0.45 P/m^2)	
	Speed (m/s)	Flow (P/h)	Speed (m/s)	Flow (P/h)
All shoppers	1.3	936	1.0	1620

Table 2.3 Mall stairway capacity

The table shows the pedestrian flows rates in persons per hour (P/h) and pedestrian speeds in metres per second (m/s) for uncrowded conditions of 0.4 persons per square metre (P/m^2) and for crowded conditions of 0.8 persons per square metre (P/m^2) per metre width of mall.

Traffic type	Speed (m/s)	Uncrowded (0.4 P/m^2)	Crowded (0.8 P/m^2)
Men	0.8	960	1920
Women	0.7	840	1680
Elderly men	0.5	600	1200
Elderly women	0.6	720	1440
Children	0.8	960	1920
Push chairs	0.5	600	1200

Table 2.4 Actual mall escalator handling capacity

The table shows the likely handling capacity of mall escalators in persons per hour (P/h) for uncrowded and crowded conditions.

Speed	Step width 600 mm		Step width 800 mm		Step width 1000 mm	
	Uncrowded	Crowded	Uncrowded	Crowded	Uncrowded	Crowded
0.50	1500	3000	2250	4500	3000	6000

CHAPTER FOUR

Principles of Lift Traffic Design

4.1 THE NEED FOR LIFTS

Lifts are installed into buildings to satisfy the vertical transportation needs of their occupants and visitors. They are necessary to provide a comfortable means of transportation to the different levels in a building. Some of these requirements are written into statutory regulations.

The transportation capacity of the lift group in a building is a major factor in the success or failure of a building as a place to work, live or receive a service. Like toilets, lifts should be available and easy to use without a second thought. Unfortunately this is not always the case and speculative building often results in the installation of an imperfect lift system.

In offices and other commercial buildings, lifts are installed to aid the efficient movement of the occupants around the building, when performing their work tasks. This has the benefit of saving time, and hence money.

These financial considerations do not apply for residential property; quite the opposite, money is saved by not providing a lift, and statutory regulations have been framed to ensure suitable lifts are installed. In Britain, for example, it is recommended that a lift be installed in all residences where there are four or more storeys, and that two lifts be installed where a building contains more than six storeys.

The increase in the numbers of high and medium rise buildings since the Second World War has been a challenge to the lift industry. The four decades between 1945 – 85 have seen the acceptance of automatic cars, the introduction of better traffic and control systems, and the inclusion of the digital computer in equipment. Improvements have also occurred in the engineering design and engineering installation of lift systems. The acceptance of traffic design methods has been slower and has only really become accepted since the early 1970s.

4.2 FUNDAMENTAL DESIGN CONSTRAINTS

The planning and selection of transportation equipment is a very involved subject. Although the basic calculations are relatively simple, the theory on which they are based is complex. The results obtained need to be tempered with a great deal of working experience of existing buildings, in order to ensure satisfactory design results.

When sizing a lift system for a new building, the major building dimensions should be known. Unfortunately it is often the case that the architect responsible for the building conception will not have taken professional advice from a lift specialist and may well have fixed the building's core dimensions, thus limiting the space available for the lift system or, even worse, may have defined the number of shafts, their dimensions and travel. This removes one very important degree of freedom from the lift traffic designer. Building circulation, both horizontal and vertical, is the lifeblood of any building, and hence if a successful building is to be designed it is essential that the architect take expert advice at conception. This does not imply that the lift designer will take over the core design, but simply that by means of a team approach various aesthetic and conceptual ideas can be considered early on and design and optimal solutions offered.

Often the result of a team and professional approach will be a better sized lift system design, possibly with less shafts or fewer shafts travelling the whole height of the building, or a rearrangement of service floors and main terminals. The net effect should be a building properly configured for good access with sufficient handling capacity to serve the proposed population and its circulation needs.

Of course, at the low end of the market, there may be only one lift in a building, or its dimensions may be fixed to conform with statutory regulations, or to accommodate the carriage of furniture, etc. But as the lift system moves "up-market", initial design decisions become more important.

When redesigning for the modernisation of an existing lift installation, the fundamental constraints mentioned above cannot be altered (or not very much) as the building actually exists. However, there is often the advantage that the building population to be served is already known.

4.3 HUMAN CONSTRAINTS

A lift system has to be acceptable to the travelling public. The most important requirement the public demands is safety. These aspects are covered by the safety standards promulgated at national, continental and world wide levels. This requirement is most important so that passengers may feel confident about the way they are handled. However, passengers are human and are subject to constraints, which fall into two categories: physiological and psychological, the body and mind.

4.3.1 Physiological Constraints

The physiological constraints (the effects of movement on the body) limit the manner in which a passenger may be moved in the vertical plane. The human body is uncomfortable if its internal organs are caused to move within the body frame. This occurs when the body is subjected to acceleration or deceleration, the well known g effect. The magnitude of the effect on an individual depends on an individual's age, physical and mental health, and whether the individual is prepared for the experience of a sudden movement. It is not clearly established what the level of acceleration is at which permanent harm may be caused to the human body, but it is known, by experience, the levels of acceleration or deceleration which have been found to be generally acceptable, when riding in a lift. These are shown in Figure 4.1.

Note that there is no limit to the velocity at which a passenger may travel in an enclosed lift car, as speed is not noticeable to the passenger. But the values of acceleration/deceleration (rate of change of velocity) should be limited to about one eighth of g_n[1] or 1.5 m/s^2 and the values of jerk (rate of change of acceleration) to 2.0 m/s^3. The affect of an acceleration of one eighth of g_n on a body weighing 80 kg travelling in an upward direction is that it then weighs 90 kg. Likewise the same body subjected to a deceleration, while travelling in an upward direction, would weigh 70 kg.

---

[1] g_n is the acceleration of a body due to gravity, numerically equal to 9.81 m/s^2.

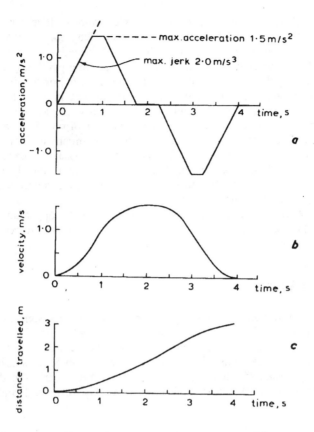

Figure 4.1 Ideal acceleration, velocity and distance travelled curves for a single floor jump

    (a) Acceleration profile: note maximum jerk 2.0 m/s³ and maximum acceleration 1.5 m/s².
    (b) Velocity profile: note maximum speed 1.5 m/s.
    (c) Distance travelled: note total distance 3.0 m.

It is the jerk values (not a very scientific sounding name, sometimes called *shock*), which cause the most discomfort. If the value of jerk is allowed to exceed 2 m/s³ for any length of time (tenths of seconds), discomfort will be experienced. Whereas velocity and acceleration/deceleration profiles can be specified and controlled in drive systems, jerk cannot. Constant values of jerk require that the acceleration/deceleration profile increase/decrease at a constant rate, and this is not always possible.

It is perhaps fortunate that these human constraints do exist as they ease the design of lift drive systems considerably !

4.3.2 Psychological Constraints

As would be expected, psychological constraints are more subtle. A passenger expects a good service from a lift system. But an individual passenger expects a different grade of service at different times of the day and at different locations. For example, an office worker will not be too annoyed if delayed when travelling up a building to work, but will become very annoyed if delays occur when leaving at night. In contrast, the same office worker would not expect the same grade of service from a lift in a residential block. This constraint can be categorised as the passenger's waiting time constraint. In general, the average waiting time in an office block should not exceed 30 s and in the residential block it should not exceed 60 s. Waiting time is the prime psychological constraint.

A secondary psychological constraint is the transit time, or travel time, in the car after the passenger boards. Here the passenger is dependent on the fellow passengers in the car and other passengers on the landings making calls. A passenger travelling high up a building becomes intolerant of stops after about 90 s of travel. Again the tolerance level depends on whether the passenger is travelling in company of friends or colleagues and on the other passengers' behaviour. For instance, one passenger boarding or alighting is obviously more "selfish" than two or three transferring at a time. This psychological constraint has been summed up by Strakosch (1967) as "a person will not be required to ride a car longer than a reasonable time".

There are other psychological effects, such as aesthetic appearance and "gentle" doors, which add to a passenger's confidence in a lift system and overcome the fears of some persons who are afraid of such machines.

4.4 TRAFFIC PATTERNS

As the users of lift systems, the passengers impose on the lift system the need for it to respond to different traffic patterns. Consider Figure 4.2, this shows the passenger demand in an office building as represented by the number of individual calls, aggregated for up and down call directions. This office building is subject to a strict time regime of fixed starting, break and leaving times. It illustrates clearly the different traffic patterns of morning up-peak, evening down-peak, midday traffic and random (balanced) interfloor traffic.

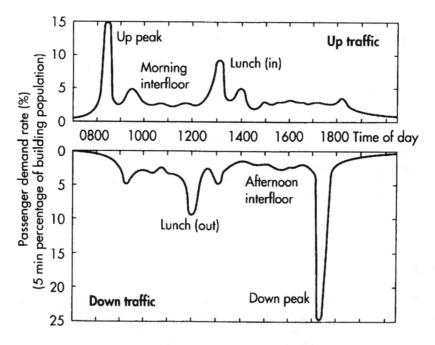

Figure 4.2 Passenger demand for an office building

At the start of the day there is a larger than average number of up-hall calls. This is due to the building's occupants arriving to start work. This traffic pattern is called the morning uppeak.

Late in the day there is a larger than average number of down-hall calls. These are due to the building's population leaving the building at the end of the working day. This traffic pattern is called the evening down peak.

In the middle of the day there are two separate sets of uppeaks and two down peaks. This represents a situation where the occupants of the building take two distinct lunch periods (ie: 12.00 to 13.00 and 13.00 to 14.00). This pattern is sometimes called two-way traffic.

During the rest of the day the numbers of up-hall and down-hall calls are similar in size and over a period are equal. This traffic pattern is called interfloor traffic, sometimes qualified as balanced interfloor traffic.

In practice this pattern may not be observed exactly as shown, as many companies have adopted a "flexitime" attendance regime. It does, however, serve as a model for discussion.

4.4.1 Uppeak traffic

This traffic condition is shown diagrammatically in Figure 4.3.

Definition 4.1: An uppeak traffic condition exists when the dominant, or only, traffic flow is in an upward direction, with all, or the majority of, passengers entering the lift system at the main terminal[1] of the building.

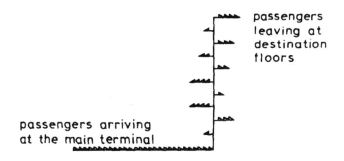

Figure 4.3 Uppeak traffic

Uppeak occurs in considerable strength in the morning when prospective lift passengers enter a building intent on travelling to destinations on the upper floors of the building. To a lesser extent an uppeak occurs at the end of the midday break. It is

---

[1] The term main terminal is used throughout this book to avoid confusion and is synonymous with ground floor (Britain), first floor (USA), lobby, foyer, main arrival floor and building entrance floor.

considered that, if a lift system can cope efficiently with the morning uppeak, then it will cope with other patterns of traffic, such as down peak and random interfloor traffic.

The uppeak condition results from employers requiring their employees to arrive at work by a specific starting time. Human nature then exacerbates the condition as the majority of employees feel that in conscience all they must do is to be in a building before the defined starting time and that the employer then has the responsibility to transport them to their work station.

The modern trend to FLEXITIME[1] working will go some way to alleviate the uppeak situation, but unfortunately it is not being applied to all classes of employment. It can also mean that other traffic conditions may become relatively more severe and if a building designed for FLEXITIME becomes one with a fixed time regime then the lift system could be seriously undersized.

The arrival rate profile for the morning uppeak thus takes a form as shown in Figure 4.4. Here the envelope of the curve describes the arrival profile in terms of the instantaneous passenger arrival rate in calls per hour for a period of one hour. Figure 4.4 reveals that the uppeak traffic profile presents a gradual build-up prior to the official starting time and then a more rapid decay afterwards. The lift installation must be able to handle the peak, if a satisfactory service is to be provided.

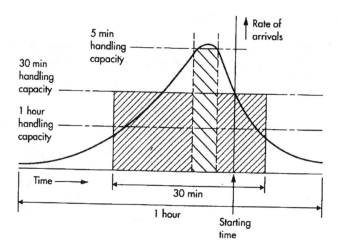

Figure 4.4 Detail of uppeak traffic profile

The profile of Figure 4.4 is often idealised by designers in terms of a 5-minute peak value taken as a percentage of the building population (the wide hatched area of Figure 4.4). The industry practice is to size a lift installation to handle the number of passengers requesting service during the heaviest 5 minutes of the uppeak traffic condition. This is a sound recommendation. To size the lift system to handle the actual peak would require too large a system, which would be very expensive and much of the equipment would be under utilised during large periods of the working day.

---

[1] Flexitime working allows workers to arrive, leave and take refreshment breaks within wide time bands. They are expected to be present during specified periods (core time) and to work a specified number of hours.

A second design practice sometimes used for the uppeak profile is to state the percentage of the building population that arrives over 30 minutes of peak activity (the close hatched area of Figure 4.4). This practice will generally result in a totally inadequate installation, not only for uppeak traffic, but also for the other traffic conditions.

The uppeak traffic condition is often detected by the traffic supervisor so specific control actions may be taken. Common detection systems determine when a predefined number of cars leave the main terminal loaded to a predefined level. The duration of the uppeak period detected in this way does not necessarily exist for precisely 5 minutes.

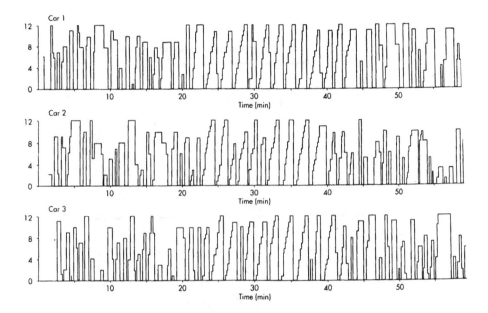

Figure 4.5 Screen shot of the spatial movements of lift cars during uppeak traffic

Figure 4.5 is taken from a computer simulation[1] of a peak morning hour. It shows the spatial movements of a number of lifts. Note the increased number of stops during the peak 5-minute period.

4.4.2 Down peak traffic
The traffic condition is shown diagrammatically in Figure 4.6.

Definition 4.2: A down peak traffic condition exists when the dominant or only traffic flow is in a downward direction with all, or the majority of, passengers leaving the lift system at the main terminal of the building.

---

[1] Figure 4.5 has been obtained by copying the display that is produced during a computer simulation of a lift system to a computer file and then printing it. Details of lift traffic design using digital computer simulation methods are discussed fully in Chapter 16.

To some extent, down peak is the reverse of the morning uppeak occurring at the end of the working day, and to a lesser extent at the start of the midday break. The evening down peak is usually more intense than the morning uppeak with higher demands and with durations of up to 10 minutes. Figure 4.7 illustrates these effects.

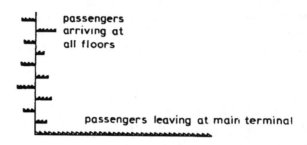

Figure 4.6 Down peak traffic

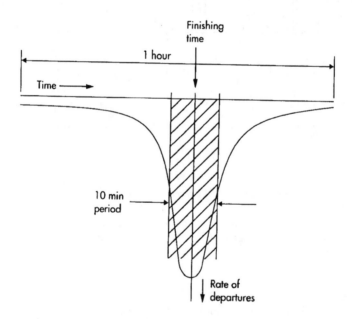

Figure 4.7 Detail of down peak profile

Figure 4.7 details the down peak traffic profile, showing the larger size and longer duration of the traffic demand. Fortunately a lift system can be shown to possess 50% more handling capacity during down peak than during uppeak. (This is because during down peak a lift car fills at three, four or five floors and then makes an express run to the main terminal. This reduction in the number of stops results in a shorter round trip time and hence a greater handling capacity during down peak.)

The detection of the onset and duration of the down peak traffic condition is usually achieved by similar methods to those used to detect uppeak. Figure 4.8 is taken from a computer simulation of a peak evening hour, again showing the spatial movement of the lifts. Note the smaller number of stops and the express runs to the main terminal floor during the peak 10-minute period.

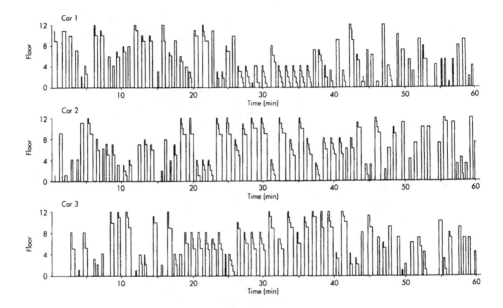

Figure 4.8 Spatial movements of lift cars during down peak traffic

4.4.3 Two Way and Mid Day (Lunch Time) Traffic

The two way traffic condition may not be easily detectable in most buildings. It can arise from the presence of a refreshment floor, which at certain times of the day attracts a significant number of stops and calls. Two way traffic could thus occur during the mid-morning and mid-afternoon refreshment breaks.

Definition 4.3: A two way traffic condition exists when the dominant traffic flow is to and from one specific floor, which maybe is the main terminal.

The mid day lunch period often presents the heaviest demand on a lift system owing to the simultaneous up, down and interfloor traffic to several floors.

Definition 4.4: A mid day (lunch time) traffic condition occurs in the middle of the day and exhibits a dominant traffic flow to and from one or more specific floors, one of which may be the main terminal.

4.4.4 Random Interfloor Traffic

This traffic condition is the most common traffic situation and exists for the majority of the working day in office buildings.

Definition 4.5: Random interfloor traffic can be said to exist when no discernable pattern of calls can be detected.

Uppeak probably exists for 5 minutes and down peak for 10 minutes, and two and four way traffic, if they occur at all, can be considered to be severe cases of unbalanced interfloor traffic. Interfloor traffic is caused by the normal circulation of people around a building during the course of their business. Sometimes this traffic is called balanced two way traffic as it involves both up and down trips, and it is balanced because passengers usually return to their original floor after moving about the building.

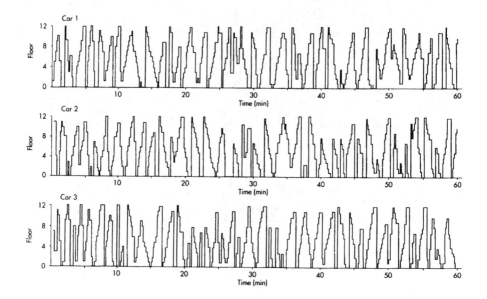

Figure 4.9 Spatial movements of lift cars during balanced interfloor traffic

Figure 4.9 is taken from a computer simulation of an hour of office activity. Note that the figure can be reversed or inverted and still no discernable pattern can be seen in the spatial activity of the lifts

4.4.5 Other traffic situations

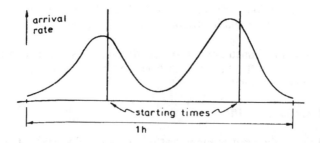

Figure 4.10 Another arrival profile for morning uppeak with two starting times

It is possible to find office buildings where no dominant traffic flows occur, especially where FLEXITIME working is used. Sometimes the uppeak situation occurs twice, as in Figure 4.10, but at a lower intensity. And obviously traffic patterns are different in institutional and residential buildings; but often dominant patterns similar to those defined above do emerge and hence ease design procedures. The effect on a lift system of applying a non-smoking regime in a building, where smoking is not permitted inside the building and smokers have to go outside, can have a significant effect. Even today 24% of the people smoke and might crave one smoke per hour.

4.4.6 Summary of traffic conditions

Traffic conditions may be summarised as follows:

- the duration of the uppeak traffic condition is about 5 minutes
- the duration of the down peak condition is about 10 minutes
- the two-way traffic condition may exist for one to two hours dependent on the arrangements for the midday break
- the interfloor traffic condition exists for most of the working day and therefore is very important.

The distinctive "fingerprints" of uppeak, down peak and balanced interfloor traffic patterns, as represented by the spatial movements of lifts, are:

Uppeak: The lifts arrive at the main terminal, load with passengers, and move up the building stopping often until the last stop when they express return to the main terminal. During the peak 5-minutes there is a "staircase" pattern.

Down peak: The lifts stop at a few floors in the building, loading with passengers and then express to the main terminal. After unloading the passengers the lifts make an express run back up the building. There is a small "staircase" pattern in the reverse direction to the uppeak case.

Interfloor: There is no discernable pattern for a balanced interfloor traffic condition.

4.5 TRAFFIC DESIGN

4.5.1 Introduction

Why is there a need for traffic design ? This could be answered as follows:

- To size a lift installation to serve a traffic requirement or meet a capital/recurrent financial requirement.
- To compare competitive tenders.

It is extremely difficult to compare competitive tenders where no standardised methods of specification or common design procedures are used. Each manufacturer and lift consultant often use different methods, and are not keen to explain their approach. Many methods that are published are often sketchy and some are inaccurate. In addition, the use today of modern control systems radically alters some of the design assumptions. An easy to use, acceptable and standard design method will be presented here. This should enable a prospective designer to gain a better understanding of the design procedure and be able to use it better.

Little theoretical or analytical work was carried out into traffic design and control, until recently. Simulation techniques were used by earlier workers (Browne and Kelly, 1968), who considered their use essential to investigate better design methods and to develop new traffic supervisory control techniques. Simulation is a tool generally used only when a sufficient mathematical definition does not exist.

By the 1970s, a recognised method of calculation had evolved, for the uppeak traffic sizing, based on the mathematical determination of average highest reversal floor (H) by Schroeder (1955), average number of stops (S) by Basset Jones (1923) and average number of passengers (P). The next lesson looks at statistical modelling theory, which provide a sound mathematical base.

The formulae by Barney and Dos Santos (1977) for the calculation of the passenger handling performance of lift systems are now universally accepted. Lift makers often use tables specific to their product range to estimate round trip times, interval, handling capacity, etc.

The problem in sizing lift systems is to match the demands for transportation from the building's occupants with the handling capacity of the installed lift system. This procedure should also result in an economic solution. The conventional procedure used in the traffic design of lift systems is to determine the handling capacity for the uppeak traffic situation. This approach is sensible as the uppeak traffic condition does yield to analytic techniques, although some of the assumptions making this possible are difficult to justify in the real life situation.

4.5.2 Some Definitions

What is uppeak or incoming traffic? This has been defined in Section 4.4.1 as Definition 4.1. The idealised profile of Figure 4.4 extends Definition 4.1 to allow a 5 minute uppeak passenger arrival rate to be defined.

Definition 4.6: The uppeak percentage arrival rate (%*POP*) is the number of passengers who arrive, at the main terminal of a building, for transportation to the upper floors over the worst 5 minute period expressed as a percentage of the total building population.

A lift system is thus expected to respond to the peak demand in such a way as to quickly and efficiently transport passengers to their destinations without excessive passenger waiting times occurring or unwieldy queues developing. This implies that the handling capacity of the lift system should be sufficient to carry all those passengers demanding service. So what is handling capacity ?

Definition 4.7: The handling capacity (*UPPHC*) of a lift system is the total number of passengers that it can transport in a period of 5 minutes during the uppeak traffic condition with a specified average car loading.

Examination of Figure 4.4 shows that the required instantaneous handling capacity varies from relatively low levels each side of the defined starting time to a high level prior to the starting time. Over a period of one hour the average arrival rate is low and can be handled by a small number of lifts. However, as soon as the arrival rate exceeds the one hour handling capacity, large queues build up and waiting times become excessive. Only when the arrival rate again falls below the one hour handling capacity can the queues reduce, and even then it will be some time before the queues disappear and the handling capacity again exceeds demand (see Figure 4.11). Thus it is not satisfactory to size a lift system to handle a one hour average rate of arrival. Conversely, high instantaneous demands obviously cannot be met except by a large and expensive system. Thus a compromise is necessary where intending passengers are required to wait a reasonable time for service during the peak demand periods.

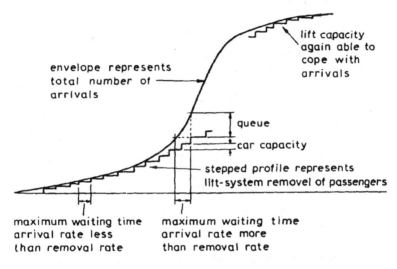

Figure 4.11 Passenger arrival rate and lift system handling capacity

A realistic approach then is to define handling capacity for a period of time less than one hour, but longer than a reasonable waiting time. A period of 5 minutes for the handling capacity definition has achieved general acceptance. From an analytical point of view it lies between one hour and a reasonable average waiting time, typically 30 s; allowing the smoothing out of short-term transients, but defining a period over which conditions remain reasonably fixed.

Thus if it is possible to equate the passenger demand as expressed by the 5 minute percentage peak arrival rate with the handling capacity of a lift system, then a suitable configuration will have been designed.

The question now arises of how to calculate handling capacity. This can be answered by considering how a single lift car services the incoming passengers during uppeak. The events are: the lift car comes to the ground floor, picks up passengers and then transports them to their destinations in the upper parts of the building; when all passengers have alighted and the car is empty it returns 'express' to the main arrival floor. The concepts of a round trip and round trip time emerge.

Definition 4.8: The round trip time (*RTT*) is the time in seconds for a single car trip around a building from the time the car doors open at the main terminal, until the doors reopen, when the car has returned to the main terminal floor, after its trip around the building.

A round trip time should not usually exceed two to three minutes (except in very tall buildings) as the majority of this time can represent the journey time for some passengers with destinations on the top floors of a building, which is undesirable.

Now if it is known how many round trips a single lift car can complete during the peak 5-minute period, then the uppeak handling capacity can be defined.

$$\text{number of round trips for a single car} = \frac{5\text{-minutes}}{RTT} = \frac{300}{RTT} \tag{4.1}$$

Therefore the 5-minute handling capacity (*UPPHC*) for a single car is:

$$UPPHC = \frac{300}{RTT} \times \text{average number of passengers per trip} \tag{4.2}$$

Equation (4.2) is one of the most important equations used in lift traffic design. Before it can be evaluated it is necessary to determine the number of passengers carried per trip.

As each car has a defined rated car capacity (*CC*) that it can accommodate, why not use this value ? Traditionally, and as a result of experience, the number of passengers assumed to be carried on each trip is taken as 80% of rated car capacity. This does not mean cars are assumed to fill only to 80% of rated car capacity each trip but that the average load is 80% of rated car capacity. The validity of this 80% diversity factor will be discussed later in Sections 5.2, 5.13, 6.6.3 and 7.8. Throughout this book the term "Car Loading" will sometimes be used in preference to "Capacity Factor" favoured by some authors (Peters, 1990).

Thus Equation (4.2) for a single car becomes:

$$UPPHC = \frac{300}{RTT} \times \frac{80}{100} \times CC \tag{4.3}$$

If the number of passengers carried is defined as *P* then Equation (4.3) becomes:

$$UPPHC = \frac{300}{RTT} \times P \tag{4.4}$$

In installations with more than one car, Equations (4.3) and (4.4) become:

$$UPPHC = \frac{300}{RTT} \times \frac{80}{100} \times CC \times L \tag{4.5}$$

$$UPPHC = \frac{300}{RTT} \times P \times L \tag{4.6}$$

where *L* is the number of lift cars.

The handling capacity of a lift installation indicates the *quantity of service* a lift system can provide. Passengers are concerned also with *quality of service,* which is how long they must wait. To some extent the frequency of lift arrivals at the main terminal floor gives an indication of quality. With a single car the interval between successive arrivals is the round trip time. However, where a lift system contains more than one car the interval becomes:

$$\text{interval} = \frac{RTT}{L} \tag{4.7}$$

Figure 4.12 illustrates the relationships between round trip and interval.

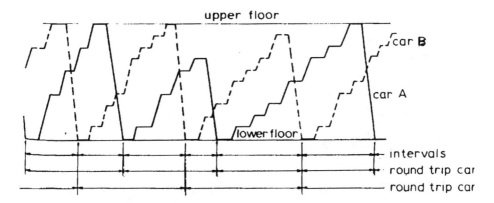

Figure 4.12 Relationship between round trip time and interval

Definition 4.9: Interval (*INT*) is the average time between successive lift car arrivals at the main terminal floor with cars loaded to any level.

It should be noted that sometimes the term waiting interval is used instead of interval. This is an attempt to create the idea that the inter-arrival period between cars can define the waiting time of a passenger. To this end it is confusing.

Definition 4.10: Uppeak interval (*UPPINT*) is the average time between successive lift car arrivals at the main terminal floor with cars loaded to 80% of rated car capacity during uppeak traffic conditions.

Equations (4.5) and (4.6) can now be rearranged, using Equation (4.7), to give the handling capacity of a group of *L* lifts:

$$UPPHC = \frac{300}{UPPINT} \times P = \frac{300}{UPPINT} \times \frac{80}{100} \times CC \qquad (4.8)$$

Another useful operating performance parameter is to determine the percentage of a building's population handled in the peak 5-minutes. This is called percentage population served.

Definition 4.11: The population served in the uppeak 5-minutes is defined as the ratio of the uppeak handling capacity and the building population given as a percentage.

$$\%POP = \frac{UPPHC}{\text{building population}} \times 100 \qquad (4.9)$$

As measures of a lift system's operational performance the three parameters, uppeak handling capacity, uppeak interval and uppeak percentage population served, are the parameters most often quoted by a lift supplier.

Example 4.1

A building is served by three lifts with a round trip time of 150 s. The building population is 400 persons and each car has a rated car capacity of 10 passengers. Calculate the uppeak interval, uppeak handling capacity and percentage population served.

From Equation (4.7):

$$\text{interval} = \frac{150}{3} = 50 \text{ s}$$

From Equation (4.8):

$$UPPHC = \frac{300}{50} \times \frac{80}{100} \times 10 = 48 \text{ persons/5 minutes}$$

From Equation (4.9):

$$\%POP = \frac{48}{400} \times 100 = 12\%$$

4.6 DERIVATION OF THE ROUND TRIP TIME OF A SINGLE CAR

Consider the way in which a single lift car circulates around a building during the uppeak traffic condition. The car opens its doors at the main terminal floor and passengers board the car; the doors close. The car then runs to the first stopping floor going through periods of acceleration, travelling at rated speed, deceleration and levelling. (Travel at rated speed may not occur if the interfloor distance is too small.) At the first stopping floor, the doors open and one or more passengers alight; the doors close. This sequence continues until the highest stopping floor is reached. After the doors have closed, the car is considered to make an express run to the main terminal, thus completing the round trip. This description indicates, and Figure 4.13 illustrates, that a round trip consists of a number of elements.

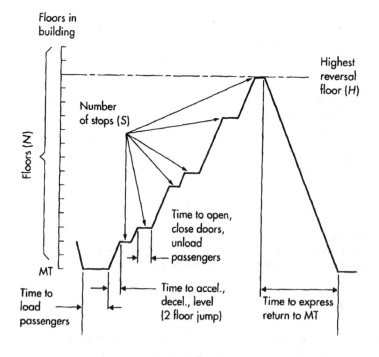

Figure 4.13 The elements of a round trip time

The elements are:

(a) passenger loading time t_l

(b) passenger unloading time t_u

(c) door closing and opening times t_c and t_o

(d) interfloor jump time (for a single floor assuming rated speed reached) $t_f(1)$

(e) time to travel in the upward direction at rated speed for jumps greater than a single floor t_a

(f) time to travel from the highest floor to the main terminal t_e.

Items (a), (b), and (c) may be considered to be standing times and (d), (e) and (f) as running times. (See Table 4.1 for full definitions of the time parameters.)

The travel sequence of the lift is much affected by the average number of stops made (S), the average highest floor reached (H) and the aveage number of passengers (P) carried. It is now possible to deduce an expression for round trip time as:

$$RTT = \quad Pt_l + Pt_u \qquad + (S+1)(t_c+t_o) \qquad + (S+1)t_f(1)$$

passenger transfer time   door operating time   time to accelerate, decelerate, level, etc.

$$+ (H-S)t_v \qquad\qquad + (H-1)t_v \qquad\qquad\qquad (4.10)$$

time to travel remaining floors   time to express to main terminal floor

Refer to Table 4.1 for the definitions of the parameters. The term (S+1) occurs to account for the stop at the main terminal floor.

Combining and simplifying:

$$RTT = 2Ht_v + (S+1)t_s + 2Pt_p \qquad\qquad\qquad (4.11)$$

The stopping t_s is an artificial time developed as a mathematical simplification and cannot be measured directly. It is:

t_s = door operating times (t_d) + single floor flight time ($t_f(1)$)

   − single floor transit time (t_v)

$$t_d + t_f(1) - t_v \qquad\qquad\qquad (4.12)$$

Another time, the performance time (T), can be more easily measured and is very useful in determining the performance of a lift.

T = door operating times (t_d) + single floor flight time ($t_f(1)$)

$$= t_d + t_f(1) \qquad\qquad\qquad (4.13)$$

which gives:

$$t_s \quad = T - t_v \qquad\qquad\qquad (4.14)$$

Equation (4.11) can be modified to show the time parameters in more detail by including Definition 4.21 and Equation (4.14), but at a loss of symmetry, ie:

$$RTT = 2Hd_f/v + (S+1)(T-t_v) + 2Pt_p \qquad (4.11\text{bis})$$

The expressions for round trip time presented here may differ from expressions derived by others authors owing to the manner in which mathematical simplifications have been made and because of the way in which the various parameters have been defined. Equation (4.11), however, is neat, symmetrical, simple and easy to remember.

Example 4.2

A lift system comprising 4 cars of rated speed 1.6 m/s and rated car capacity of 10 persons have door opening times of 3.0 s and door closing times of 4.0 s. The flight time between adjacent floors of interfloor distance 3.5 m is 4.5 s. Assuming passengers can enter/exit at 1.2 s (average time), calculate the round trip time. Assume that the highest floor reached is 10 and the number of stops is 9. Then consider the effect on the lift system, if the speed is increased to 2.5 m/s. What are the changes to the three elements of the round trip equation ?

Considering the slower rated speed. Data:

$t_v = 3.5/1.6 = 2.2$ s $\qquad$ [see Equation (4.10)]
$t_s = 3.0 + 4.0 + 4.5 - 2.2 = 9.3$ s $\quad$ [Equation (4.12)]
$t_p = 1.2$ s $\qquad$ [given]

Using Equation (4.11):

$$RTT = 2 \times 10 \times 2.2 + (9 + 1)9.3 + 2 \times 10 \times 0.8 \times 1.2$$

$$= 44 + 93 + 19.2 = 156.2 \text{ s}$$

Considering the higher rated speed. Data:

$t_v = 3.5/2.5 = 1.4$ s
$t_s = 3.0 + 4.0 + 4.5 - 1.4 = 10.1$ s
$t_p = 1.2$ s

Again using Equation (4.11):

$$RTT = 2 \times 10 \times 1.4 + (9 + 1)10.1 + 2 \times 10 \times 0.8 \times 1.2$$

$$= 28 + 101 + 19.2 = 148.2 \text{ s}$$

The first term gets smaller and the middle term gets larger and the round trip time is 5% smaller. There was not a significant difference.

To properly evaluate Equation (4.11), values for *S*, *H* and *P* must be determined, they were assumed in this example. Their derivation will be dealt with in the next chapter. Sometimes it is suggested that a faster lift will provide better service. Example 4.2 demonstrates this is not always so as there was only a small change.

Table 4.1 Further definitions of terms

No.	Time period	Symbol	Description
4.12	Passenger loading time	t_l	The average time for a single passenger to enter a car (boarding time, entry time)
4.13	Passenger unloading time	t_u	The average time for a single passenger to leave a car (alighting time, exit time)
4.14	Passenger transfer time	t_p	The average time for a single passenger to enter or leave a car, ie: $t_p = t_l + t_u/2$
4.15	Door closing time	t_c	A period of time measured from the instant that the car doors start to close until the doors are locked
4.16	Door opening time	t_o	A period of time measured from the instant that the car doors start to open until they are open 800 mm
4.17	Door operating time	t_d	The sum of the door opening and closing times, ie: $t_d = t_c + t_o$
4.18	Car call dwell time	t_{cd}	The period of time that the car doors remain open at a stop in response to a car call, provided no passengers cross the threshold
4.19	Landing call dwell time	t_{ld}	The period of time that the car doors remain open at a stop in response to a landing call, provided no passengers cross the threshold
4.20	Single floor flight time	$t_f(1)$	The period of time measured from the instant that the car doors are locked until the lift is level at the next adjacent floor
	Multi-floor flight time for a jump of n floors	$t_f(n)$	The period of time measured from the instant that the car doors are locked until the lift is level at the nth adjacent floor
4.21	Single floor transit time	t_v	The period of time for a lift to travel past two adjacent floors at rated speed, ie: $t_v = d_f/v$ where d_f is the interfloor distance and v is the rated speed
4.22	Stopping time	t_s	A composite time associated with each stop, ie: $t_s = t_f(1) + t_c + t_o - t_v$
4.23	Performance time	T	The period of time between the instant the car doors start to close and the instant that the car doors are open 800 mm at the next adjacent floor
4.24	Cycle time	t_{cyc}	The period of time between the instant the car doors begin to close until the instant that the car doors begin to close again at the next adjacent floor provided no passengers have crossed the threshold

CHAPTER FIVE

Evaluating the Round Trip Time Equation

5.1 DATA SETS

To calculate the performance of a lift system, Equation (4.11) for round trip time must be evaluated. This equation depends on three data sets concerning the building, the lift system and the passengers.

5.1.1 Building Data Set

The building data set comprises the following:

(a) number of floors
(b) interfloor distance.
(c) express jump

Values for items (a), (b) and (c) are generally decided early in a building's design or are already fixed in an existing building. Sometimes the lift designer will be able to suggest the number of floors in each building zone, in order to design a more effective lift installation. Item (c), the express jump, may be nothing more than a few extra metres between the main terminal floor and the first served floor or could be some distance if the first served floor above the lobby is an express zone terminal floor high in the building. More often than not, items (a), (b) and (c) will literally be "set in stone".

5.1.2 Lift System Data Set

The lift system data set comprises the following:

(a) number of cars
(b) rated car capacity
(c) rated speed
(d) flight times between floors
(e) door opening times
(f) door closing times
(g) miscellaneous times (such as car and landing door dwell times, etc.)
(h) traffic control system.

Items (a) and (b) are design variables to meet the required passenger traffic demands and the quality of service. Items (c) to (g) can be established from the lift manufacturer. The effect of (h) cannot always be analysed in precise mathematical terms, but may be assessed empirically.

5.1.3 Passenger Data Set

The passenger data set comprises the following:

(a) number of passengers boarding from specific floors
(b) number of passengers alighting at specific floors
(c) traffic mode, ie: unidirectional or multidirectional, etc.
(d) transfer times for passengers entering and leaving cars
(e) passenger actions.

Items (a) and (b) are dependent on floor populations. During the uppeak traffic condition, item (a) determines the 5-minute handling capacity when all passengers board at one terminal floor and, similarly, item (b) defines the activity during down peak traffic condition. It is these two items which determine the level of duty for a lift system ie: the number of starts an hour it will be required to make. Items (c) and (d) are dependent on human behaviour and are not easily predictable. Item (e) is included to cover passenger misbehaviour (door holding, excessive operation of pushbuttons, etc.).

This data set is the least well defined and is subject to considerable error in its estimation.

5.1.4 Numerical Values

The round trip time equation, Equation (4.11), is:

$$RTT = 2Ht_v + (S+1)t_s + 2Pt_p \tag{4.11}$$

It will be noted that the equation comprises three parts. Each part comprises a time independent variable (H, S, P) and a time dependant variable (t_v, t_s, t_p). It is these six parameters which must be evaluated in order to determine the round trip time equation. Each variable will now be taken in turn in the following sections and typical values indicated. Installed systems may have different values.

5.2 DETERMINATION OF P

To evaluate Equation (4.11) a value for the average number of passengers (P) carried during each trip needs to be determined. If passengers arrived efficiently then as each lift arrived there would be a number of passengers waiting (P) equal to the rated car capacity (CC) of the lift ready to board. Unfortunately life is not like that and passengers arrive in a random fashion.

To avoid passengers being left behind (a queue) to wait for the next lift, it is necessary to assume a lower than 100% utilisation factor for car occupancy. This assumption arises as statistical theory implies that, as the utilisation of a facility increases towards its maximum, the probability of immediate use of that facility reduces. Thus to achieve maximum utilisation of a facility it is necessary to have a queue of applicants waiting (like an airport). This is not considered satisfactory for a lift system. Therefore the design utilisation has to be lower to allow for statistical variations to be accommodated. How much lower than 100% should P be set below CC ?

The probability of the immediate use of a facility is shown diagrammatically in Figure 5.1 with respect to system utilisation. As system utilisation increases, then the probability of a passenger being left behind increases, until at 100% utilisation there is a high probability of being left behind to queue. The shape of the curve has been shown to apply to such diverse facilities, access to a telephone line, availability of a lavatory, a free bank teller, etc.

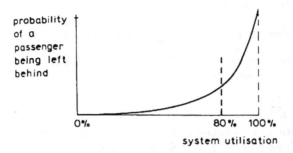

Figure 5.1 System utilisation

Looking at the curve of Figure 5.1 again, it can be seen that above 70% utilisation the change in slope increases significantly and at 90% and above it is very rapidly increasing. The ratio of the slope at 90% compared to the slope at 50% is some 25:1. Usually the 80% point is considered to be the "knee" of the curve for most system utilisation judgements and this value is selected for lifts also. Therefore, for lift systems it is reasonable to consider:

$$P = 0.8 \times CC \tag{5.1}$$

Early explanations justified the 80% value by saying that passengers never loaded a car above 80% car capacity even when queues existed, thus showing remarkable restraint. Other theories to explain the 80% figure proposed are either: circulation difficulties (passengers at the back of the car always want to get out at the first stop); or operational problems (passengers obstructing door closing), both of which have the effect of increasing the round trip time. The 80% derating factor appears to have been arrived at by intuition and experience, rather than theory.

5.3 DETERMINATION OF S

During a round trip a lift car stops at a number of floors for passengers to alight. The round trip time is affected by the number of stops made. Each trip will be different, but what is the average number of stops that can be expected? The term "probable number of stops" has been avoided as statisticians understand[1] a parameter labelled "probable"

---

[1] It is not sensible to proceed further without defining some statistical terms. The *probability* of a variable is defined by a numeric value between zero and unity. Probability then is simply a measure of the likelihood of some occurrence. For example, the probability of stopping at a floor may be 0.1. However, the actual number of stops is called the *expected* number of stops. The expectation is the *mean* or *average* value of some random variable, often termed the *expectance*. It is often of interest to know the spread of possible outcomes about the mean. This variation is essentially measured by the statistical characteristic termed *variance*. If the square root of a value of variance is taken, the value obtained is termed *standard deviation*. If large values of variance occur with respect to values obtained for expectance of some variable, then that variable has wide deviations of value from its mean. Statistical analysis and probability theory are highly mathematical fields and are best dealt with in specialised texts.

will yield a number between zero and unity. Thus the more acceptable term which will be used is *expected number of stops*. Basset Jones in 1923 published a method of calculating the expected number of stops for floors with equal populations. Consider a building with N floors above the main terminal. Assume that each floor is equally likely as a destination for passengers.

The probability that one passenger will leave the lift at any particular floor is $1/N$.

The probability that one passenger will NOT leave the lift at any particular floor is the complement of this probability, *viz*:

$$1 - \frac{1}{N} = \frac{N-1}{N} \tag{5.2}$$

Since each passenger is assumed to be independent of all others, the product law of probability gives the probability that NO passengers from a lift containing P passengers will leave the lift at any particular floor as:

$$\left[\frac{N-1}{N} \right] \times \left[\frac{N-1}{N} \right] \cdots \left[\frac{N-1}{N} \right] = \left[\frac{N-1}{N} \right]^P \tag{5.3}$$

Note there are P terms. Hence the probability that a stop will be made at any particular floor is:

$$1 - \left[\frac{N-1}{N} \right]^P \tag{5.4}$$

The expected or average number of stops (S) for N floors will then be:

$$S = N \left[1 - \left[\frac{N-1}{N} \right]^P \right] \tag{5.5}$$

Values for S can be calculated each time using the formulae. As there are a finite set of rated capacities offered by lift manufacturers, eg: those indicated in BS EN81 and BS ISO4190, it is possible to tabulate these values as shown in Table 5.1.

5.4 DETERMINATION OF H

Some design procedures assume H to be N, or in tall buildings, $N-1$. Arbitrary rules are offered (Strakosch, 1998[1]). It is possible, however, to deduce an expression for H with respect to N and P using probability theory. It is not clear who first derived an expression for H or when, although Schroeder (1955) writing in German could have been the first. Using the same assumptions and definitions as in Section 5.3, assume a passenger is equally likely to travel to any floor.

The probability that one passenger will leave the lift at any given floor is $1/N$ and so the probability that one passenger will NOT leave the car at a given floor is:

$$1 - \frac{1}{N} \tag{5.6}$$

---

[1] Page 82.

The probability that none of the P passengers will leave the car at a given floor is:

$$\left(1-\frac{1}{N}\right)^P \tag{5.7}$$

The probability of the car travelling no higher than the ith floor is equal to the probability that no one leaves the lift at the Nth, $(N-1)$th, $(N-2)$th, ... and $(i+1)$th, floors is:

$$\left[1-\frac{1}{N}\right]^P \left[1-\frac{1}{N-1}\right]^P \left[1-\frac{1}{N-2}\right]^P \left[1-\frac{1}{N-3}\right]^P \cdots \left[1-\frac{1}{i+1}\right]^P \tag{5.8}$$

Expanding and simplifying produces:

$$\left[\frac{i}{N}\right]^P \tag{5.9}$$

It is now possible to propose that the {probability that i is the highest floor attained} is equal to the {probability that a lift travels no higher than the ith floor} minus the {probability that the lift travels no higher than the $(i-1)$th floor}, *viz*:

$$= \left[\frac{i}{N}\right]^P - \left[\frac{i-1}{N}\right]^P \tag{5.10}$$

Then the average (or mean) highest floor H is:

$$H = \sum_{i=1}^{N-1} i \left[\left[\frac{i}{N}\right]^P - \left[\frac{i-1}{N}\right]^P \right] \tag{5.11}$$

Expanding and simplifying, the expected or average highest reversal floor (H) for N floors will then be:

$$H = N - \sum_{i=1}^{N-1} \left[\frac{i}{N}\right]^P \tag{5.12}$$

Values for H can be calculated each time using the formulae. As there are a finite set of rated capacities offered by lift manufacturers, eg: those indicated in BS EN81 and BS ISO4190, it is possible to tabulate these values as in Table 5.1. This table shows that H is indeed approximately equal to N, where a large capacity lift is installed, but grossly in error, where a small capacity lift is installed, especially where many floors are served (see Section 7.3.3 for discussion).

5.5 EXAMPLE 5.1

Calculate the round trip time using the data below and Equation (4.11).

$N = 10$, $d_f = 3.0$ m, $CC = 8$ persons, $v = 1.5$ m/s, $t_f(1) = 5.0$ s, $t_d = 4.5$ s, $t_p = 1.2$ s.

The data above gives:

$t_v = 3.0/1.5 = 2.0$ s, $t_s = 5.0 + 4.5 - 2.0 = 7.5$ s, $t_p = 1.2$ s, $P = 8 \times 0.8 = 6.4$ persons.

Table 5.1 Values of H and S for EN81 rated capacities *

Floors	6 (4.8)		8 (6.4)		10(8.0)		13(10.4)		16(12.8)		21(16.8)		26(20.8)		33 (26.4)	
N	H	S	H	S	H	S	H	S	H	S	H	S	H	S	H	S
5	4.6	3.3	4.7	3.8	4.8	4.2	4.9	4.5	4.9	4.7	5.0	4.9	5.0	5.0	5.0	5.0
6	5.4	3.5	5.6	4.1	5.7	4.6	5.8	5.1	5.9	5.4	6.0	5.7	6.0	5.9	6.0	6.0
7	6.2	3.7	6.5	4.4	6.6	5.0	6.8	5.6	6.8	6.0	6.9	6.5	7.0	6.7	7.0	6.9
8	7.1	3.8	7.4	4.6	7.5	5.3	7.7	6.0	7.8	6.6	7.9	7.2	7.9	7.5	8.0	7.8
9	7.9	3.9	8.2	4.8	8.4	5.5	8.6	6.4	8.7	7.0	8.8	7.8	8.9	8.2	9.0	8.6
10	8.7	4.0	9.1	4.9	9.3	5.7	9.5	6.7	9.7	7.4	9.8	8.3	9.9	8.9	9.9	9.4
11	9.6	4.0	10.0	5.0	10.2	5.9	10.5	6.9	10.6	7.8	10.8	8.8	10.8	9.5	10.9	10.1
12	10.4	4.1	10.8	5.1	11.1	6.0	11.4	7.1	11.5	8.1	11.7	9.2	11.8	10.0	11.9	10.8
13	11.2	4.1	11.7	5.2	12.0	6.1	12.3	7.3	12.5	8.3	12.7	9.6	12.8	10.5	12.9	11.4
14	12.1	4.2	12.6	5.3	12.9	6.3	13.2	7.5	13.4	8.6	13.6	10.0	13.7	11.0	13.8	12.0
15	12.9	4.2	13.4	5.4	13.8	6.4	14.1	7.7	14.3	8.8	14.6	10.3	14.7	11.4	14.8	12.6
16	13.7	4.3	14.3	5.4	14.7	6.5	15.0	7.8	15.3	9.0	15.5	10.6	15.7	11.8	15.8	13.1
17	14.5	4.3	15.3	5.5	15.6	6.5	16.0	8.0	16.2	9.2	16.5	10.9	16.6	12.2	16.8	13.6
18	15.4	4.3	16.0	5.5	16.6	6.6	16.9	8.1	17.1	9.3	17.4	11.1	17.6	12.5	17.7	14.0
19	16.2	4.3	16.9	5.6	17.4	6.7	17.8	8.2	18.1	9.5	18.4	11.3	18.5	12.8	18.7	14.4
20	17.0	4.4	17.8	5.6	18.2	6.7	18.7	8.3	19.0	9.6	19.3	11.6	19.5	13.1	19.7	14.8
21	17.9	4.4	18.6	5.6	19.1	6.8	19.6	8.4	19.9	9.8	20.3	11.7	20.5	13.4	20.6	15.2
22	18.7	4.4	19.5	5.7	20.0	6.8	20.5	8.4	20.9	9.9	21.2	11.9	21.4	13.6	21.6	15.6
23	19.5	4.4	20.4	5.7	20.9	6.9	21.4	8.5	21.8	10.0	22.1	12.1	22.4	13.9	22.6	15.9
24	20.3	4.4	21.2	5.7	21.8	6.9	22.4	8.6	22.7	10.1	23.1	12.3	23.3	14.1	23.5	16.2

* 80% capacity shown in parentheses. N is the number of floors above the main terminal.

To determine S, use Equation (5.5):

$$S = 10 \left[1 - \left[\frac{10-1}{10} \right]^{6.4} \right] = 10(1 - 0.51) = 4.9$$

To determine H, use Equation (5.12):

$$H = 10 - \left[\frac{1}{10} \right]^{6.4} + \left[\frac{2}{10} \right]^{6.4} + \left[\frac{3}{10} \right]^{6.4} + \left[\frac{4}{10} \right]^{6.4} + \left[\frac{5}{10} \right]^{6.4} + \left[\frac{6}{10} \right]^{6.4}$$

$$+ \left[\frac{7}{10} \right]^{6.4} + \left[\frac{8}{10} \right]^{6.4} + \left[\frac{9}{10} \right]^{6.4}$$

$$= 10 - (0.0000004 + 0.00003 + 0.0006 + 0.0028 + 0.012 + 0.049$$

$$+ 0.095 + 0.24 + 0.51)$$

$$= 10 - 0.91 = 9.09$$

Using Equation (4.11):

$$RTT = 2 \times 9.09 \times 2.0 + (4.9 + 1) \times 7.5 + 2 \times 6.4 \times 1.2$$

$$= 36.4 + 44.2 + 15.4 = 96 \text{ s.}$$

Compare the calculated values of S and H to those given in Table 5.1.

5.6 EFFECT OF PASSENGER ARRIVAL PROCESS

5.6.1 General

The derivation of the formulae for H and S assumed that passengers arrived at a constant inter-arrival interval (according to the particular level of arrivals existing) and that the lift arrival at a constant interval to take the intending passengers to their destinations. This effect is illustrated in Figure 5.2(a) where there is a lift system with a rated car capacity of 6 persons and an interval of 20 s; and 6 passengers arrive every 20 s.

No account has been taken of the way in which passengers arrive in a building or the randomness of their destinations. In practice passengers do not conveniently arrive in batches equal to 80% of the rated car capacity nor do they register the same number of destinations during each trip. The effect of this randomness is to cause the lifts to take different times to carry out a round trip and they become unevenly spaced. This effect is called bunching. (Buses in the street seem to do this. Although they may be on a 20 minute timetabled frequency they only appear every hour in threes !). This effect is illustrated in Figure 5.2(b) for the same conditions as Figure 5.2(a). Note the overall passenger average waiting times have increased and queues develop.

Note that these "snapshots" of lift behaviour are a very simple minded representation of an extremely complex process and does not bear close scrutiny. This is why it is best to set up statistical models to represent the process and then to draw general and averaged conclusions from them.

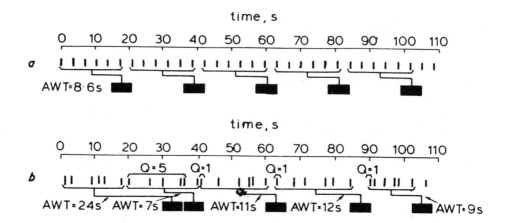

Figure 5.2 A simple representation of passenger arrival and lift car departures
(a)   Constant passenger arrivals: constant lift departures.
      Overall average passenger waiting time 8.6 s.
(b)   Random passenger arrivals: irregular lift departures.
      Overall average passenger waiting time 12.6 s.

So if passengers do not obey the constant (sometimes called rectangular) arrival process used to derive Equations (5.5) and (5.12) what process do they obey ?

It is generally accepted that passengers arrive into a lift system according to the Poisson probability process. This probability distribution function has been used to describe other phenomena such as: the generation of radioactive particles and telephone calls; failures of electronic equipment; and the demands on digital computer central servers. Although this arrival process is not proven with respect to lift systems, work by Alexandris (1977) did go some way to confirm it. Using observers, Alexandris surveyed three buildings with widely differing lift and other physical characteristics. He came to certain conclusions:

(a)   Comparison of the observed and theoretical values calculated for the mean and variance showed a Poisson fit to be reasonable.
(b)   The chi-squared goodness-of-fit tests gave evidence that a Poisson arrival rate assumption at least cannot be rejected.
(c)   Although there may be other theoretical distributions which might better accommodate the data, the Poisson distribution must be considered as a good approximation to the actual empirical distribution.

5.6.2 Formulae for S and H using the Poisson probability distribution function

Accepting the Poisson probability distribution function (pdf) as the best representation of the passenger arrival process, what effect does this have on the evaluation of the round trip time equation ? Assume that the probability of n calls being registered in the time interval T for an average rate of arrivals λ (in calls per second) is:

$$p_r(n) = \frac{(\lambda T)^n}{n!} \times e^{-\lambda T}$$

$$(5.13)$$

Tregenza (1972) used this relationship to derive formulae for H and S (shown below with subscript, p). For a building with equal floor populations, he defined p_{r0} as the probability of no calls being registered from the main terminal to any floor above during the period of one interval (T):

$$P_{r0} = e^{-\frac{\lambda}{N}T} \tag{5.14}$$

Then by the same arguments used in developing Equations (5.5) and (5.12):

$$S_p = N(1-p_{r0}) = N\left(1-e^{-\frac{\lambda}{N}T}\right) \tag{5.15}$$

$$H_p = N - \sum_{i=1}^{N} p_{r0}i = N - \sum_{i=1}^{N}\left(e^{-\frac{\lambda}{N}T}\right)^i \tag{5.16}$$

The arrivals and departures are equal and can be defined as:

$$P_p = \lambda T \tag{5.17}$$

5.6.3 Comparison between the Rectangular and Poisson pdfs

The expressions for H_p and S_p obtained for the Poisson pdf can be shown to be almost identical to the expressions obtained for the rectangular probability distribution function.

For S this is achieved by substituting Equation (5.17) into Equation (5.15) to give:

$$S_p = N\left(1-e^{-\frac{P}{N}}\right) = N\left[1-\left(e^{-\frac{1}{N}}\right)^P\right] \tag{5.18}$$

Expanding the exponential gives:

$$e^{-\frac{1}{N}} = 1 - \frac{1}{N} + \frac{1}{2!\,N^2} - \frac{1}{2!\,N^3} \quad \text{etc.} \tag{5.19}$$

Taking the first two terms ONLY of Equation (5.19) back into Equation (5.18) produces an equivalent equation to Equation (5.5).

If the third and subsequent terms are included, the Poisson probability values using Equation (5.15) for S become smaller than the values obtained using the rectangular probability values from Equation (5.5).

For H this is achieved by substituting Equation (5.17) into Equation (5.16) and transposing the indices, *viz*:

$$H_p = N - \sum_{i=1}^{N}\left(e^{-\frac{P}{N}}\right)^i$$

$$= N - \sum_{i=1}^{N}\left(e^{-\frac{i}{N}}\right)^P \tag{5.20}$$

Expanding the exponential by series and taking the first two terms only produces:

$$\sum_{i=1}^{N}\left(e^{-\frac{1}{N}}\right)^{P} = \left[1 - \frac{1}{N}\right]^{P} + \left[1 - \frac{2}{N}\right]^{P} + \left[1 - \frac{3}{N}\right]^{P}$$
$$+ \left[1 - \frac{N-2}{N}\right]^{P} + \left[1 - \frac{N-1}{N}\right]^{P}$$

(5.21)

Equation (5.21) is identical to the expansion of the summation part of Equation (5.12). Thus as the remainder of Equations (5.12) and (5.20) are identical, then the values obtained for H will also be equal. If the third and subsequent terms are included in the expansion, the Poisson probability values obtained using Equation (5.16) for H become smaller than the rectangular probability values obtained from Equation (5.12).

5.6.4 Example 5.2

Calculate values for S and H using the formulae for the rectangular and Poisson pdfs for the following lift systems.

$N = 5;\qquad CC = 6$
$N = 24;\qquad CC = 6$
$N = 5;\qquad CC = 20$
$N = 24;\qquad CC = 20$

$N = 5;\qquad CC = 6:\qquad S = 3.3;\qquad H = 4.6;\qquad S_p = 3.1;\qquad H_p = 4.4.$
$N = 24;\qquad CC = 6:\qquad S = 4.4;\qquad H = 20.3;\qquad S_p = 4.4;\qquad H_p = 19.5.$
$N = 5;\qquad CC = 20:\qquad S = 4.9;\qquad H = 5.0;\qquad S_p = 4.8;\qquad H_p = 5.0.$
$N = 24;\qquad CC = 20:\qquad S = 11.9;\qquad H = 23.0;\qquad S_p = 11.7;\qquad H_p = 22.9.$

It should be noted that the Poisson values are always smaller or identical and the differences will not be significant in practice.

5.7 SINGLE FLOOR TRANSIT TIME t_v

The parameter t_v requires the interfloor distance (d_f) and the rated speed (v) to be known.

5.7.1 Interfloor Distance

This is the time that a car takes to travel past two adjacent floors at rated speed and is defined by Definition 4.21 as the average interfloor distance divided by the rated speed.

The average interfloor distance d_f is normally determined as the total travel to the highest served floor divided by the number of possible stopping floors above the main terminal. Domestic dwellings average about 3.0 m per floor and commercial buildings range from 3.0 m to 3.3 m for older buildings to 3.6 m to 4.2 m (or greater) for modern buildings. In the latter case the increased floor to floor distance is required to accommodate other services (eg: air conditioning, electrical supplies) and various modern technological services (eg: computer networks, telecommunications).

Commercial buildings often introduce a mixed floor pitch for a number of reasons:

■ Some floors have increased heights, such as lobby/main terminal floors, service floors, special floors (eg: those containing a restaurant, lecture room, conference room, VIP suite, etc.).

■ Some floors are sometimes unavailable for alighting during periods of the day, such as the first floor (and sometimes the second floor) above the main terminal, service floors, security floors, etc.

It is recommended that an average floor height be assumed, and the irregularities be dealt with separately, as discussed in Chapter 7.

Where a lift is serving a set of floors or zone in a building, which are not adjacent to the main terminal, an extra time to make the jump to or from the express zone must be added to Equation (4.11), ie: $2t_e$, where t_e is the time the lift takes to travel (without stopping) from the main terminal to the express zone terminal.

$$RTT = 2Ht_v + (S+1)t_s + 2Pt_p + 2t_e \qquad (5.22)$$

The long "flight time" t_e can be found using the equations in Appendix 1.

5.7.2 Rated Speed

The value of the rated speed (v) is usually supplied by the lift maker, who will select it to meet various engineering requirements (i.e. gearing, drive controllers, product line considerations, etc.) and traffic purposes. For instance, goods lifts are generally slower than passenger lifts. Speed, however, is not a dominant factor in Equation (4.11), as illustrated by Example 6.1. It does become significant if the served floors are in an upper zone, where a higher speed will permit the unserved zone to be more rapidly traversed. If a value for v is not provided it must be chosen by the traffic designer.

The appropriate value for rated speed could be that recommended in the British, European and International Standard Codes of Practice, taken together with experienced judgement. In general, the higher the building rise the faster the rated speed used. Often in a zoned building the rise from an express zone terminal may be small eg: 10 floors, but the express jump from the lower terminal to the express zone terminal may be large. It is this express jump, which largely determines the rated speed, that allows journey times to be kept at reasonable values.

Fire codes can determine a minimum value by requiring that it shall be possible to travel to the highest floor in the building from the fire control entrance level in (say) 60 s. Clearly this is not possible in very tall buildings and special arrangements must be made in these circumstances.

BS 5655: Part 6: 1990 recommends rated speeds in relation to total travel according to building usage. This can be translated into the time to travel at rated speed (without allowance for acceleration, deceleration or levelling) between the highest and lowest floors (the terminal floors), as shown in Table 5.2. This time is sometimes called the nominal travel time.

ISO 4190-6 recommends a maximum (theoretical) time of transit of between 20 s and 40 s to travel, at the rated speed, a distance equal to the total travel of the lift. The time is graded according to the likely interval at the main floor.

There is no theoretical upper limit to lift rated speed and it does not, for example, affect passenger comfort. However, it is limited by practical factors, such as the maximum sheave diameter, rope bending radius, rope wear, safety (eg: overtravel), etc.

Table 5.2 Total time required to travel between
terminal floors in different types of building

Building type	Transit time (s)
Offices	
- large	17 – 20
- small	20
Hotels	
- large	17 – 20
- small	20
Hospitals	24
Nursing and residential homes	24
Residential buildings	20 – 30
Factories and warehouses	24 – 40
Shops	24 – 40

Table 5.3 provides guidance on the selection of the speed of a lift based on the premise that the total time to travel the distance between terminal floors at rated speed should take between 20 s and 30 s. In the table the single floor flight times assume a 3.3 m interfloor distance and are slightly larger than theoretically derived values to allow for the doors to be locked and proved, the brake to lift and other start-up delays. The range of values given for acceleration is typical of those found on installed installations. Some installations nowadays limit the acceleration to about 1.2 m/s^2 in order to provide a good ride quality. This will increase the single floor flight time and the eventual handling capacity.

Table 5.3 Typical lift dynamics

Lift travel (m)	Rated speed (m/s)	Acceleration (m/s^2)	Single floor flight time (s)
<20	<1.00	0.4	10.0
20	1.00	0.4 – 0.7	7.0
32	1.60	0.7 – 0.8	6.0
50	2.50	0.8 – 0.9	5.5
63	3.15	1.0	5.0
100	5.00	1.2 – 1.5	4.5
120	6.00	1.5	4.3
>120	>6.00	1.5	4.3

The figures given in Table 5.3 apply principally to commercial buildings; speeds in residential and institutional buildings may be subject to other design regulations, and similar height and similar function buildings can be installed with a wide range of equipment, eg: old persons' homes compared to prestige flats.

5.7.3 Example 5.3

Two tenders have been received for the provision of a lift system of 10 person rated car capacity in a 15-storey office block having an interfloor height of 3.3 m. Assume a passenger transfer time of 1.2 s. Compare the two tenders. Table 5.4 gives the received tender information and Table 5.5 gives data deduced from the given data.

Table 5.4 Tender information

Parameter	Tender A	Tender B
Rated speed (m/s)	1.6	2.5
Door opening time (s)	1.0	3.0
Door closing time (s)	3.0	3.5
Flight time (s)	6.0	5.5

Table 5.5 Deduced data for Example 5.3

Parameter	Tender A	Tender B	Deduced from
Average number passengers (P)	8.0	8.0	$CC = 10$
Average highest floor (H)	13.8	13.8	Table 5.1
Average number of stops (S)	6.4	6.4	Table 5.1
Single floor transit time (t_v) (s)	2.06	1.32	Definition 4.21
Stopping time (t_s) (s)	7.94	10.68	Definition 4.22
Passenger transfer time (t_p) (s)	1.2	1.2	Design brief
Performance time (T) (s)	10.0	12.0	Definition 4.23

Calculation of the round trip time for Tender A is as follows:

$$RTT = (2 \times 13.8 \times 2.06) + 7.4(6.0 + 1.0 + 3.0 - 2.06) + (2 \times 8 \times 1.2)$$
$$= 56.9 + 58.8 + 19.2$$
$$= 134.9 \text{ s}$$

Calculation of the round trip time for Tender B is as follows:

$$RTT = (2 \times 13.8 \times 1.32) + 7.4(5.5 + 3.0 + 3.5 - 1.32) + (2 \times 8 \times 1.2)$$
$$= 36.4 + 79.0 + 19.2$$
$$= 134.6 \text{ s}$$

Although the speed was different between Tender A and Tender B, other changes altered the values obtained from each of the component parts of the round trip time equation. The result was very similar traffic handling systems.

5.8 TIME CONSUMED WHEN STOPPING (t_s)

The parameter t_s involves evaluation of flight and door times. The time consumed when stopping is given by Definition 4.22 as

$$t_s = t_f(1) + t_c + t_o - t_v \tag{5.23}$$

which can be expressed in terms of Equation (4.14) as:

$$t_s = T - t_v \tag{4.14}$$

The lift performance time (T) has the most significant effect on the round trip time. It is defined as the period of time between the instant (when a stationary) lift starts to close its doors and the instant when its doors are open by 800 mm at the next adjacent floor (Definition 4.23). The time to transit two adjacent floors at rated speed (t_v) was dealt with in the previous section.

5.8.1 Cycle Time (Definition 4.24) and Other Times

There can be some confusion regarding the meaning of flight time, performance time and cycle time and also when door opening time is considered to be finished.

Figure 5.3(a) is the usual operating cycle, without advanced door opening. This begins with passengers transferring into the car, the doors closing, the interlocks making up and the car moving (brake to brake time). When the car stops moving the doors open and the passengers transfer out of the car and some passengers may transfer into the car.

Figure 5.3(b) illustrates what happens when the lift is supplied with the advance door opening feature. Here some of the time of the door opening is shared with the levelling operation. That is the doors open BEFORE the lift has finished moving. This feature can reduce the cycle and performance times by between 0.5 s to 1.7 s.

The cycle time is measured from some consistent point in the cycle, ie: when the doors are closed, or when the doors are just opening. The measurement of cycle time is made with no passengers entering or leaving the car. The cycle time thus includes any car/landing door dwell times, see Section 5.11. Figure 5.3(c) shows the operating cycle for a lift responding to a car call and Figure 5.3(d) shows the operating cycle for a lift responding to a landing call. Cycle time therefore includes time which might be wasted if a car call is registered in error and when the lift stops no one leaves; or when a landing call is registered and when the lift stops no one enters.

Performance time is the most important variable, as this can be controlled and predicted. The components of the performance time must be carefully selected to achieve the correct handling capacity for the lift installation. The lift maker should be contracted at the tender stage to provide them at the specified values and the maintenance contractor should be required to keep them at the rated values throughout the life of the installation. Failure to do so will invalidate any traffic design. The equation for the performance time is given by:

$$T = t_f(1) + t_c + t_o \tag{5.24}$$

5.8.2. The Single Floor Flight Time

The single floor flight time $t_f(1)$ is the time taken from the instant the car doors close to the instant the car is level with the next adjacent floor (Definition 4.20). It is dependent on the rated speed, the acceleration value and the jerk value. Jerk, sometimes called "shock" is the rate of change of acceleration (units in m/s^3). The relationships between distance travelled, velocity, acceleration and jerk are complex and are given in detail in Chapter 7 and Appendix 1. Flight times need to be obtained for any distance or number of floors travelled.

Fortunately for designers of lift drives there are limits on the maximum values of both acceleration and jerk. These constraints are imposed by human physiology, as described in Chapter 4. Passengers are uncomfortable when subjected to acceleration values greater than about one sixth of the acceleration due to gravity (ie: about $1.5 \ m/s^2$). Similarly, the maximum value of jerk commonly used in calculations is about $2.2 \ m/s^3$, although there is no drive control on this variable.

As a result of these limits there is little or no difference in single floor flight times for rated speeds in excess of about 5.0 m/s, as the maximum possible values of acceleration and jerk have been reached. Below 5.0 m/s, the flight times are dependent on the type of drive and drive controller, mainly owing to variations in levelling times and non-ideal acceleration deceleration profiles.

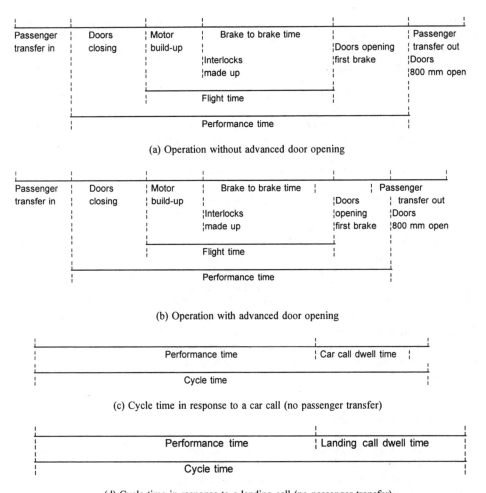

Figure 5.3 Illustration of lift system timings

Table 5.3 indicates the likely range of acceleration values and single floor flight times. The single floor flight times are slightly larger than a theoretical calculation would give to allow for start-up delays. Different manufacturers and drive systems will result in small variations in the values given. Naturally the flight times also depend on the interfloor distance. Where the higher speeds have the most effect is where a lift serves a high zone in a building and has to pass non-stop through a number of lower floors.

5.8.3 Door Operating Times

5.8.3.1 General

Door operating times comprise opening and closing times, and door dwell times. The dynamic operating (opening and closing) times are dependent on a number of factors: door panel velocity, panel arrangement, width and control.

There is a limit to door panel velocity commensurate with passenger safety. The European and US Standards require that the energy present in a moving door should not exceed 9.5 J (joules) and 10.0 J (respectively). This restriction applies mainly to closing doors. The restriction has the effect of limiting the maximum door velocities during closing to about 300 mm/s. Doors can operate faster if they are not allowed to touch a passenger. However, this involves the use of complex (and sometimes expensive) passenger detection and door control systems.

There are two basic door types: side opening and centre opening. Side opening doors have to open the whole width of the doorway, which takes more time. Here the width is taken as the clear opening width ignoring any returns or architraves. Centre opening doors open and close more quickly and the symmetrical reaction against the car frame will reduce car sway.

There are several standard widths. Narrow doors of 800 mm width are usually fitted to cars with a rated car capacity of up to 12 persons. Wider doors of 1100 mm width are fitted to lifts with a rated car capacity of over 12 persons. Doors of 1300 mm width or larger are fitted to goods lifts and hospital lifts. Obviously, the narrower the door, the faster the operation.

The control of the door operator can significantly affect door timings in respect of start-up, slow down and safety. There are a number of types usually rated according to their operating speed and method of control of the door operator (open or closed loop). The speeds are not well defined and in broad brush terms are defined as: low (about 300 mm/s), medium (about 450 mm/s) or high speed (about 750 mm/s). Generally a low speed operator will be found on a low speed lift and obviously cost less than a high speed operator. Low speed operators are often characterised by delivering the same opening and closing times.

5.8.3.2 Door closing time (t_c)

Door closing always takes place whilst the lift is stationary, and typical door closing times are given in Table 5.5. The door closing time (t_c) is the time taken from the instant the car doors start to close to the time they are locked up. Remember the energy constraints of the previous section.

5.8.3.3 Door opening time (t_o)

Door opening time is not subject to the energy constraints found in door closing and can be much faster. The doors can operate at any speed provided the trapping hazard for fingers, etc. against the door architrave or door lining is negligible. However, as the same door operator will be used for both directions of movement, opening times may not be much improved. There are two cases of door opening to be considered: with and without advanced door opening.

Where the advanced door opening time feature is not installed, the door opening time is considered to have ceased, when the passenger transfer may begin. This usually can occur when the doors are open by approximately 800 mm. Thus in this case the door opening time is taken to be the time from the instant the doors start to open until the instant the doors are open to a width of 800 mm.

Time can also be saved by advanced door opening. Once the car has entered the door zone some 200 mm from a landing, the doors can be unlocked and opened. Then an improvement in opening times can be achieved by overlapping the levelling operation with the first part of the opening of the doors called advanced door opening, or pre-opening. This is possible within the door zone, provided that the tripping hazard is negligible.

The door opening value to be used in this case in the evaluation of the parameter t_s is the time from the moment the lift is level at the landing, until the doors are open by 800 mm. This time will be less than the measured door opening time for lifts without advanced opening, viz: $t_o - t_{ad}$, where t_{ad} is the time saved by opening the doors during levelling.

Typical door opening times are given in Table 5.6 for normal and advanced opening. The table gives representative values for two door types, two door sizes, and with and without advanced door opening. These values may be used where specific values are not available.

Table 5.6 Typical door closing and opening times (s) for stated door width (mm)

Door operation	Opening (advanced)		Opening (normal)		Closing	
Door type	800	1100	800	1100	800	1100
Side	1.0	1.5	2.5	3.0	3.0	4.0
Centre	0.5	0.8	2.0	2.5	2.0	3.0

5.8.3.4 Door weight

The weight[1] of the door is determined by many factors, such as fire resistance, height, width, configuration, etc. A moving door gathers considerable kinetic energy. To protect passengers from injury, the standards require the maximum energy to be limited to 10 J, provided the safety edge is operative. If the safety edge is inoperative then the energy value must not exceed 4 J. The maximum values of energy acquisition limit the maximum door speed when closing. Typically a 150 kg door has a maximum speed of 0.23 m/s and a 500 kg door has a maximum speed of 0.13 m/s. For a particular door weight (M), the maximum speed (s) at which the doors may move to meet the energy value requirements are given in Table 5.7.

Table 5.7 Maximum door movement speeds for different weight of doors

Total door weight (kg)	Maximum speed for 10 J (m/s)	Maximum speed for 4.0 J (m/s)
150	0.36	0.23
200	0.31	0.20
250	0.28	0.18
300	0.25	0.16
350	0.23	0.15
400	0.22	0.14
450	0.21	0.13
500	0.20	0.13

Where the weight of a door is not known, the weight can be approximately estimated by assuming:

A painted hoistway door weighs 35 kg/m^2
Painted car doors weigh 24 kg/m^2
Hangers per door weigh 10 kg
Other hardware (vanes, operating arms, safe edges, etc.) per system weigh 5 kg

---

[1] Purists will say mass.

5.8.3.5 Example 5.4

A 1100 mm single panel side opening door has an area of 2.5 m². Calculate the door closing time.

Hoistway door weighs	87.5 kg
Car door weighs	60.0 kg
Hangers weigh	20.0 kg
Other hardware weighs	5.0 kg
	172.5 kg

From Table 5.7 (by interpolation) the maximum door speed is 0.34 m/s.

It is the practice to measure door closing times from a point 50 mm from the open and closing jambs. This allows for the acceleration and deceleration time periods at the extremes of travel. Thus the door will move the 1.0 m in 2.94 s.

The actual door closing time will be longer than this to allow for the acceleration and deceleration of the door panels. Add 1.0 s to account for this giving a door closing time of some 4.0 s.

5.9 PASSENGER TRANSFER TIMES (t_p)

The passenger transfer time is the time a single passenger takes to enter or leave a car (Definitions 4.12 and 4.13). This parameter is the vaguest of all the components of Equation (4.15), principally because it is dependent on human behaviour. The passenger loading times and the passengers unloading times are not necessarily equal. The transfer time is affected by the direction of the transfer (in or out), the width of the doors, whether a lobby or car is crowded, the shape of the car, the size and type of car entrance, the building type (eg: commercial, institutional, residential, etc.) and the characteristics of the passengers (eg: age, agility, purpose, etc.).

Phillips (1973)[1] provides graphs for loading and unloading. These were mathematically described by Jones (1971) as two equations (see Section 8.3.6). Strakosch (1967) offers tables of transfer times related to standard car sizes, which he updates in the 1998 edition to include disability needs. He suggests that the round trip time might include an inefficiency factor to account for passenger transfers.

General rules can be offered. If the car door width is 1000 mm or less it may be assumed that passengers enter or exit in single file. For door widths of 1000 mm and above it may be assumed that the first six passengers enter or exit in single file and the remainder in double file.

The average passenger transfer time (entry or exit) may be taken as 1.2 s (each way). This time could be reduced for single cars, but increased for groups. The time might be increased for small door openings and reduced for large door openings. For situations where passengers are elderly, or have no reason to rush, the transfer times should be increased to about 2.0 s. ISO 4190-6 considers a passenger transfer time of 1.75 s suitable for residential buildings.

---

[1] The first edition of this book is 1938.

5.10 EXAMPLE 5.5

A speculative office block of 16 floors above the main terminal is to be built. The interfloor distance is 3.3 m with no express jump. A single 16 person lift with 1100 mm two panel centre opening (2PCO) doors is to be installed. What will be its characteristics? Assume the passenger transfer time is 1.2 s.

Basic data:

$N = 16$, $CC = 16$, so $P = 12.8$, $H = 15.3$, $S = 9.0$ (from Table 5.1)

The travel will be $16 \times 3.3 = 52.8$ m; so from Table 5.3 the rated speed should be 3.15 m/s. However, remember that it is a speculative building so use 2.5 m/s. Hence:

$t_v = 3.3/2.5 = 1.32$ s

From Table 5.3 the single floor flight time could be 5.5 s.

From Table 5.6 (advanced opening not considered) the door times could be:

$t_o = 2.5$ s, $t_c = 3.0$ s

Thus:

$t_s = 5.5 + 2.5 + 3.0 - 1.32 = 9.68$ s

Then solving the RTT Equation (4.11):

$$RTT = 2Ht_v + (S+1)t_s + 2Pt_p \qquad (4.11)$$

$$RTT = 2 \times 15.3 \times 1.32 + (9.0 + 1)9.68 + 2 \times 12.8 \times 1.2$$

$$= 40.4 + 96.8 + 30.7 = 167.9 \text{ s}$$

Therefore:

$INT = 167.9$ s

and hence:

$UPPHC = 300/167.9 \times 12.8 = 22.9$ persons/5-minutes

5.11 OTHER TIMES

Many lifts are fitted with a number of features that can affect the evaluation of the round trip time. These are:

(a)   car call dwell time
(b)   landing call dwell time
(c)   differential door time
(d)   lobby dwell time.

When a lift arrives at a landing it opens its doors. With lifts fitted with mechanical safe edges the doors are held open for a period of time, called the dwell time, to allow passengers to enter or leave the car before the doors start to close. In an office building it is usually sufficient to keep the door open for a stop in response to a landing call for 3.0 s dependent on the lobby arrangement. And for a stop in response to a car call, the door hold open time can be shorter (as passengers are prepared to leave) at 2.0 s. In a residential building, or a shopping centre, these times might be set to 7.0 s to allow prams and bicycles to be transported. Where a lift is designated for disabled use the dwell times will be considerably extended from the 2 s to 3 s acceptable by the able bodied. Regulations may require dwell times of from 5 s to 7 s to accommodate disabled persons.

The car and landing call dwell times can be reduced if passenger collision detectors are fitted. Once a passenger crosses the threshold, the dwell time can be reduced to 0.5 s and the passenger detection system can be used to keep the doors open whilst passengers are present in the threshold. This feature is called differential door timing. Many lift controllers also cause the dwell times to be reduced to about 0.5 s whenever a new car call is registered, or when an existing car call is registered again on the car operating panel.

The effect of these dwell times can be significant. For example, if the dwell times are set too long, and differential timing is not available, the passenger transfer may well be shorter than the dwell times. Thus the third term in the round trip equation becomes redundant and extra time would need to be added to the performance time to compensate, thus increasing the size of the middle term of the round trip equation. If, however, differential door timers are in operation, then the third term of the round trip time equation is required.

Lobby dwell time operates during uppeak periods of traffic and causes a car to remain at the main terminal floor for a period of time after the first car call has been registered. This period of time is often set to allow a car to load to 80% of its rated car capacity and prevents a car leaving the main terminal with a small number of passengers. The lobby dwell time should not affect the calculation during uppeak. But if it is not turned off during other traffic conditions it will delay the movement of the lift.

5.11.1 Example 5.6

To illustrate the effects of these times, suppose that in Example 5.5 the lobby dwell time was set to 8.0 s, the car call dwell time was set to 2.0 s, the landing call dwell time was set to 5.0 s and advanced door opening reduced the door opening time by 0.8 s. What effect would this have ? What would be the values for performance time, cycle time (in response to a car call) and cycle time (in response to a landing call) ? Consider each in turn.

Lobby dwell time
This is set at 8.0 s. With 1100 mm doors it can be assumed that the first six passengers enter single file and the remainder (6.8) enter double file. The total passenger transfer time would therefore be:

$$6 \times 1.2 + 3.4 \times 1.2 = 11.3 \text{ s}$$

This time is larger than the lobby dwell time of 8.0 s and therefore has no effect.

Car call dwell time

This is set at 2.0 s. At each floor an average number of passengers will leave. There are 9 stops and 12.8 passengers. This gives an average time of 12.8/9 ie: 1.42 passengers. They will take $1.42 \times 1.2 = 1.7$ s to leave. Thus the car call dwell time is longer (on average) than the average time for passengers to leave the car. It might be prudent to add 0.3 s to t_s.

Landing call dwell time

This is set to 5.0 s. As the calculation is for an uppeak traffic condition this time will not operate. However, it is rather long for other traffic conditions and a single installed lift. It should be reduced to about 3.0 s.

Performance time

In the case of no advanced opening it is:

$$T = 5.5 + 2.5 + 3.0 = 11.0 \text{ s}$$

In the case of advanced opening it is:

$$T = 5.5 + 2.5 + 3.0 - 0.8 = 10.2 \text{ s}$$

Cycle time

No advanced door opening and responding to a car call:

$$T_{cyc} = 11.0 + 2.0 = 13.0 \text{ s}$$

No advanced door opening and responding to a landing call:

$$T_{cyc} = 11.0 + 5.0 = 16.0 \text{ s}$$

With advanced door opening and responding to a car call:

$$T_{cyc} = 11.0 - 0.8 + 2.0 = 12.2 \text{ s}$$

With advanced door opening and responding to a landing call:

$$T_{cyc} = 11.0 - 0.8 + 5.0 = 15.2 \text{ s}$$

Remember that the formulae all use average values for all the evaluated variables, hence the non-integer values (eg: 1.42 passengers). Clearly there can only be an integer number of real passengers.

5.12 FACTORS AFFECTING THE VALUE OF THE ROUND TRIP EQUATION

It was noted in Section 5.1.4 that the round trip equation comprised six parameters. Three parameters were time independent variables, ie: H, S and P, and three parameters were time dependant variables, ie: t_v, t_s and t_p. The round trip time is dependent on the values of these six parameters. The smaller they are, the smaller the resulting value of the round trip time and the higher the handling capacity of the lift system. How can these six parameters be manipulated to achieve this ?

5.12.1 The Time Independent Variables

A lift is provided to transport passengers as represented by the variable P, so little can be done to reduce this parameter. Section 12.8 and Case Study CS17 indicate how values of the variable H and S can be reduced by either limiting the number of floors a lift serves or by advanced traffic control systems.

5.12.2 The Time Dependent Variables

Examination of the answer to Example 5.5 shows that the first and last (third) terms of the round trip equation are significantly less than the middle (second) term. This is generally so and means that for each second saved or added to the middle term reduces or increases the value of the round trip time. This can be developed into a general rule of thumb that for a one second change in the performance time, there is a consequential change in the handling capacity of about 5%.

5.12.3 Five Lift Systems

Consider the following five lift systems, which could be considered to be the "four corners of the lift world and its centre of gravity":

$N = 10$: $CC = 10$: $v = 1.6$ m/s
$N = 10$: $CC = 24$: $v = 1.6$ m/s
$N = 16$: $CC = 16$: $v = 2.5$ m/s
$N = 20$: $CC = 10$: $v = 3.15$ m/s
$N = 20$: $CC = 24$: $v = 3.15$ m/s

All the systems have:

■ a performance time of 10.0 s
■ an assumed passenger transfer time of 1.0 s
■ an interfloor distance of 3.3 m.

For the five lift systems determine the effect of:

■ increasing the performance time by +1.0 s
■ decreasing the rated speed by 10%
■ increasing the passenger transfer time by +0.2 s.

Table 5.8 tabulates the effects.

Table 5.8 Effect of changing the time dependent variables

N	CC	v (m/s)	RTT (s)	$T + 1.0$ s	$v - 10\%$	$t_p + 0.2$ s
10	10	1.6	108	114 (6%)	111 (3%)	111 (3%)
10	24	1.6	156	166 (6%)	158 (1%)	164 (5%)
16	16	2.5	153	163 (6%)	158 (3%)	158 (3%)
20	10	3.15	123	131 (7%)	127 (3%)	127 (3%)
20	24	3.15	200	214 (7%)	203 (2%)	208 (4%)

All figures in parenthesis are percentage increases. All figures are rounded.

There are a number of conclusions that can be drawn from Table 5.8. Changing the speed by 10% or the passenger transfer time by 0.2 s has a smaller effect than changing the performance time by 1.0 s. There is little that a designer can do to hasten passengers into or out of a lift, except to provide good signalling and good circulation areas, so the emphasis must be on specifying the performance time as low as possible and keeping it at the specified value by good maintenance.

5.13 SUMMARY

Statistical theory has been used to derive two of the parameters of the round trip time equation S and H. Two different probability distribution functions have been considered. The first based on a simple statistical process, and used as long ago as 1923 by Basset Jones, assumes a rectangular input distribution with random arrival times. The second uses the Poisson process to describe arrivals. It is interesting to note that the older procedure (the rectangular pdf) will always produce a more pessimistic result than the Poisson pdf. Thus if a designer wishes to be a little conservative in a traffic design, then the conventional (rectangular pdf) expressions should be used.

The other parameter in the RTT expression P is usually assumed to be 80% of rated car capacity by designers. The conventional design assumes the arrival of passengers at constant intervals served by lifts arriving at constant intervals as shown in Figure 5.2(a). In practice neither situation occurs (illustrated by Figure 5.2(b) where both arrivals and lift departures are random). The problem with randomness in a lift system is that both the random passenger processes will add together to produce queues and large waiting times, if an attempt is made to realise 100% handling capacity. To counteract this additive process, it is sensible to derate the lift system to allow for periodic overloads. Hence the use of the 80% factor for determining a value for P, when calculating handling capacities.

This chapter (and most design procedures) use the uppeak traffic condition to size a lift installation as it is simple and easily modelled. It is possible to extend the statistical techniques to derive expressions for the more general case of interfloor traffic. The full significance of quality of service (represented by system handling capacity) and car load versus quality of service (represented by average passenger waiting time) is considered in detail in Chapter 6.

CHAPTER SIX

Determination of Passenger Demand

6.1 INTRODUCTION

The difficulty in planning a lift installation is not in calculating its probable performance, but in estimating the likely passenger demand. Quite often the building has yet to be built and estimates have to be based on the experience gained with previous similar structures. Existing buildings can be surveyed, by observation, or by means of an attached data logger, to determine the current activity. However, even this is prone to error, as the building's population may have adapted to poor (or good) lift performance.

It is essential, therefore, that all the parties involved in the planning of a lift installation have a clear understanding of the basis for the planning. For example, it is important that the architect or planner establishes the lift system required at a very early stage and not after the rest of the building has been designed, as often happens.

It is important to remember that the distribution and size of the population of any large building changes regularly. Thus a tightly planned design may prove inadequate once a building has been occupied for a year or more. To understand the effect of these changes on a design, it is essential to document the criteria and decisions taken at all stages of a design.

6.2 QUALITY AND QUANTITY OF SERVICE

There are two key factors affecting the demand that a building's occupants will make on a lift system, as indicated in Section 4.5.2. These are:

■ Quantity of Service

■ Quality of Service.

The quantity of service factor, ie: how many people will use the lift system over a defined period of time, is represented by the handling capacity. The quality of service factor, ie: how well must the lift system deal with its passengers, is represented by passenger waiting time. Both factors are interrelated. Both factors depend, among other things, on the type of building and its use and the type of occupier. This makes the design task very difficult for buildings of a speculative nature.

The following sections indicate methods to facilitate the design task by looking at building populations, the likely demand on the lift installation and the Quality of Service factor. The analysis is relevant mainly to commercial office type buildings. Other buildings will be discussed briefly in this chapter.

The traffic period for evaluation when sizing an office building is usually a morning up-peak, 5-minute segment. During this period of time, little or no traffic is moving interfloor or down in the building. The lifts are loading passengers at the main lobby, distributing those passengers to various upper floors and then making an express trip back to the main lobby for the next load.

Therefore, studies are based upon "one-way traffic" in the up direction with no stops at the intervening floors in the down direction. It is possible to consider some down travelling passengers, but there is no consensus as to how big this flow should be. Thus for a method which can be used to compare any designers results it is best to stick to a pure uppeak traffic flow.

In general, if the uppeak traffic pattern is sized correctly all other traffic patterns will also be adequately served. There are exceptions to this comment. For example: in hotels at meal times; in hospitals at visiting times; in buildings with trading floors (insurance and stock markets), which open at specified times and at lunch time in all buildings.

6.3 REPRISE

The sizing of lift systems to serve the demands of a building's population has interested the lift community since the 1920s. The methods used then were somewhat rough and ready. The problem is to match the demands for transportation from a building's occupants with the handling capacity of the installed lift system. This procedure should also result in an economic solution.

As Chapter 5 indicated, the first attempts to size a lift system to meet passenger demands occurred in the 1920s when Basset Jones (1923) derived a formula for the average number of stops that a lift would make under uppeak traffic conditions. The derivation of a formula for the probable average highest floor for the same condition was not made until 1955 by Schroeder (published in German in *Fordern und Haben*). He also produced a three term formula for the round trip time. Barney and Dos Santos (1975) independently derived a slightly different three term formula (in English).

Thus, by the 1970s a recognised method of calculation had evolved, for the uppeak traffic sizing, based on the mathematical determination of the H, S and P parameters (average highest reversal floor, average number of stops and average number of passengers).

The formulae for the calculation of the passenger handling performance of lift systems are now a universally accepted method of analysis. Lift makers are able to use tables applicable to their product range, based on the formulae, to estimate round trip times, intervals, handling capacity, etc.

During the 1970s digital computer calculation and simulation packages evolved, which allowed other traffic conditions to be examined. These computer techniques utilise the proven mathematical methods.

6.4 ESTIMATION OF POPULATION

6.4.1 Passenger Data Sets

In Section 5.1.3 the passenger data set was found to comprise:

(a) the number of passengers boarding from specific floors
(b) the number of passengers alighting at specific floors
(c) the traffic mode ie. unidirectional or multidirectional
(d) the transfer times for passengers entering and leaving cars
(e) passenger actions.

During uppeak traffic these data are simplified.

Point (a): passengers only load at the lobby.
Point (b): passengers never alight at the lobby.
Point (c): the traffic is unidirectional ie: travel is up the building.
Point (d): the passenger transfer times will generally be "brisk".
Point (e): there is little opportunity for passengers to misbehave.

To determine the number of passengers, who will board and what their demand will be, depends on the building population. The following sections indicate the factors, which need to be considered, when estimating the building population.

6.4.2 Purpose of a Building

The size of the intended population should be obtained from the building owner or proposed occupier, if possible (and in writing). However, it may be that the population size is not available, or the building is a speculative one, when an estimation must be made.

The number of occupants will vary according to:

(a) the purpose of the building (residential, commercial or institutional)
(b) the quality of the accommodation
(c) the type of occupancy (in the case of office buildings, the type of tenancy).

With regard to point (a), the buildings are generally defined as:

- residential: where people live, eg: blocks of flats
- commercial: where people work, eg: offices
- institutional: where people receive a service, eg: a hospital.

Point (b) implies that the more prestigious the building, eg: a head office, the more space is available to each occupant. Point (c) is more complex. There are three main types of tenancy:

(i) Diversified
(ii) Mixed
(iii) Single.

Definition 6.1: A diversified tenancy is a building occupancy condition, where no single tenant occupies more than a single floor and no more than one quarter of the tenants of the building are engaged in the same type of business activity.

Definition 6.2: A mixed tenancy anticipates the possibility of multi-floor occupancy by a single tenant or multiple tenants with the same business activity.

Definition 6.3: A single tenancy is a building occupancy condition where a single tenant occupies a substantial portion or zone of the building (say 80%).

The single tenancy situation can present a severe traffic design condition. The group handling capacity with such occupancy can be high (about 14%) for calculation purposes. And some single-tenant insurance companies, government entities, or utilities, with large numbers of clerical workers can have handling capacity requirements of substantially more than 15% of the population in five minutes, if they operate a fixed starting time regime. In these cases, it would be important to establish this demand from the prospective building owner before carrying out any calculations.

In Tables 6.1, 6.2 and 6.3 diversified and mixed tenancies are grouped as multiple tenancy.

6.4.3 Main Terminal Population

The main terminal population is not normally included in the design population for the following reasons:

(a) The main terminal floor is the bottom terminal for the lift group and passengers on this level walk to their work places.
(b) The main terminal floor is occupied in total by a bank or is a retail space. Employees of these types of businesses generally start work later than businesses occupying the majority of the building and, therefore, do not affect the major morning office building traffic peak.
(c) The main terminal in a building with subterranean or off-site/street parking is served by a separate bank of lifts. In this situation, the persons working on the main terminal floor and parking in the subterranean level would ride this separate bank of lifts to that floor. In the case of traffic from off-site parking or ground level drop-off traffic (taxis, etc.), they would walk directly into the main terminal floor and on to their work place.

The main terminal population may be included in the building population in the following situation. There are subterranean parking levels served by the same group of lifts that serve the upper floors of the building. Thus persons who work on the main terminal floor and who park in the subterranean levels would use this group of lifts.

6.4.4 Usable Area

Most population estimates start from a knowledge of the net usable area, ie: the area which can be usefully occupied and which excludes circulation space (stairs, corridors, waiting areas), structural intrusions (steelwork, space heating, architectural features, etc.), toilet facilities, cleaners' areas, etc.

The American National Standard ANSI Z65.1-1980 "Standard Method for Measuring Floor Area in Office Buildings" gives a useful guide to calculating areas in office buildings. It defines two important terms: rentable area and usable area.

Definition 6.4: The rentable area is determined by measuring between the inside finished surfaces and/or dominant parts of permanent outside walls, excluding any major vertical penetrations (stairs, lift shafts, flues, pipe shafts, ducts and their enclosing walls). No deductions are made for any columns and projections necessary to the building.

Rentable area generally remains fixed for the life of the building and is used to calculate rents.

Definition 6.5: The usable area is determined by measuring between the finished surfaces of the office side of corridors and/or other permanent walls and/or the dominant parts of outside permanent walls and/or the centre of partitions within the rentable area. No deductions are made for any columns and projections necessary to the building.

Usable area indicates the actual occupiable area and is important in lift traffic design calculations. Usable area can vary during the life of the building as corridors, partitions, etc. are moved.

In most traffic design cases, it is necessary to calculate and project the building population from imprecise data. It is necessary always to calculate the usable area and then divide that area by the area allocated per person (in m²) to derive the estimated population.

Where architectural drawings are too schematic to make an accurate estimate of areas, one of the following approximate rule of thumb relationships may be used, when the gross area is known:

Rentable area = 90−95% of gross area
Usable area   = 75−80% of gross area

or the relationship below if the rentable area is known:

Usable area   = 80−85% of rentable area

Sometime the term Net Internal Area (NIA) is used. This is basically the area from the inside surfaces of the external walls. No concession is given for penetrations.

Whenever traffic calculations are made it is important and advisable to indicate (in writing) which estimations have been utilised and that a check review of the initial study is made once the architectural drawings are developed to a point where an accurate usable area calculation can be made.

In some cases, the building population may be dictated by the Owner/Client. This is particularly true if the building is being designed for a known occupant.

6.4.5 Example 6.1

Using rules of thumb above, what are the rentable and usable areas of (a) a tall/slender building and (b) a low/squat building, each having a gross area 5000 m² ?

(a) This will have a large core compared to the footprint, but the occupants will always be close to a lift.

Rentable area = 90% of gross area, ie: 4500 m²
Usable area   = 75% of gross area, ie: 3750 m²

(b) This will have a small core compared to the footprint, and the occupants may be far from a lift.

Rentable area = 95% of gross area, ie: 4750 m²
Usable area   = 80% of gross area, ie: 4000 m²

6.4.6 Practical Population Estimations

Table 6.1 gives guidance for a variety of buildings based on surveys and experience of the population to be accommodated.

6.5 ESTIMATION OF ARRIVAL RATE

It is necessary now to determine the percentage of a building's population who will require transportation to the higher floors of a building during the morning five minute uppeak. This peak will vary due to effects such as:

Table 6.1 Estimation of population

Building type	Population estimate
Hotel	1.5 – 1.9 persons/room
Flats	1.5 – 1.9 persons/bedroom
Hospital	3.0 persons/bedspace[*]
School	0.8 – 1.2 m² net area/pupil
Office (multiple tenancy):	
regular	10 – 12 m² net area/person
prestige	15 – 25 m² net area/person
Office (single tenancy):	
regular	8 – 10 m² net area/person
prestige	12 – 20 m² net area/person

[*] Patient plus three others (doctors, nurses, porters, etc.).

(a) the type of building occupancy (different business interests or single tenant);
(b) the starting regime (unified or flexitime);
(c) the location of bulk transit facilities such as bus, train, etc. (distant alighting places will result in a spread of arrivals owing to different walking speeds).

The arrival rate is expressed as a percentage of a building's total population. It is unlikely in many buildings that all the total population is present on any day. The Greater London Council (date unknown) assumed and attendance of 84% (GLC 25). The effective population considered during the uppeak period can be reduced to account for:

(a) persons away on holiday
(b) persons away sick
(c) persons away on company business
(d) vacant posts
(e) persons who arrive before or after the peak hour of incoming traffic.

Thus the total building population could be reduced by 15% to 20% to account for these factors. Table 6.2 gives guidance of probable peak arrival rates of the remaining occupants.

6.6 QUALITY OF SERVICE

6.6.1 Passenger Average Waiting Time

The first passenger to arrive at a landing registers a call and then waits for a lift to arrive. Other passengers arrive and wait shorter periods of time.

Definition 6.6: The passenger average waiting time is the average period of time, in seconds, that an average passenger spends waiting for a lift, measured from the instant that the passenger registers a landing call (or arrives at the landing), until the instant the passenger can enter the lift.[1]

---

[1] This must be the first lift to arrive, otherwise queues will develop.

Table 6.2 Percentage arrival rates

Building type	Arrival rate
Hotel	10 – 15%
Flats	5 – 7%
Hospital	8 – 10%
School	15 – 25%
Office (multiple tenancy):	
regular	11 – 15%
prestige	17%
Office (single tenancy):	
regular	15%
prestige	17 – 25%

The instant a passenger can enter a car (assuming no passengers are exiting) is when the doors have opened 800 mm. This is because a person feels confident they can pass through a "gap" of 800 mm, which is about 200 mm wider than the body ellipse of Figure 1.1.

Actual passenger waiting time would be the best indicator of the Quality of Service that an installed lift system could provide, ie: the shorter the time the better the service. Passengers tend to be upset if they are made to wait too long, ie: over 30 s !

However, passenger waiting times cannot be easily measured. Some designers, therefore, use the interval of car arrivals at the main terminal as an indication of service quality. Interval, however, is part of the evaluation of handling capacity, which simply determines the Quantity of Service. In general terms, interval can be used to indicate the probable Quality of Service, when considering office buildings as an interval of:

- 20 s or less would indicate an excellent system
- 25 s would indicate a good system
- 30 s would indicate a satisfactory system
- 40 s would indicate a poor system
- 50 s or greater would indicate an unacceptable system.

Table 6.3 gives guidance for values of suitable intervals for office and other types of buildings. These values are better than those given in some early standards, eg: BS5655, Part 6: 1990. As the expectations of passengers, particularly in major city centre offices, have increased, these recommendations have been updated in the current BS5655: Part 6: 2002.

6.6.2 Example 6.2

A speculative, regular 10 floor (above the main terminal) building is to be built. Each floor is 1200 m² of usable space. What is the basic specification of the lift system ?

Table 6.1 indicates that 10 – 12 m²/person should be considered.
Assume 12 m², this gives 100 persons per floor.

The total population will then be 10 × 100 = 1000.

Assume 80% daily occupancy, ie: this gives a design population of 800.

Table 6.3 Uppeak intervals

Building type	Interval (s)
Hotel	30–50
Flats	40–90
Hospital	30–50
School	30–50
Office (multiple tenancy):	
regular	25–30
prestige	20–25
Office (single tenancy):	
regular	25–30
prestige	20–25

Table 6.2 indicates 11–15% arrival rate; assume 12.5%.

Then peak arrival will be 12.5% of 800, ie: 100 persons.

Table 6.3 indicates an interval of 25–30 s. This is a speculative building so to save capital expenditure assume 30 s.

The lift system should be sized to be able to handle 100 persons with a 30 s interval.

6.6.3 Uppeak performance

Caution must be exercised when using interval as a Quality of Service indicator as passenger waiting time depends on car load. A simple rule of thumb has been to assume that the average passenger waiting time is half the interval. This would be so if passengers were to arrive with equal time spacing, ie: a rectangular probability distribution function. This rule delivers an imprecise result as it is only accurate when the cars load to less than half full. Lightly loaded cars are an unlikely situation for an uppeak traffic condition.

It has been shown (Barney and Dos Santos, 1977) that a theoretical relationship exists between interval and passenger average waiting time dependent on the actual percentage car load (passengers in the car). This relationship is depicted in Figure 6.1 and tabulated in Table 6.4. The figure shows the average car load as a percentage of rated car capacity versus performance represented by passenger average waiting time (AWT) divided by average interval (INT) at that car load. The relationship AWT/INT normalises the results.

To a first approximation the relationship can be used to indicate the probable Quality of Service of a lift installation. At the conventional assumed car loading of 80%, average passenger waiting time is 85% of the calculated interval. But at a 90% car loading, the average passenger waiting time has extended to 130% of the calculated interval. For loadings greater than 90%, the average passenger waiting time increases rapidly and in theory at 100% would be infinite.

For car loads between 50% and 80% it is possible to develop an approximate equation for the AWT as:

$$AWT = [0.4 + (1.8\ P/RC - 0.77)^2]INT \tag{6.1}$$

where RC is the rated car capacity.

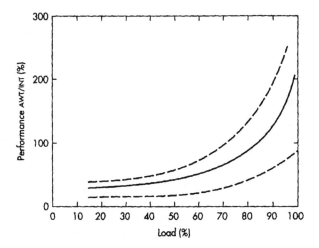

Figure 6.1 Uppeak performance − graphical representation

Table 6.4 Uppeak performance − numerical values

Car load (%)	*AWT/INT* (%)	Car load (%)	*AWT/INT* (%)
30	0.32	75	0.74
40	0.35	80	0.85
50	0.40	85	1.01
60	0.50	90	1.30
70	0.65	95	1.65

6.6.4 Example 6.3

Example 6.2 requires a lift system to be sized to be able to handle 100 persons with a 30 s interval. Design such a system and estimate the average passenger waiting time.

If the interval is 30 s, then a lift group must provide 10 trips over five minutes.

Each car must load with 10 passengers to handle 100 persons in five minutes.

If a 13 person car is used, the percentage car load is 10/13, ie: 77%.

From Figure 6.1 or Table 6.4 a car load of 77% indicates that AWT/INT will be 78.4%.

The estimated AWT will be 0.784 × 30 = 23.5 s. (The formula gives 23.4 s.)

6.6.5 Average (lift) System Response Time (*ASRT*)

The average passenger waiting time is calculated by adding all the individual waiting times together and dividing by their number. At present it is not possible to easily measure these times, unless a squad of observers are employed or sophisticated (expensive) scanning equipment is installed. It is possible, however, to determine average passenger waiting times from computer simulations and models. Another way of measuring Quality of Service is to measure the average system response time.

Definition 6.7: The (lift) system response time is the period of time that it takes for a group of lifts to respond to the first registered landing call at a floor.

Individual system response times can always be measured, unlike the average passenger waiting time, using external data loggers or the in-built monitoring systems provided in most traffic controllers. Each individual system response time, however, only indicates the maximum waiting time that the first passenger has to wait. Later passengers will wait for less time.

The average system response time (*ASRT*) is the average of a number of individual response times. Care needs to be exercised when determining the average, as some in-built data loggers record system response times over very long periods, eg: since the lift system was installed. In this case, or even over a 24 hour period, the *ASRT* value will be very low, as it would include all periods of time, when the lift system is hardly used, eg: at night, weekends and holidays. The best systems will provide averages over five minute periods ie: over the comparable period used for the calculation of handling capacities.

Definition 6.7 is applicable whether there is a single car or up to eight cars in the group. Strictly, the system response time is measured from the time that the first passenger at a floor registers a landing call until the car doors of the lift servicing that call has opened its doors to a width of 800 mm. In practice the measurement will be made until the landing call registration is cancelled by the traffic controller. This can be from 5 s to 10 s before the doors are 800 mm open and this time error must be added to the system response time that is measured to find the actual response times. When presented with data from a logger showing average system response time, it is important to ensure that it is not in error by this value.

It is possible for office buildings to establish criteria for the grade of service provided by an installed lift system. These are best expressed as either the percentage of calls answered in specified time intervals, or the time to answer a specified percentage of calls. That is to serve the majority of users as well as possible and to some extent ignore the tail of the service distribution. This tail to the distribution has been called (*sic*) the "forgotten man" problem. Thus 100% satisfaction should never be quoted.

Table 6.5 Office Building Average System Response Time Performance

Grade of Service	Percentage of calls answered in		Time to answer calls (s)	
	30 s	60 s	50%	90%
Excellent	>75	>98	20	45
Good	>70	>95	22.5	50
Fair	>65	>92	25	55
Poor/unacceptable	<65	<92	>25	>55

Table 6.5 indicates the percentage and time values for several grades of service over one hour of peak activity in an office building. An hour of peak activity is taken in order to obtain sensible and realisable results. It should be possible to obtain the grades of service indicated in the table during the worst hour of activity. This might occur during the mid day break rather than during the intense, but shorter, uppeak and down peak periods at the beginning and end of the working day. To allow for this during a peak fifteen minute period, say down peak, the next lower grade of service should be possible.

During a peak five minute period, say during uppeak, two grades of service lower should be possible.

To illustrate the use of these criteria, suppose the building being considered has a good system installed. Then over one hour the system should provide the figures given in the Good row, in down peak the figures given in the Fair row and during uppeak the figure given in the Poor/unacceptable row. This illustrates that passenger expectations change according to the purpose of their travel.

6.7 OTHER USEFUL DESIGN PARAMETERS

6.7.1 Passenger Average Travel Time to Destination (*ATT*)

It is useful to know the average time it would take for a passenger to reach their destination floor (assumed to be half way up the building zone being served) after their allocated lift is ready for boarding, ie: opened its doors. This extra knowledge would help to evaluate the suitability of a planned lift group. It is, however, a secondary Quality of Service design consideration after average passenger waiting time. This is because passengers travelling to the upper floors of a building zone become annoyed if a lift takes too long to reach their floor. Strakosch (1998) states[1] that for most people 100 s is a tolerable travel time which can be further tolerated to some 150 s of travel time if two people exit at each stop. He regards 180 s as the absolute limit. These should be considered maximum values with average values being about half of these.

Definition 6.8: The passenger average travel time (*ATT*) is the average period of time, in seconds, which an average passenger takes to travel from the main terminal floor to the requested destination floor, measured from the time the passenger enters the lift until alighting at the destination floor.

A quick rule of thumb, which has been used to evaluate this time, is to use the formula of: adding one half of the uppeak interval for a group of lifts to one quarter of the uppeak round trip time for the individual lift in the group *viz*:

$$ATT = 0.5 \times UPPINT + 0.25 \times UPPRTT \tag{6.2}$$

The figure obtained considerably understates the likely *ATT* as it "forgets" how quickly a car expresses back to the main terminal floor after the last passenger has alighted. A better rule of thumb, found by comparison to calculations, is to add one half of the uppeak interval to one half of the uppeak round trip time, *viz*:

$$ATT = 0.5 \times UPPINT + 0.5 \times UPPRTT \tag{6.3}$$

A more accurate estimate of how long it takes the average passenger to reach their destination is to modify the round trip time in Equation (4.11) and calculate *ATT* to the midpoint of the local travel for any group of lifts. This means travel for a distance of *H*/2 with the number of stops being *S*/2 and a transfer of *P*/2 passengers boarding the lift and *P*/2 passengers alighting.

---

[1] 3rd edition, page 64.

The resulting formula is given below:

$$ATT = 0.5Ht_v + 0.5St_s + Pt_p \tag{6.4}$$

If there is any express travel through a number of floors (E_j), the formula becomes:

$$ATT = 0.5Ht_v + 0.5St_s + Pt_p + t_e \tag{6.5}$$

Thus Equation (6.5) calculates *ATT* to the midpoint of the local and express travel for any group of lifts. This will be to a point halfway between the lobby and the high call reversal floor (H). Also the equation takes account of the passenger transfer times and the express travel. To illustrate this consider a lift carrying eight passengers serving a building with 22 floors above the main terminal. What is the position of the average destination floor ?

Using Table 5.1, the column for a 10 person rated load allows 8 passengers in the car. Following this column down to the line corresponding to 22 floors shows that the highest reversal floor is Floor 20. The average destination floor is thus Floor 10.

6.7.2 Passenger Average Journey Time (*AJT*)

The primary consideration of passenger average waiting time (*AWT*) can be combined with the secondary consideration of passenger average travel time (*ATT*) to give a passenger average journey time (*AJT*).

Definition 6.9: The passenger average journey time is the average period of time, in seconds, measured from the instant an average passenger first registers a landing call (or arrives at the landing), until alighting at the destination floor.

Thus the passenger average journey time is the sum of the average passenger travel time (*ATT*) and the average passenger waiting time (*AWT*). The average passenger travel time is simple to calculate, but the average passenger waiting time depends on car loading, which can only be determined after the car size has been selected. The passenger average travel time plus one half of the uppeak interval will give a close approximation for evaluation purposes.

$$AJT = 0.5Ht_v + 0.5St_s + 1.0Pt_p + t_e + 0.5INT \tag{6.6}$$

The passenger average journey time is more accurately obtained by adding the average passenger waiting time to Equation (6.5) and is given by:

$$AJT = 0.5Ht_v + 0.5St_s + 1.0Pt_p + t_e + AWT \tag{6.7}$$

The average waiting time should be estimated from Table 6.4 according to the car loading.

6.7.3 Summary of *AWT*, *ATT* and *AJT*

The Quality of Service is particularly important for office buildings. The values given in Table 6.6 indicate the performance times to aim for in a traffic design and the maximum acceptable values.

Table 6.6 Summary of times

Time	Aim for	Poor
AWT	<20 s	>25 s
ATT	<60 s	>70 s
AJT	<80 s	>90 s

Where a passenger uses a shuttle lift (a two stop lift serving the main entry level and an upper lobby) to first reach the upper terminal floor and then uses another group of lifts to reach their final destination floor, the values obtained for *AWT, ATT* and *AJT* should be calculated separately for each journey. For example, in a building with a shuttle service to floor 40 and a transfer to a group of lifts serving an upper zone of 20 floors, it is necessary to calculate the shuttle and upper zone time values separately.

6.8 EXAMPLE 6.4

A speculative building, of no great prestige, with 10 floors above the main terminal floor is to be built. Each floor has 1500 m² of gross space. The interfloor height is a regular 3.3 m and assume the passenger transfer time is 1.2 s.

6.8.1 Given data

$N = 10$, $d_f = 3.3$ m, gross floor area 1500 m², $t_p = 1.2$ s.

A speculative building could be occupied by one tenant, ie: single tenancy.

The usable area could be 80% of gross, ie: 1200 m².

Table 6.1 indicates that the density of occupation for a regular building is in the range 10−12 m² per person. As a speculative building assume worst case, ie: 10 m²/person. The population will be:

    1200/10 = 120 persons/floor

Total population will be:

    120 × 10 = 1200 persons

Assume 80% daily attendance (see Section 6.5) and the design population becomes 960 persons.

Table 6.2 indicates that 11−15% of the population will arrive in the busiest 5-minute period. Assume the worst case is 15%, then the peak arrival rate will be:

    960 × 0.15 = 144 persons

Table 6.3 gives a suitable interval as 25−30 s. The building is speculative, so assume 30 s.

Design the lift system to handle 144 people with an interval of 30 s.

6.8.2 Initial sizing

There will be 10 trips in 5-minutes (300 seconds), ie: a 30 s interval. This means that 14.4 passengers (on average) are transported on each trip. If this represents 80% car occupancy, then the rated load would need to be:

$14.4/0.8 = 18$ persons

The nearest car (BS ISO4190) size is 1600 kg (21 persons).

6.8.3 Calculation

It is now necessary to evaluate the round trip time equation (Equation 4.11).

$$RTT = 2Ht_v + (S+1)t_s + 2Pt_v$$

The total travel is 10×3.3 ie: 33 m. From Table 5.2 this suggests a rated speed of 1.6 m/s. Therefore the single floor transit time is:

$t_v = 3.3/1.6 = 2.1$ s

Table 5.3 suggests the likely single floor flight time $(t_f(1))$ could be 6.0 s.

From Table 5.6 select 1100 mm centre opening doors, which gives a door opening time of 0.8 s (advanced opening) and a door closing time of 3.0 s. Thus a value for T can be obtained as:

$T = 6.0 + 0.8 + 3.0 = 9.8$ s.

$t_s = 9.8 - 2.1 = 7.7$

A 21 person rated load car has been selected, hence from Table 5.1 values for H and S can be obtained:

$P = 16.8; H = 9.8; S = 8.3$

Using Equation (4.11):

$$RTT = 2 \times 9.8 \times 2.1 + (8.3+1)(7.7) + 2 \times 16.8 \times 1.2$$
$$= 41.2 + 71.6 + 40.3$$
$$= 153.1 \text{ s}$$

To achieve an uppeak interval of 30 s (or thereabouts) would require 5 cars.

$UPPINT \quad = 153.1/5 = 30.6$ s

The uppeak handling capacity will then be:

$UPPHC \quad = (300 \times 16.8)/30.6 = 165$ persons/5-minutes

The design installation would comprise 5 cars of 21 person rated car capacity. This would deliver an uppeak interval of 30.6 s and an uppeak handling capacity of 165 person per 5 minutes. There would be too much handling capacity, but an acceptable interval.

6.9 EXAMPLE 6.5

Suppose the building of Example 6.4 were now to be a prestigious building, what system would then be required ?

The assumed floor population will now be 12 m^2/person, giving a total population of 1000 persons. At 80% attendance the design population becomes 800 persons. Again, assuming the worst case of a 15% peak, the arrival rate will be 120 persons. The interval required will be 25 s.

Design the lift system to handle 120 people with an interval of 25 s.

With 12 trips per 5 minutes the car size could be 1000 kg (13 person). A car of this size would be considered too small for a prestige office building. A better size would be a 1275 kg (16 person) lift. This gives values for $P = 12.8$; $H = 9.7$; $S = 7.4$.

Keeping the lift dynamic times the same and using Equation (4.11):

$$RTT = 2 \times 9.7 \times 2.1 + (7.4 + 1)(7.7) + 2 \times 12.8 \times 1.2$$
$$= 40.7 + 64.7 + 30.7$$
$$= 136.1 \text{ s}$$

To achieve an uppeak interval of 25 s (or thereabouts) would require 5 cars.

$$UPPINT \quad = 136.1/5 = 27.2 \text{ s}$$

The uppeak handling capacity will then be:

$$UPPHC \quad = (300 \times 12.8)/27.2 = 141 \text{ persons/5-minutes}$$

The design installation would comprise 5 cars of 16 person rated car capacity. This would deliver an uppeak interval 27.2 s, which is longer than specified and an uppeak handling capacity of 141 person per 5 minutes, which is larger than specified.

6.10 AN IMPROVED DESIGN PROCEDURE

6.10.1 The Iterative Balance Method

In the previous two examples the handling capacity was larger than that required and in Example 6.5 the required interval was not achieved. What should the designer do ? Clearly in Example 6.5 there was too much handling capacity and a poorer interval than desired. In effect there were not enough intending passengers to use the capability of the installed system.

Should the designer modify the component parts of the *RTT* expression to achieve a balance ? The first term can only be altered by changing the rated speed and the effect would be small. The second term can be altered by changing the single floor flight time or door timings. Thus a lower specification door gear could produce a matching handling capacity. The third term can be altered by changing the rated car capacity, but this will alter *S* and *H* and can be counterproductive. Experienced designers will use intuitive procedures incorporating combinations of the above to establish a suitable design to cater for the desired handling capacity.

Using the conventional design method above, a designer would propose initial values for the dynamic parameters and estimate the rated car capacity of the lifts based on experience. It is then assumed that the average number of passengers (*P*) carried per trip is 80% of the rated car capacity and values for the expected number of stops (*S*) and the average highest reversal floor (*H*) are evaluated. Hence the round trip time (*RTT*), interval (*INT*) and handling capacity (*HC*) are calculated.

At this stage, the designer compares the calculated value of *HC* with the number of passengers to be moved in the peak five minutes. If *HC* is greater than or equal to this number of passengers, then the designer is satisfied that the system will cope with the traffic. The configuration will be trimmed if the handling capacity is too large, and should it be smaller than the required value, then the designer must repeat the evaluation for more and (or) bigger and (or) faster cars. However, the values calculated for *RTT*, *INT* and *HC* are exact only if there is a perfect match between the arrival rate and handling capacity.

The procedure suggested by Tregenza (1972) as Equation (5.17) is significant as he presents the idea of matching the lift handling capacity to the desired handling capacity exactly. This is achieved by not rigidly fixing *P* as a percentage of rated car capacity. *P* is allowed to take the most appropriate value. From Tregenza, *P* is equal to λINT.

A new design procedure, the Iterative Balance Method (Barney and Dos Santos, 1975), can now be proposed, which can be used with either the conventional or Poisson formulae. For simplicity, it is presented as a series of steps (Table 6.7).

To obtain values for *H* and *S* for the number of passengers to be carried in the car, Table 5.1 could be used by interpolating between the 80% values shown in parenthesis. These values follow a nonlinear sequence and make it difficult to estimate intermediate values. To assist, Table 6.8 is provided which gives values for integer values of *P* from five to 20 persons.

The Iterative Balance Method is a classical two point boundary problem, where the start and end results are known − in this case the balancing interval. To arrive at an answer, a two point boundary problem has to iterate, ie: change the start value to converge on the end value. To do this a suitable algorithm has to be chosen. Pick the wrong algorithm and the start and end values diverge rapidly, leading to infinite values, ie: no solution. The algorithm chosen in step (7), which is simply "twice the new value minus the old value", does converge.

In step (8) it is important to remember that, where an average car load is much greater than 80%, poor passenger service might result, ie: long waiting times and queues. The designer must select a suitable car size to meet desired economic and operating conditions. It is important not to make any simplifications either for the sake of arithmetic ease or to meet a preconceived idea of a suitable lift installation at any stage of the calculation until step (8) is reached. Always calculate primary and secondary values as precisely as possible.

The procedure outlined in Table 6.7 allows the *RTT* to be calculated for any values of *P*, *H* and *S*. Only when the initial (estimated) interval matches the final (calculated) interval are any decisions made.

Table 6.7 New design procedure

Step	Procedure
1	Decide on λ rate of passenger arrivals over 5 minutes
2	Obtain or decide upon lift system data: N Number of floors t_v the interfloor time t_s the operating time t_p the passenger transfer time
3	Estimate an appropriate interval
4	Obtain: P average car load H average reversal floor S expected number of stops
5	Calculate *RTT*
6	Select L, the number of lifts to produce an interval close to that estimated in step (3)
7	Compare the estimated interval in step (3) with the calculated interval in step (6) and if significantly different, estimate another value for the interval and then iterate from step (4). A possible new trial could be: New $INT = INT$ (step (6)) + [INT (step (6)) − INT (step (3))]
8	Select a suitable standard car capacity, which allows approximately 80% average car load

6.10.2 Example 6.6

In Example 6.5 there was too much handling capacity and a poorer interval than desired. Using the Iterative Balance Method indicated in Table 6.7, the following is obtained:

(1) λ = 120 persons/5-minutes = 120/300 = 0.4 persons/s

(2) $N = 10$
$t_v = 3.3/1.6 = 2.1$ s
$t_s = 7.7$ s
$t_p = 1.2$ s

(3) Let $INT = 25$ s

(4) $P = \lambda INT = 0.4 \times 25 = 10$ persons
$H = 9.5$ (calculated from Equations (5.12) and (5.5)
$S = 6.5$ or from interpolation in Table 6.8)

(5) $RTT = 2 \times 9.5 \times 2.1 + 7.5 \times 7.7 + 2 \times 10 \times 1.2 = 121.7$ s

(6) Let $L = 5$, then $INT = 121.7/5 = 24.3$ s

Table 6.8a Values of H and S for different values of P from 5 to 12 persons

Floors	5		6		7		8		9		10		11		12	
N	H	S	H	S	H	S	H	S	H	S	H	S	H	S	H	S
5	4.6	3.4	4.7	3.7	4.8	4.0	4.8	4.2	4.9	4.3	4.9	4.5	4.9	4.6	4.9	4.7
6	5.4	3.6	5.6	4.0	5.7	4.3	5.7	4.6	5.8	4.8	5.8	5.0	5.9	5.2	5.9	5.3
7	6.3	3.8	6.4	4.2	6.5	4.6	6.6	5.0	6.7	5.3	6.7	5.5	6.8	5.7	6.8	5.9
8	7.1	3.9	7.3	4.4	7.4	4.9	7.5	5.3	7.6	5.6	7.7	5.9	7.7	6.2	7.8	6.4
9	8.0	4.0	8.2	4.6	8.3	5.1	8.4	5.5	8.5	5.9	8.6	6.2	8.7	6.5	8.7	6.8
10	8.8	4.1	9.0	4.7	9.2	5.2	9.3	5.7	9.4	6.1	9.5	6.5	9.6	6.9	9.6	7.2
11	9.6	4.2	9.9	4.8	10.1	5.4	10.2	5.9	10.3	6.3	10.4	6.8	10.5	7.1	10.6	7.5
12	10.5	4.2	10.7	4.9	11.0	5.5	11.1	6.0	11.2	6.5	11.3	7.0	11.4	7.4	11.5	7.8
13	11.3	4.3	11.6	5.0	11.8	5.6	12.0	6.1	12.1	6.7	12.3	7.2	12.3	7.6	12.4	8.0
14	12.1	4.3	12.5	5.0	12.7	5.7	12.9	6.3	13.0	6.8	13.2	7.3	13.3	7.8	13.4	8.2
15	13.0	4.4	13.3	5.1	13.6	5.7	13.8	6.4	14.0	6.9	14.1	7.5	14.2	8.0	14.3	8.4
16	13.8	4.4	14.2	5.1	14.5	5.8	14.7	6.5	14.9	7.0	15.0	7.6	15.1	8.1	15.2	8.6
17	14.6	4.4	15.0	5.2	15.3	5.9	15.6	6.5	15.8	7.1	15.9	7.7	16.0	8.3	16.1	8.8
18	15.5	4.5	15.9	5.2	16.2	5.9	16.5	6.6	16.7	7.2	16.8	7.8	16.9	8.4	17.1	8.9
19	16.3	4.5	16.8	5.3	17.1	6.0	17.4	6.7	17.6	7.3	17.7	7.9	17.9	8.5	18.0	9.1
20	17.1	4.5	17.6	5.3	18.0	6.0	18.2	6.7	18.5	7.4	18.6	8.0	18.8	8.6	18.9	9.2
21	18.0	4.5	18.5	5.3	18.8	6.1	19.1	6.8	19.4	7.5	19.6	8.1	19.7	8.7	19.8	9.3
22	18.8	4.6	19.3	5.4	19.7	6.1	20.0	6.8	20.3	7.5	20.5	8.2	20.6	8.8	20.8	9.4
23	19.6	4.6	20.2	5.4	20.6	6.2	20.9	6.9	21.2	7.6	21.4	8.3	21.5	8.9	21.7	9.5
24	20.5	4.6	21.1	5.4	21.5	6.2	21.8	6.9	22.1	7.6	22.3	8.3	22.5	9.0	22.6	9.6

N is the number of floors above the main terminal.

Table 6.8b Values of H and S for different values of P from 13 to 20 persons

Floors	13		14		15		16		17		18		19		20	
N	H	S	H	S	H	S	H	S	H	S	H	S	H	S	H	S
5	4.9	4.7	5.0	4.8	5.0	4.8	5.0	4.9	5.0	4.9	5.0	4.9	5.0	4.9	5.0	4.9
6	5.9	5.4	5.9	5.5	5.9	5.6	5.9	5.7	6.0	5.7	6.0	5.8	6.0	5.8	6.0	5.8
7	6.9	6.1	6.9	6.2	6.9	6.3	6.9	6.4	6.9	6.5	6.9	6.6	6.9	6.6	7.0	6.7
8	7.8	6.6	7.8	6.8	7.9	6.9	7.9	7.1	7.9	7.2	7.9	7.3	7.9	7.4	7.9	7.4
9	8.7	7.1	8.8	7.3	8.8	7.5	8.8	7.6	8.8	7.8	8.9	7.9	8.9	8.0	8.9	8.1
10	9.7	7.5	9.7	7.7	9.8	7.9	9.8	8.1	9.8	8.3	9.8	8.5	9.8	8.6	9.9	8.8
11	10.6	7.8	10.7	8.1	10.7	8.4	10.7	8.6	10.8	8.8	10.8	9.0	10.8	9.2	10.8	9.4
12	11.6	8.1	11.6	8.5	11.6	8.7	11.7	9.0	11.7	9.3	11.7	9.5	11.8	9.7	11.8	9.9
13	12.5	8.4	12.5	8.8	12.6	9.1	12.6	9.4	12.7	9.7	12.7	9.9	12.7	10.2	12.8	10.4
14	13.4	8.7	13.5	9.0	13.5	9.4	13.6	9.7	13.6	10.0	13.7	10.3	13.7	10.6	13.7	10.8
15	14.4	8.9	14.4	9.3	14.5	9.7	14.5	10.0	14.6	10.4	14.6	10.7	14.6	11.0	14.7	11.2
16	15.3	9.1	15.4	9.5	15.4	9.9	15.5	10.3	15.5	10.7	15.6	11.0	15.6	11.3	15.6	11.6
17	16.2	9.3	16.3	9.7	16.4	10.2	16.4	10.6	16.5	10.9	16.5	11.3	16.6	11.6	16.6	11.9
18	17.2	9.4	17.2	9.9	17.3	10.4	17.4	10.8	17.4	11.2	17.5	11.6	17.5	11.9	17.6	12.3
19	18.1	9.6	18.2	10.1	18.2	10.6	18.3	11.0	18.4	11.4	18.4	11.8	18.5	12.2	18.5	12.6
20	19.0	9.7	19.1	10.2	19.2	10.7	19.3	11.2	19.3	11.6	19.4	12.1	19.4	12.5	19.5	12.8
21	19.9	9.9	20.0	10.4	20.1	10.9	20.2	11.4	20.3	11.8	20.3	12.3	20.4	12.7	20.4	13.1
22	20.9	10.0	21.0	10.5	21.1	11.1	21.1	11.5	21.2	12.0	21.3	12.5	21.3	12.9	21.4	13.3
23	21.8	10.1	21.9	10.7	22.0	11.2	22.1	11.7	22.2	12.2	22.2	12.7	22.3	13.1	22.3	13.5
24	22.7	10.2	22.9	10.8	22.9	11.3	23.0	11.9	23.1	12.4	23.2	12.8	23.2	13.3	23.3	13.8

N is the number of floors above the main terminal.

(7) The value calculated of 24.3 s does not closely match the start value of 25 s. Try a new value *viz*:

$$24.3 + (24.3 - 25) = 23.6 \text{ s}$$

(4′) $P = \lambda INT = 0.4 \times 23.6 = 9.4$ persons
$H = 9.4$
$S = 6.3$

(5′) $RTT = 2 \times 9.4 \times 2.1 + 7.3 \times 7.7 + 2 \times 9.4 \times 1.2 = 118.3$ s

(6′) Let $L = 5$, then $INT = 118.3/5 = 23.7$ s

(7′) This is a close enough match (error <0.1 s)

(8) Standard rated car capacities are 10, 13 and 16. The 10 passenger car size would be too small and 16 passenger car size too large. Try 13 person cars.

$$9.4/13 \times 100 = 72\%$$

Examining the result above, it can be seen that the arrival rate and the lift handling capacity exactly balance. This was achieved by the number of passengers in the car changing until the resulting system handling capacity was equal to the demand represented by the passenger arrival rate. The interval is less than the specified 25 s, but not a lot smaller. This is an improvement.

6.10.3 Example 6.7

As the Iterative Balance Method is important, a further example is presented.

The arrival rate is 110 calls/5-minutes and an interval of 30 s is required.

Relevant data are: $N=16$, $t_v=1.3$ s, $t_s=6.4$ s, $t_p=1.2$ s.

(1) $\lambda = 110$ persons/5-minutes $= 110/300 = 0.37$ persons/s

(2) $N = 16$
$t_v = 3.3/2.5 = 1.32$ s
$t_s = (4.3 - 1.32) + 3.4 = 6.4$ s
$t_p = 1.2$ s

(3) Let $INT = 30$ s

(4) $P = \lambda INT = 0.37 \times 30 = 11$ persons
$H = 15.1$ (calculated from Equations (5.12) and (5.5)
$S = 8.1$ or from interpolation in Table 6.8)

(5) $RTT = 2 \times 15.1 \times 1.32 + 9.1 \times 6.4 + 2 \times 11 \times 1.2 = 124.5$ s

(6) Let $L = 4$, then $INT = 124.5/4 = 31.1$ s

(7) The value calculated of 31.1 s does not match the start value of 30 s. Try a new value *viz*:

31.1 + (31.1 − 30) = 32.2 s

(4′) $P = \lambda INT = 0.37 \times 32.2 = 11.8$ persons
     $H = 15.2$
     $S = 8.5$

(5′) $RTT = 2 \times 15.2 \times 1.32 + 9.5 \times 6.4 + 2 \times 11.8 \times 1.2 = 129.2$ s

(6′) Let $L = 4$, then $INT = 129.2/4 = 32.3$ s

(7′) This is a close enough match (error <0.1 s)

(8) Standard rated car capacities are 13, 16 and 21. The 13 passenger car size would be too small and 21 passenger car size too large. Try 16 person cars.

11.8/16 × 100 = 74%

This is a satisfactory system. Remember that the most suitable lift configuration chosen would be based on economic grounds, the cost of the system and performance.

6.11 EPILOGUE

In this chapter a number of definitions have been proposed. It is important that each design be prepared against a set of mutually understood definition of terms. Unfortunately most designers and lift companies have their own set. A set of over 50 terms, widely used in the USA, are those produced by the NEII (2000). Another set of nearly 2000 terms and cross references can be found in Barney *et al.* (2001).

CHAPTER SEVEN

Uppeak Traffic Calculations: Limitations and Assumptions

7.1 SOME ASSUMPTIONS IN THE DERIVATION OF THE ROUND TRIP EQUATION

The standard traffic design uses the "pure" uppeak calculation, ie: with only an up flow of passengers, to determine the likely performance of a lift system. The calculation of a value for the round trip time using Equation (4.11) relies on a number of assumptions. These are that:

(1) the traffic profile is ideal
(2) all floors are equally populated or present equal attraction
(3) rated speed is achieved for a single floor jump
(4) interfloor heights are assumed constant
(5) the traffic supervisory system is assumed ideal
(6) various lost times, such as passenger disturbance, despatch intervals, loading intervals, etc. are negligible
(7) passengers arrive uniformly in time (rectangular probability function).

It is proposed to deal now with each of these assumptions and their effects on the value obtained for the round trip time (*RTT*).

7.2 THE TRAFFIC PROFILE IS IDEAL

Figure 4.4 in Chapter 4 indicated that the 5-minute peak traffic profile would be considered as a rectangle of height equal to the peak arrival rate and of width 5-minutes. That is, it had a square top and nothing happened before or after the peak 5-minutes. During the peak 5-minutes the lifts are assumed to be evenly distributed about the building, will pick up equal numbers of passengers and make an equal number of stops on each round trip.

In practice the shape of the arrival profile is as shown by the curved thick line of Figure 4.4. So there is not at any time a constant arrival rate of passengers. Also the lifts will have been carrying out service around the building at the start of the peak 5-minutes and can be in any spatial positions.

Undoubtedly the past history of the lift positions and the varying arrival rate will make some affect on the calculation, but what ?

7.3 UNEQUAL FLOOR POPULATIONS OR DEMAND

In the derivations of the highest reversal floor (*H*) and the expected number of stops (*S*), it was assumed all floors held equal attraction for calls originating at the main terminal. In real buildings this is not always so. Floor populations can vary and result in unequal demand. Alternatively, floor populations may be substantially equal, but work routines may be such as to cause unequal rates of population arrival during the peak period. The effect of unequal floor demands is to modify the derivations of H and S.

7.3.1 Number of Stops (S) for Unequal Demand

Consider a building with:

- N floors above the main terminal
- P the average number of passengers present in a lift as it leaves the main terminal
- U the total building population above the main terminal
- U_i the population of floor i.

Using a similar approach to that of Section 5.3, the probability that one passenger will leave the lift at any particular floor i is:

$$\frac{U_i}{U} \tag{7.1}$$

The variable U_i has been represented as the population: it could equally represent the demand for floor i.

The probability that one passenger will leave the lift at the first floor is then:

$$\frac{U_1}{U} \tag{7.1}$$

Assuming that the passengers are independent of each other, the probability that one passenger will not leave the lift at the first floor is:

$$1 - \frac{U_1}{U} \tag{7.2}$$

The probability that none of the P passengers in the lift will leave the lift at the first floor is:

$$\left[1 - \frac{U_1}{U}\right]^P \tag{7.3}$$

Thus the probability that no passengers will leave the lift for the first i floors is:

$$\left[1 - \frac{U_1}{U}\right]^P + \left[1 - \frac{U_2}{U}\right]^P + \dots \left[1 - \frac{U_i}{U}\right]^P \tag{7.4}$$

This is synonymous to the lift not stopping at the first i floors.

Then the probability that stops will be made at the first i floors is:

$$1 - \left[1 - \frac{U_1}{U}\right]^P + \left[1 - \frac{U_2}{U}\right]^P + \dots \left[1 - \frac{U_i}{U}\right]^P$$

$$= 1 - \sum_{j=1}^{i}\left[1 - \frac{U_i}{U}\right]^P \tag{7.5}$$

Hence it can be shown (after some algebraic manipulation) that the expected number of stops for N floors is:

$$S = N\left[1 - \frac{1}{N}\sum_{i=1}^{N}\left[1 - \frac{U_i}{U}\right]^P\right] \tag{7.6}$$

This is the general case of Equation (5.5) for equal floor demands. Equation (7.6) becomes that of Equation (5.5) when $U_1 = U_2 = U_3 = \ldots = U_i = U/N$.

Hence:

$$S = N - \left[N\left[1 - \frac{1}{N}\right]^P\right] \tag{7.7}$$

$$= N\left[1 - \left[\frac{N-1}{N}\right]^P\right] \tag{5.5}$$

The evaluation by hand of Equation (7.6) for S is obviously tedious, but it is easily calculated by digital computer. Simplifications can be made for cases where there is no population on some floors and equal populations or demand on others, and where groups of floors have the same populations or demand, as for example in "stepped" buildings.

7.3.2 Examples 7.1 and 7.2

To illustrate some easy and often common population distributions/demands, consider the following two examples.

(a) Example 7.1

What is a suitable expression, where there is no population on some floors, or the lift is not permitted to stop at some floors and there is equal population on other floors ?

Consider that the lift is not to stop at floor 1 and there is no population on floor 3.

This is best dealt with using Equation (5.5) rather than the general Equation (7.6). Quite simply the number of served floors is $N - 2$. Hence:

$$S = (N-2)\left[1 - \left[\frac{N-3}{N-2}\right]^P\right] \tag{7.8}$$

Thus it is possible to suggest that instead of thinking of N as the number of floors above the main terminal, it can be considered as the number of floors served above the main terminal; in which case equations like Equation (7.7) are not appropriate and Equation (5.5) or Equation (7.6) should always be used instead. Thus to determine the number of served floors N, simply deduct the number of floors not being served from the number of possible floors in a building zone.

(b) Example 7.2

What happens where there are groups of floors with similar population densities or demands ?

The lower part of a building comprising B floors has a population of U_B per floor and the upper part of the building of T floors has a population of U_T per floor. Formulate an expression to cover this case.

The number of floors served is $N = B + T$. It is useful to use Equation (7.7) which was an intermediate step to obtain Equation (4.5).

$$S = N - \left[B\left[1 - \frac{U_B}{U}\right]^P + T\left[1 - \frac{U_T}{U}\right]^P \right] \tag{7.9}$$

This shows the reduction of N by the deduction of the two probabilities of no stops at floors in the two parts of the building to obtain a value for S.

7.3.3 Highest Reversal Floor (H) for Unequal Demand

Using the previous definitions in Section 7.3.1, the procedure is similar to the approach in Section 5.4. The probability that one passenger will leave the lift at a given floor i is:

$$\frac{U_i}{U} \tag{7.10}$$

which becomes the probability of the lift travelling no higher than the ith floor obtained by extension and subsequent algebraic simplification:

$$\left[\sum_{i=1}^{j} \left[\frac{U_i}{U}\right]^P \right] \tag{7.11}$$

Using the same procedure as in Section 5.4 the expectance[1] H is:

$$H = \sum_{j=1}^{N} j \left[\left[\sum_{i=1}^{j} \frac{U_i}{U}\right]^P - \left[\sum_{i=1}^{j} \frac{U_{i-1}}{U}\right]^P \right] \tag{7.12}$$

which becomes:

$$H = N - \sum_{j=1}^{N-1} \left[\sum_{i=1}^{j} \frac{U_i}{U}\right]^P \tag{7.13}$$

---

[1] The remarks made in the footnote to Section 5.3 regarding variance, expectance and mean also apply to the above derivations of S and H.

Evaluation of Equation (7.13) by manual methods is again somewhat tedious, which may account for some designers guessing H to be equal to N or $N - 1$. The error caused by guessing can be large, eg: for a 6 person lift serving 24 floors H is 20.3 (15% error) giving an error in round trip times of perhaps 5% – 10%. Figure 7.1 illustrates the error (vertical axis) for a different number of floors (horizontal axis) and different lift sizes.

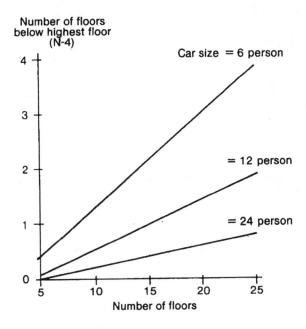

Figure 7.1 Error in the value of H when guessed as equal to N

Little simplification of Equation (7.13) is possible as the summation is a series of different terms. Only when floors have zero population will terms repeat, giving some simplification in the calculation.

7.3.4 Examples 7.3 and 7.4

Two examples are provided to illustrate the calculation of both H and S.

(a) Example 7.3

Consider a building has 10 floors above the main terminal with the following distribution of population:

Floor	1	2	3	4	5	6	7	8	9	10
Population	5	10	25	25	50	50	100	100	100	100

Using a slightly modified Equation (7.6) for S:

$$S = N - \sum_{i=1}^{N} \left[1 - \frac{U_i}{U} \right]^P$$

$$= 10 - \left[\left[1 - \frac{5}{565} \right]^8 + \left[1 - \frac{10}{565} \right]^8 + 2 \left[1 - \frac{25}{565} \right]^8 + 2 \left[1 - \frac{50}{565} \right]^8 + 4 \left[1 - \frac{100}{565} \right]^8 \right]$$

$$= 10 - \frac{1}{565^8} [560^8 + 555^8 + 2 \times 540^8 + 2 \times 515^8 + 4 \times 454^8]$$

$$= 10 - 4.986 = 5.014$$

Using Equation (7.13) for *H*:

$$H = N - \sum_{j=1}^{N-1} \left[\sum_{i=1}^{j} \frac{U_i}{U} \right]^P$$

$$= 10 - \frac{1}{565^8} [5^8 + 15^8 + 40^8 + 65^8 + 115^8 + 265^8 + 365^8 + 465^8]$$

$$= 10 - 0.245 = 9.755$$

For comparison, a building with 10 floors serviced by 10 person lifts serving floors with equal demand would give:

$$S = 5.7; \; H = 9.3$$

With the unequal populations in this example: *S* is smaller at 5.0, but *H* is larger at 9.8.

(b) Example 7.4

Consider a building has 10 floors above the main terminal with the distribution of the populations reversed to those given in Example (7.3):

Floor	1	2	3	4	5	6	7	8	9	10
Population	100	100	100	100	50	50	25	25	10	5

Calculate values for *H* and *S*.

Using Equation (7.6) for *S*:

$$= 10 - \left[4 \left[1 - \frac{100}{565} \right]^8 + 2 \left[1 - \frac{50}{565} \right]^8 + 2 \left[1 - \frac{25}{565} \right]^8 + \left[1 - \frac{10}{565} \right]^8 + \left[1 - \frac{5}{565} \right]^8 \right]$$

This is the same expression as that obtained in Example 7.3 but order of the tems is reversed. Hence $S = 5.014$

Using Equation (7.13) for *H*:

$$H = 10 - \frac{1}{565^8} [100^8 + 200^8 + 300^8 + 400^8 + 450^8 + 500^8 + 525^8 + 550^8 + 560^8]$$

$$H = 10 - 2.99 = 7.01$$

Comparing again with a building with equal floor populations where $S = 5.7$ and $H = 9.3$. With the unequal populations in this example, both *S* and *H* are smaller.

(c) Comments on Examples 7.3 and 7.4

H and *S* can be seen to be dependent on the relative floor demands in a building. Also it can be seen that *S* is independent of the distribution of demand, but that *H* is extremely sensitive to it. Also the unequal populations will make the *RTT* smaller as the dominant term containing *S* is always smaller.

7.4 INTERFLOOR FLIGHT TIMES AND UNEQUAL FLOOR HEIGHTS

7.4.1 Interfloor Flight Times

In modern fast lift systems it will take a travel (flight) of several floors for the rated speed to be reached. Hunt (1975) presents an example of speed/time curves for 5.0 m/s rated speed, 1.5 m/s^2 acceleration, where rated speed is only reached during a flight of seven floors or further. The values below illustrate this point:

Floors jumped	1	2	3	4	5	6	7	8
	$t_f(1)$	$t_f(2)$	$t_f(3)$	$t_f(4)$	$t_f(5)$	$t_f(6)$	$t_f(7)$	$t_f(8)$
Flight time (s)	3.6	4.8	5.7	6.5	7.1	7.7	8.3	8.9
Incremental time (s)		1.2	0.9	0.8	0.6	0.6	0.6	0.6

The times above show that only after a flight of between 3 and 4 floors do the incremental times[1] increase by t_v. This implies that, in this case, the acceleration or deceleration process takes some 3.5 floors, invalidating the assumption that jumps of greater than one floor are travelled at rated speed.

The flight times to travel various distances can be measured on an actual system, when calculating installed performance. To obtain estimates of flight times for proposed systems, either a manufacturer's time distance graph should be used or a calculation can be preformed. The calculation is complex as not only does speed have to be considered, but also values of acceleration/deceleration and values of the rate of change of acceleration/deceleration (jerk) have to be known. A calculation method for lift dynamics is given in Appendix 1.

It is also possible for interfloor flight times to vary according to the load being carried and the direction of travel. This occurs particularly in low cost systems such as those installed in residential accommodation. Some allowance may need to be made for this variation.

Thus a problem exists as to the numeric value to attribute to t_s in Equation (4.11), where it was assumed that the rated speed was attained during a single floor jump. This difficulty is exacerbated as it is not known how many single, double, triple, etc. floor jumps are made during each round trip. Phillips (1973) evaluates acceleration and deceleration periods and running time at rated speed as a function of the distance to accelerate from rest to rated speed, and presents typical time/distance curves, but does not indicate how to cater for rated speed not being attained. Jones (1971) assumes rated speed is always reached, and Petigny (1972) in his detailed work does not refer to the problem.

---

[1] Interfloor distance 3.0 m; rated speed 5.0 m/s; then $t_v = 0.6$ s.

Strakosch (1967) takes fixed standard values, chosen according to the rated speed. Gaver and Powell (1971), in order to linearise their expressions for the round trip time, use an approximated solution. They assume that the lift does an express flight to the lowest stopping floor (lowest floor of zone being served by the group of lifts), does all but one of the expected stops, and finally does a direct flight to the highest stopping floor.

Tregenza (1972) modifies Equation (4.11) for *RTT* when the distance required for acceleration and retardation are each greater than half the interfloor distance. He decomposes the term $(S+1)t_s$, considering that from the main terminal a fraction p_s of all flights produce jumps of a single floor and the remaining $(1 - p_s)$ produce jumps which attain the rated speed. Figure 7.2 illustrates the assumption of rated speed being reached after a jump of one floor.

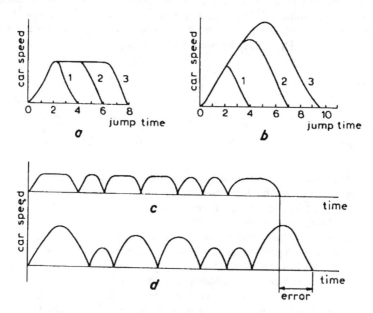

Figure 7.2 Nonlinear flight times for a 13 floor building with 8 stops and jumps of 3 × 1, 2 × 2 and 2 × 3:
(a) interfloor flight times, assuming rated speed is reached in a one floor jump
(b) interfloor flight times, no assumptions
(c) illustration of flight pattern with 8 stops using interfloor flight times as (a)
(d) illustration of flight pattern with 8 stops using interfloors flight times as (b).
Note time error between (c) and (d).

It is possible (Roschier and Kaakinen, 1980) by using probability theory, and knowing the population or demand on each floor, to determine the relative numbers of 1, 2, 3, etc. floor jumps. However, the expression is complex.

Another simplistic approach is to determine the most likely number of floors jumped by evaluating the expression *H/S* and then to approximate to the nearest integer below (designated *LJ*). Some further knowledge of the building, such as any heavily populated or service floors, should enable a designer to estimate whether there will be many jumps in excess of *LJ*. Of course if the rated speed has been reached for jumps equal to or less than *LJ*, all other floors will be travelled at rated speed. If *LJ* is greater than two, some estimate of the number of these jumps must be made.

7.4.2 Unequal Interfloor Distances

All the floors in a building may not be of equal height. For example the entrance floor is often higher to give a spacious aspect. Other floors may be service floors or may accommodate public halls, again causing the interfloor distance to vary. The main effect of unequal floor heights is to increase the travelling time and to further alter interfloor flight times.

It is unlikely that floor heights will vary in a gross fashion, and so probably the easiest method to account for the additional times involved is to adopt the following procedure. Add together the incremental floor distances, extra to the standard interfloor height, multiply by two to account for both directions of travel, and divide by the rated speed to obtain the additional time. The special case of an express jump can also be dealt with in this way.

As an example consider a building where the interfloor distance for most floors is 3.0 m, but four floors are 5.0 m. Then the extra distance will be 16.0 m. If the rated speed were to be 4.0 m/s, then the additional time would be 4.0 s.

7.4.3 Taking Account of Both Interfloor Flight Time Variations and Unequal Interfloor Distances

A more accurate way to determine the effect of both interfloor flight time variations and uneven interfloor distances is to calculate the average jump for an average floor height. This can be achieved by taking the actual distance to travel to the high call reversal floor H and dividing it by the probable number of stops S. The procedure could be:

(1) Determine H/S to give average interfloor jump.

(2) Determine height of building (d_H) to floor H and divide by H to give average interfloor height.

(3) Calculate (1) × (2) to give average distance travelled .

(4) Look up on a manufacturer's time/distance graph[1] the time to travel the distance found in (3), or calculate it with a dynamics program (see Appendix 1).

(5) Calculate the assumed time to travel the distance calculated in (3).

(6) Calculate the difference between the time obtained in (4) and (5).

(7) Add the time obtained to t_s, when calculating the round trip time.

Peters (1997b) carried out many round trip time calculations (RTT) using the flight time to travel the average interfloor distance and compared the results to using the flight time for a standard interfloor height. He found an error of only a few per cent. This was not an unexpected result.

---

[1] An example of a time/distance graph can be seen in Phillips (1973), Figure 2.6.

7.4.4 Example 7.5

Use the method in Section 7.4.3 to estimate the additional time to be added to the *RTT* for a lift system, where floors 4, 8, 12, 16 and 20 are 5.0 m high. The other data are:

$H = 22.2$, $S = 8.0$, $d_f = 3.0$ m, $v = 5.0$ m/s,
$t_f(1) = 3.6$ s, $t_f(2) = 4.7$ s, $t_f(3) = 5.6$ s, $t_f(4) = 6.4$ s.

(1) Average jump: 22.2/8.0 = 2.77 floors.

(2) Average interfloor height: (22.2 × 3.0 + 5 × 2.0)/22.2 = 3.45 m.

(3) Average distance: 3.45 × 2.77 = 9.6 m.

(4) Travel time: for 3.2 floors (9.6/3.0).
    By interpolation from the time values for $t_f(3)$ and $t_f(4)$, the time to jump 9.6 m would be about 5.8 s.

(5) Assumed time: for $t_f(1)$ was 3.6 s and for $t_f(3.2)$ will be 3.6 + 2.2 × 0.6 = 4.9 s. (0.6 s is from d_f/v, ie: 3.0/5.0; and 2.2 is from 3.2 − 1)

(6) Difference time: 5.8 − 4.9 = 0.9 s.

(7) Additional time: add 0.9 s to t_s.

7.5 EFFECT OF TRAFFIC SUPERVISORY SYSTEM

The conventional calculation procedure based on the *RTT* expression requires that lifts present themselves at the main terminal evenly spaced by a period of time equal to one interval. Two factors can upset this situation.

First the uppeak five minutes is part of a continuous process, as shown in Figure 4.12. Hence lifts, in practice, will already be transporting passengers at the onset of the uppeak. As a result lifts will tend to be randomly dispersed around the building. A severe condition occurs when the distribution of lifts becomes so disturbed that lifts bunch together and move round the building together. The effect of bunching is to reduce the quality of service by making most passengers wait longer for lifts.

Secondly, the traffic supervisory control system during most of the day is arranged to deal with interfloor traffic. The peak periods, uppeak, down peak and lunch time two way traffic are generally supervised by special algorithms, which must be switched on when required. The changeover from interfloor to uppeak or down peak control is achieved by monitoring (say) car load, and when this exceeds a predetermined value the appropriate control algorithm is selected. Thus the uppeak algorithm must be active just before the peak five minutes starts, if it is to be effective.

It is assumed in the round trip calculation that during uppeak all lifts are express to the main terminal after depositing the last passenger at the highest floor. This is all the uppeak traffic supervisory algorithm can do for uppeak service. However the time when the uppeak control is switched in is important. If it is too late, only those cars with completed trips will be travelling to the main terminal. Those lifts just starting, or part way through, a round trip can only become available at the main terminal some two or three minutes later, halfway through the uppeak period ! Some may still be dealing with interfloor or down direction traffic.

It is essential therefore to detect the uppeak condition well before it "takes off" to ensure the full lift system handling capacity is available at the lobby. Some authors (Strakosch, 1967; Tregenza, 1972; Phillips, 1973; Morley, 1962) suggest the *RTT* should be increased to account for running out of schedule by (say) 10%. This so called "inefficiency factor" should not form a part of the design calculation, but be considered after a design has been completed. Some uppeak control systems allow down direction and other traffic to be catered for on a timed basis (at least three minute wait). If this feature is incorporated then the uppeak handling capacity is reduced for the time the delegated lift is not serving up peak.

Various techniques are employed to improve uppeak performance, such as uppeak subzoning, sectoring and hall call allocation. These techniques can still be analysed by calculation and will be discussed in Chapter 12.

7.6 VARIOUS LOST TIMES

The first two terms of the round trip time, Equation (3.11), have been calculated as idealised values. This is not unreasonable as they are basically dependent on electromechanical machinery. The third term relating to passenger transfer times is far more speculative as it is dependent on human behaviour, which is unpredictable at the best of times.

The round trip time can be considerably increased by passengers unthinkingly or maliciously preventing lift cars from continuing on their journey. For example, a passenger may hold a door whilst finishing a conversation; or a tea trolley may be loaded thereby reducing the car capacity; or, maliciously, a person may enter additional car calls. The effect of all these disruptions is an increase in *RTT*.

It is difficult to quantify these disruptions. Some account might be taken by following Phillips (1973) and Strakosch (1967) and adding 10% to the *RTT*, or it might be better to increase the passenger transfer times as the effect is due to passenger behaviour and is probably proportional to their numbers. It is always worth bearing in mind possible disruption when selecting a lift system. It could be sensible to select one with capacity in hand, in order to counteract this unpredictable effect.

There are four control features which can affect the design calculation. First, some lift control systems cause a lift to remain at the main terminal for a fixed time interval (Dispatch Interval). Secondly, others hold a lift at the main terminal for a fixed time from the registration of the first car call (Loading Interval). These two control features should not affect the running of the lifts provided:

dispatch interval < time to load 80% of the rated car capacity
loading interval < time to load 80% of the rated car capacity

Thirdly, other control systems hold a lift at the main terminal floor with its doors closed until the previous lift is full, or is ready to leave (Next Car). This feature simply delays the departure of a lift from the main terminal floor.

Finally, some lift systems cause the lift doors to be held open for a fixed time at each stop (Door Holding Interval, or Door Dwell Time) before attempting to reclose the doors. This is provided to allow the movement of passengers into and out of the lift without collision with the doors. This time is often extended to accommodate the movement of push chairs, bicycles, and the elderly and the disabled to enter safely. Typically the door dwell time will be longer for a stop in response to a landing call than for a car call, on the assumption that passengers may alight as well as board the lift. Also the door dwell time, in systems without dispatch, or loading interval features, may be longer at the main terminal than at other floors. If door dwell times are longer than the time to off load the average number of passengers per stop than the passenger transfer time t_p should be set to zero.

7.7 PASSENGERS ARRIVE UNIFORMLY IN TIME

7.7.1 Passenger Arrival Process

A number of assumptions in the derivations of the round trip time expression have been discussed in previous sections. These assumptions were concerned with the nonlinear effects of the building and lift system and also the static population distributions. No account has been taken of the way in which passengers arrive in a building or the randomness of their destinations. In practice passengers do not conveniently arrive at a building such that 80% of the rated car capacity of a lift is available to step into a lift and be transported to their destinations. Also passengers do not request the same number of destinations on each trip. The effect of random arrival rates and destinations is to cause cars to become unequally spaced and, in extreme circumstances, to bunch. If large numbers of passengers arrive, queues can also develop. Throughout this chapter values for H and S have been described as average values. It is using these values that the average RTT is calculated. Obviously, if the RTT has wide deviations from the average, a poor quality can result. How then does the RTT vary ?

7.7.2 Effect of Randomness of Passenger Destinations

Returning to statistical terminology,[1] in Section 5.3 it has been indicated that the probability of a variable is defined by a numeric value between zero and unity. Probability then is simply a measure of the likelihood of some occurrence. For example, the probability of stopping at a floor may be 0.1. However, the actual number of stops is called the expected number of stops. The expectation is the mean or average value of some random variable, often termed the expectance.

It is often of interest to know the spread of possible outcomes about the mean. This variation is essentially measured by the statistical characteristic termed variance. If the square root of a value of variance is taken, the value then obtained is termed standard deviation. If large values of variance occur with respect to values obtained for expectance of some variable then that variable has wide deviations of value from its mean.

A mathematical study has been made by Gaver and Powell (1971). They considered an uppeak situation, with no interfloor traffic, and assumed that all lifts depart from the main terminal with a constant number of passengers. No analysis was made with car load variations. The expression they deduced allows for lifts to be expressed from the main terminal to a certain floor and then to service that floor and the floors above. They also made some assumptions to try to overcome the problem that the rated speed may not be reached for a flight of one floor. Both cases of equal and unequal floor population are considered. The expression is a linear function of the number of stops and the reversal floor, so the evaluation of the variance of the round trip time was reduced to the

---

[1] Statistical analysis and probability theory are highly mathematical fields and are best dealt with in specialised texts (Papoulis, 1965)

calculation of the variance of the two variables (number of stops and reversal floor). The final expression is complex and is best illustrated by examples.

Consider Table 7.1 for a group of lifts serving floors 15 to 26 (12 floor zone) in a building, the lift rated speed being 4.0 m/s.

Table 7.1 Statistical data

Passengers in lift	15	17	19	21
Expectance of RTT(s)	149	157	164	170
Variance of RTT(s^2)	68	66	62	57
Standard deviation(s)	8.25	8.15	7.89	7.55

The variance decreases when the number of passengers increases, since there is less randomness in the number of stops. However, two points must be made: the number of floors served is modest and smaller than the number of passengers in the lift; and the group serves an express zone, which means that a significant part of the round trip time, corresponding to the express flights in both directions, is not subjected to variance of round trip time. If a group of lifts with the same characteristics but serving the 12 floors next to the main terminal is considered, the variance will be the same but the round trip time will be decreased by the travelling time of two flights of 15 floors. This, for an average interfloor distance and a rated speed of 4.0 m/s, is about 25 s. For such a group of lifts next to the main terminal the variance becomes more significant.

The figures above assume equal floor populations; for unequal floor populations the variance increases. An example given by Gaver and Powell with slightly uneven populations gives an RTT expectance of 148 s and a variance of 74 s^2 for 15 passengers in a lift (compare with Table 7.1).

Thus significant variance of RTT does exist for equal and slightly unequal floor populations. Gaver and Powell conclude "if the distribution varies considerably from uniform we must formulate a new representation of round trip time". How these variations affect a lift system is not quantified. In the long term little effect on the carrying capacity will be noticed, but in the short term queues can build-up reducing the quality of service. It would appear, therefore, that as far as conventional calculations are concerned randomness of destination need not be considered.

7.7.3 Effect of Variations in Arrival Rate

Passengers do not arrive uniformly in time. It is generally accepted that the arrival of people at a building obeys a Poisson process, ie: the probability of n calls being registered in the time interval T for an average rate of arrivals λ is:

$$p_r(n) = \frac{(\lambda T)^n}{n!} e^{-\lambda T} \qquad (7.14)$$

The assumption of passenger arrival by the Poisson process is not completely proven, and its validation would require extensive data logging on a wide range of lift systems (see Section 5.6.1).

Petigny (1972) and Tregenza (1972), following different methods of analysis but both considering a Poisson arrival distribution, have related the system behaviour with the arrival of passengers. Petigny's study is purely mathematical and considers a single lift during an uppeak situation. By using probability theory he deduces Equations (7.6) and (7.13) for S and H for the case of defined floor populations. Also by setting up a complex mathematical study involving Markov chain theory, he deduces an expression for the average number of passengers per trip. Although an interesting study, Petigny's work relates to a single lift only.

Tregenza uses a more practical approach and produces an expression similar to Equation (4.11) for RTT, which he writes as:

$$RTT = E_h t_1 + E_s t_s' + E_p t_3 \tag{7.15}$$

where $t_1 = 2t_v$ and $t_3 = 2t_p$ and E symbolises expectance.

Rewriting Equation (7.15) in the style of Equation (4.11):

$$RTT = 2H_p t_v + (S_p+1)t_s + 2P_p t_p \tag{7.16}$$

[The subscript p indicates Poisson pdf.]

Instead of taking P_T as the car capacity or a fraction of it, as in conventional design, he relates the parameters H_T, S_T and P_T as universal functions of an arrival rate parameter p_{ro}. He defines p_{ro} as the probability of no calls being registered from the main terminal floor to any floor above during the period of one interval. For equal floor populations he produces four equations. These have already been derived as Equations (5.14), (5.15), (5.16) and (5.17).

Section 5.6.3 showed that if the first two terms of the expansion of the equations for H_p and S_p were taken, then H, using a rectangular probability function and Equation (5.12), was identical to H_p and likewise for S using Equation (5.5) and S_p.

Figures 7.3 and 7.4 show the equations for H and S plotted for both the rectangular and Poisson pdfs. These graphs confirm that if the rectangular pdf is used to derive the H and S equations, their values will always be larger than if the Poisson pdf is used. Thus the resulting value for the RTT will always be larger and a design based on it will be slightly conservative. As the equations using the rectangular pdf are simpler to calculate, they are generally used in design calculations.

The above analysis assumes equal floor populations/demand. Using statistical techniques, expressions can be obtained (Alexandris, 1976) for unequal floor populations. They are:

$$S = N - \sum_{i=1}^{N} e^{-\lambda INT \frac{U_i}{U}} \tag{7.15}$$

$$H = N - \sum_{j=1}^{N} \prod_{i=N-j+1}^{N} e^{-\lambda INT \frac{U_i}{U}} \tag{7.16}$$

(The mathematical symbol $\prod$ means "multiply all terms like ...".)

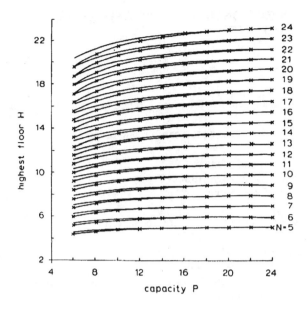

Figure 7.3 Mean highest reversal floor *H* for different numbers of floors and car capacities

solid line: conventional calculations, rectangular probability distribution
line with x: calculation using Poisson probability distribution

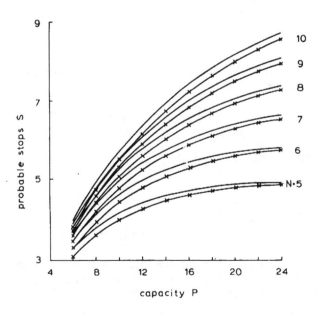

Figure 7.4 Mean number of stops *S* for different numbers of floors and car capacities

solid line: conventional calculations, rectangular probability distribution
line with x: calculation using Poisson probability distribution

7.8 COMPARISON OF NONLINEARITIES AND THEIR TOTAL EFFECT ON THE VALUE OF THE *RTT*

7.8.1 Categories

In Section 7.1 seven assumptions were indicated and now can be categorised as follows:

(a) There are three assumptions, which can be calculated:

(2) Unequal floor populations.
(3) Nonlinear interfloor flight time.
(4) Unequal interfloor heights.

(b) There are three assumptions, which are not determinate:

(1) The traffic profile is ideal.
(5) The traffic supervisory system is assumed ideal.
(6) Various lost times, such as passenger disturbance, dispatch intervals, loading intervals etc. are negligible.

(c) There is one assumption, which does not matter, or has little effect:

(7) Passengers arrive uniformly in time.

It is now worth considering the magnitude of their effects by looking at an example.

7.8.2 Example 7.6

Consider the building used in Example 7.3. It has 10 floors above the main terminal and the floor populations are:

Floor	1	2	3	4	5	6	7	8	9	10
Population	5	10	25	25	50	50	100	100	100	100

The rated speed is 1.5 m/s and the interfloor height 3.3 m.

Flight times are:

Jump	1	2	3	4 ... 10
Time(s)	5.1	7.5	9.7	at rated speed

The height of the main terminal, fourth and tenth floors is 5.0 m. Door closing time is 4.1 s and door opening is 1.0 s. Assume passenger transfer time is 1.2 s per passenger. Calculate the round trip time (*RTT*) and any corrections.

(a) Usual procedure
Values for H and S can be obtained from Table 5.1: $S = 5.7$ and $H = 9.3$.

$$t_v = 3.3/1.5 = 2.2 \text{ s}$$

$t_s = (5.1 - 2.2) + 4.1 + 1.0 = 8.0$ s
$t_p = 1.2$ s

Then using Equation (4.11):

$RTT = 2Htv + (S+1)t_s + 2Pt_p$
$= 2 \times 9.3 \times 2.2 + 6.7 \times 8.0 + 2 \times 8 \times 1.2$
$= 40.9 + 53.6 + 19.2$
$= 113.7$ s

(b) Effect of unequal populations

From Example 7.3: $S = 5.1$, $H = 9.8$. Then using Equation (4.11):

$RTT = 2 \times 9.8 \times 2.2 + 6.1 \times 8.0 + 8 \times 2.4$
$= 43.1 + 48.8 + 19.2$
$= 111.1$ s

This is 2.6 s smaller, or 2.3% smaller than the value obtained in (a).

(c) Effect of nonlinear flight times and unequal floor heights.

Using the procedure in Section 7.4.3 and the values for S and H using Equation (4.11) gives:

(1) Average jump: 9.8/5.1 = 1.9 m.

(2) Average interfloor distance: $(9.8 \times 3.3 + 2 \times (5 - 3.3))/9.8 = 3.7$ m.
Floor 10 is not considered as it is above H.

(3) Average distance: $3.7 \times 1.9 = 7.0$ m.

(4) Travel time: for 2.1 floors (7.0/3.3).
By interpolation between $t_f(2)$ and $t_f(3)$, flight time will be: 7.7 s.

(5) Assumed time: for $t_f(1)$ was 5.1 s and for $t_f(2.1)$ will be $5.1 + 1.1 \times 2.2 = 7.5$ s.
(1.1 s is d_f/v, ie: 3.3/1.5).

(6) Difference time: 7.7 - 7.5 = 0.2 s.

(7) Additional time: add 0.2 s to t_s.

The value of RTT now becomes:

$= 111.1 + 6.1 \times 0.2 = 112.3$

This is 1.4 s smaller or 1.2% smaller than the value obtained in (a).

(d) Discussion

The corrections do not make much difference in this case. No account has been taken of lost times or the effect of the control system. The above example illustrates the procedure to determine the additional times resulting from nonlinearities. The examples do not show very large changes, but each selected system should be considered in this way as it may exhibit significant extra times. In general, however, the effects are likely to be of secondary importance.

It is important to remember that the values of round trip times obtained are average values and they are the sum of three average values for H, S and P. All average values have some variance, or likely range of values. In particular cases, therefore, the RTT may deviate greatly from the average. This deviation may be smaller, or greater, than the effects discussed above. However, it is still prudent to determine the magnitude of the nonlinearities.

7.9 NUMBER OF PASSENGERS AND CAR LOAD REVISITED

7.9.1 Preamble

Industry experts and consultants (Strakosch, 1983, 1998[1]) state that in practice the lifts are not observed to fill with passengers to the numbers permitted by the rating plate, but a lower value between 60% and 70% of rated car capacity, particularly in larger lifts. These observations of lower occupancy values are supported by work done by Day (2001a, 2001b), who found that the comfort of passengers was very important and that many reasons exist as to why passengers will not fill a lift even to the Design Capacity.

It was hinted in Chapters 1 and 2 that it is not possible to carry the number of passengers suggested by the standards (eg: EN81) in the lifts of 1000 kg and larger. There is a necessary safety requirement limiting the load (mass) that a lift can carry and there is no problem with this concept. However, it is very difficult to fill the larger lifts with the number of passengers that this mass represents. When sizing a lift system to handle the passenger traffic this is an important factor.

7.9.2 Definitions

What, therefore, are the actual and design car capacities ?

Definition 7.1: The Rated Car Load (RL), in kg, is the load that the lift is rated to carry.

The Rated Load is the maximum safe load a lift can carry and this value must not be exceeded. EN81 (1998) indicates in its Table 1.1, the rated loads a lift may carry related to maximum platform size. The way in which this rated load is obtained in the standards is based on a formula of unknown provenance. EN81 calculates the number of passengers the lift may carry by either dividing the rated load by 75 or by using its Table 1.2, which is related to minimum platform size.

---

[1] 2nd edition, page 74; 3rd edition, page 84.

Definition 7.2: The Rated Car Capacity (*CC*), in persons, is the rated car load (in kg) divided by 75.

The number of passengers (Rated Car Capacity) is thus related to the available car area, but in a nonlinear way. A 450 kg lift, which has a platform area of 1.3 m^2, would have a Rated Car Capacity of six passengers; and a 2500 kg lift, which has a platform area of 5.0 m^2, would have a Rated Car Capacity of 33 passengers. This allows each passenger in a 450 kg lift, 0.21 m^2 of platform space, whereas each passenger in a 2500 kg lift is only allowed a space of 0.15 m^2 !

Another smaller consideration is that the whole of the platform area may not be available for passenger occupation, but may be reduced by architectural finishes (panels, handrails, mirrors, etc.). The US A17.1 Code (1984) assumes a 5% reduction for this factor.

The Rated Car Capacity should never be used in traffic calculations; the lifts would physically be too small !

Allowing 0.21 m^2 per passenger, as recommended in Section 1.2.2, the EN81 lift sizes lead to the Actual Car Capacities given in Table 7.2. These figures confirm the industry observations by Day (2001a, 2001b) of lower occupancies particularly in larger lifts.

Definition 7.3: The Actual Car Capacity (*AC*), in persons is the platform area, in m^2, divided by 0.21.

The Actual Car Capacity is a realistic value for the maximum number of passengers a lift can carry per trip allowing for all sizes of person, clothing, hand baggage and crowd psychology. This leads to the need to define the recommended Design Car Capacity as a lower value than that given in the EN81 standards for the rated capacity.

Definition 7.4: The Design Car Capacity (*DC*), in persons, is 80% of the Actual Car Capacity.

The Design Car Capacity (*DC*) is the average number of persons a lift is assumed to carry during an uppeak traffic period. Setting this parameter at 80% allows for the statistics of passenger arrivals and the effects of lift bunching arising from the group control system. The Design Car Capacity is the number of passengers that the lift is assumed to carry on average, when calculating traffic performance.

Table 7.2 gives a full list of values of these important parameters. The table shows the rated load, the maximum platform area, the Rated Car Capacity, Actual Car Capacity and Design Car Capacity. Strakosch (1983, 1998) offers a similar table (Table 4.4).

It is interesting to note that the actual load column in Table 7.2 is very close to the permitted loading of a hydraulic lift according to Table 1.1A of EN81-2: 1998.

7.9.3 Example 7.7

What are the Rated, Actual and Design Car Capacities for a lift of 2500 kg rated load ?

Rated Car Capacity is:	33 persons
Actual Car Capacity is:	23.8 persons
Design Car Capacity is:	19.0 persons
Capacity factor is:	80%

Table 7.2 Car loading and car capacity

Rated load (kg) (RL)	Max area (m²) (CA)	Rated capacity (persons) (CC)	Actual capacity (persons) (AC)	Design capacity (persons) (DC)	Capacity factor (%) (CF)	Actual load (kg) (AL)
320	0.95	4	4.5	3.6	90	338
450	1.30	6	6.2	5.0	82	465
630	1.66	8	7.9	6.3	79	593
800	2.00	10	9.5	7.6	76	713
1000	2.40	13	11.4	9.1	70	855
1275	2.90	16	13.8	11.0	69	1035
1600	3.56	21	16.9	13.5	64	1268
1800	3.92	24	18.6	14.9	62	1395
2000	4.20	26	20.0	16.0	62	1500
2500	5.00	33	23.8	19.0	58	1785

Figures are rounded.
Rated load (RL) values, in kg, taken from ISO 4190-1.
Maximum area values, in m², taken from EN81, Table 1.1.
Rated Car Capacity (CC) calculated by dividing the value for RL by 75 as EN81, Clause 8.2.3.
Actual Car Capacity (AC) calculated by dividing the value for maximum car platform area (CA) by 0.21.
Design Car Capacity (DC) calculated as 80% of actual car capacity (AC).
Capacity factor (CF), in per cent, calculated by dividing Actual Car Capacity (AC) by Rated Car Capacity (CC).
Actual load (AL), in kg, calculated by multiplying Actual Car Capacity (AC) by 75.

7.9.4 Example 7.8

A lift traffic design requires that the lift accommodate 10.5 passengers on average. What size of lift is indicated ?

Under the Design Car Capacity column of Table 7.2 it can be seen that a 1275 kg rated car load will accommodate up to 11.0 passengers. Thus a lift of 1250 kg rated load should be selected.

7.9.5 Other Considerations: Geographical Factors

Humans are not the same size all over the world. So far the lift sizes have been based on US and UK values. It can be said that Asia/Pacific people are smaller and that Scandinavians are the tallest people in Europe, and Europeans from the Latin countries are smaller than North Europeans, etc.

The loading for lift cars is believed to have originated in the early 20th Century US-A17 Elevator Codes. Simply put, some 80 years ago (Barney, 1996, 2000a) a male person was assumed to weigh 150 pounds, ie: 68 kg, and to stand on two feet occupying a platform area of 2.0 square feet, ie: 0.19 m². The European EN81 standards of the 1980s decided that all persons weighed 75 kg. This allowed for the fact that people had become larger over the intervening 60 years. Fruin (1971) proposed a body template, based on an ellipse of 600 mm by 450 mm, which has an area of 0.21 m². The Fruin template assumes persons can touch. Strakosch (1983)[1] said "Future development should include capacity loading based on volume rather than weight". ISO/TR 11071-2 (1996) says:

> "While the entire subject of capacity and loading has historically been treated in safety codes as one and the same, it might be more meaningful in the future writing of safety codes to cover loading as a separate issue from capacity. One refers more appropriately to the traffic handling capacity, whereas the other refers to the maximum carrying capacity which has a direct bearing on safety."

This is what Table 7.2 does.

It is interesting to note that the ratio 68/75 and 0.19/0.21 are almost the same ratio. It implies that the loading of a human being is 360 kg/m^2 (75/0.21). If a linear ratio is assumed between area occupied and weight[2] and their loading is assumed constant, then smaller, or larger people, by weight, will take up smaller, or larger areas, of the car platform. Table 7.3 can be drawn.

Table 7.3 Table of Actual Car Capacity related to weight of passengers

Available car area (m^2)	Maximum possible number of passengers of weight/occupancy			
	65 kg/0.18 m^2	68 kg/0.19 m^2	75 kg/0.21 m^2	80 kg/0.22 m^2
1.30	7	7	6	6
1.66	9	9	8	7
2.00	11	11	10	9
2.40	13	13	11	11
2.90	16	15	14	13
3.56	20	19	17	16
4.20	23	22	20	19
5.00	27	26	24	23

For example the 65 kg column might be used in the Asia/Pacific region, the 68 kg column for Latin Europeans, 75 kg column for northern Europeans and the 80 kg column for parts of Russia. The traffic designer would use the values in Table 7.3 to suit the local operating conditions. The load plate in the lift would then only indicate the "recommended number of passengers" or remain silent.

7.9.6 Other Considerations: Intended Use

The last column of Table 7.2 indicates the likely actual load a lift may be asked to carry based on the possible passenger occupancy. How does this relate to intended usage ? ISO 4190-1 (1999) indicates four classes of intended use: residential, health care, general purpose and intensive traffic, and ISO 4190-2 (2001) adds goods lifts.

---

[1] Page 108.

[2] Purists say mass.

The rated loads (kg) for each class are:

Residential   320, 450, 630, 1000
Health care   1275, 1600, 2000, 2500
General purpose 630, 800, 1000, 1275
Intensive traffic 1275, 1600, 1800, 2000
Goods   630, 1000, 1600, 2000, 2500, 3500, 5000

The range of the rated loads are intended to match their purpose of their class.

In residential buildings the range is from the smallest 320 kg, through a 450 kg that can accommodate a wheelchair to a 1000 kg lift that is both spacious and allows a wheelchair full manoeuvrability. The actual load given in Table 7.2 is close to the rated load and the intended purpose matches well.

The smallest lift (1275 kg) in the health care range allows full manoeuvrability for a wheelchair and the 2500 kg lift permits beds and operating trolleys with full attendance. The actual loads carried are likely to be much lower than the rated loads indicated in Table 7.2. This is because the passenger number are low, or they have walking aids, or they are in wheelchairs, or on trolleys, or beds. In all these cases the lift does not have to support loads of the intensity of 360 kg/m^2. In the health care range, the actual loads will be even lower than the actual loads in Table 7.2, which are already about 75% of the rated loads (column 1 and column 7). Traffic design for health care lifts is not usually required.

The general purpose lifts are intended for small offices and hotels, etc. The range is from the smallest 630 kg, which allows wheelchair access to 1275 kg, which gives full manoeuvrability. The actual loads are less than the rated, but are reasonable in case the lifts are used for goods movements. Traffic design will be important for this purpose.

The intensive traffic lifts would be used in office buildings and range from the smallest (1265 kg), which allows wheelchair full manoeuvrability. The range runs from lifts, which are probably too small for most office buildings, and there is a strong likelihood that intending passengers may be left behind, to the largest (2000 kg), which is probably too big, and where the passenger loading time becomes too long. The actual load is some 75% of the rated load. Traffic design is most important for this purpose.

The goods lifts have a wide range of rated loads to suit all intended usage. Traffic design is not appropriate, but car loading is most important. The loads may have a much higher imposed load than 360 kg/m^2, which passengers can apply. Therefore a goods lift should be able to carry the rated load value shown in Table 7.2 for safety reasons.

If rated loads were to be calculated for passenger lifts by relating them to the platform area on the global basis of 75 kg for every 0.21 m^2 there would be less confusion for lift traffic designers. However, as ISO/TR 11071-2 says:

"Experience shows that it is not realistic to assume passengers are aware of the risk of overloading by the number of passengers in the car".

Thus, unless an effective load weighing system with a back-up redundancy device is installed, the rated loads should not be reduced.

CHAPTER EIGHT

Special Situations
and their effect on the
Round Trip Time Equation

8.1 INTRODUCTION

This chapter indicates how the round trip time equation might need to be changed to deal with a number of special situations. Such situations will be discussed with respect to considerations by lift function, building form and building function. For example a firefighting lift (lift function) installed in a tall building (building form) used as a hospital (building function). The target equation is Equation (4.11):

$$RTT = 2Ht_v + (S+1)t_v + 2Pt_p \qquad (4.11)$$

8.2 CONSIDERATION BY LIFT FUNCTIONS

8.2.1 Shuttle Lifts (with sky lobbies)

Many tall buildings are divided into several zones: low zone, mid zone, high zone, etc., with service direct from the main terminal floor, situated at ground level. These are called "local" zones. This becomes impractical with very tall buildings of 70 stories or more and shuttle lifts are employed (Schroeder, 1989b) to take passengers from the ground level main lobby to a "sky lobby" (Browne and Kelly, 1968). Passengers disembark at the sky lobby and then take the local lifts to their final destination. Service is then provided to further low, mid, high zones, etc. using the sky lobby as an upper main terminal floor. The advantage is that the core efficiency is improved (Fortune, 1995, 1996), as the hoistways extend the whole height of the building (except for the intervening equipment spaces) and occupy the same hoistway "footprint". Sometimes passengers travel down from the sky lobby as well as up (Fortune, 1986, 1990). Most shuttle lifts are single deck, but there are a number of double deck installations. Schroeder (1989a) defines four basic sky lobby configurations:

1. Single deck shuttles, single deck locals, eg: World Trade Center.
2. Double deck shuttles, single deck locals, eg: Sears Tower.
3. Double deck shuttles, double deck locals, eg: Petronas Towers.
4. Single deck shuttles, single deck top/down locals, eg: none.

Configuration 4 would be difficult to engineer, as offset lobbies would be required. A configuration Schroeder did not consider should be added:

5. Double deck shuttles, single deck top/down locals, eg: UOB Plaza.

Examples of these configurations are discussed in Section 8.3.2.

Shuttle lifts are sometimes employed over shorter travel distances, such as from car parks to a main terminal and at underground railway stations, or very long distances to observation platforms. Generally shuttle lifts serve between two stops only, hence the term "shuttle", but sometimes serve three stops, ie: with two[1] sky lobbies.

Shuttle lifts are usually quite large and fast and provide an excellent service to the sky lobby. Their main disadvantage is that the passengers must change lifts mid journey, hence increasing their total journey time. When a traffic design involves a change of lift, the two journey times are best quoted separately.

There is no need to modify Equation (4.11). However, note that the value for t_v will be for the travel between the stopping floors. This could be 200 m or more.[2]

As the cars are generally large (<2000 kg) and will fill more fully than is usual for a group of lifts, the passenger transfer times (t_p) when loading and unloading will be more efficient and smaller. The reasons for this is that waiting passengers are "batched" outside a shuttle entrance expecting its arrival; the lift doors will be 1100 mm or more wide; and there may be through cars, allowing the separation of the incoming and outgoing passengers. Example 5.6 indicated that for a 16 person lift the total passenger transfer time for 12.8 passengers, each requiring a t_p of 1.2 s, was 11.3 s, ie: t_p is less than one second on average. For a 5000 kg lift, where it might be possible for 44 persons to be accommodated, the total passenger transfer could be 30.0 s, ie: the average t_p becomes 0.7 s.

The traffic design of a shuttle lift can use Equation (4.11), but as both H and S are known (usually "1"), then Equation (8.1) can be proposed:

$$RTT = 2T + 2Pt_p \qquad\qquad (8.1)$$

where T is the performance time as defined by Definition 4.23.

Example 8.1

Consider a shuttle lift transporting passengers from a main terminal to a single sky lobby ($N = 1$) during uppeak with the following data:

$CC = 66$ persons (5000 kg), $v = 8.0$ m/s,

$t_c = 3.5$ s, $t_o = 3.5$ s, $d_f = 200$ m, $t_f(1) = 30$ s.

Thus:

$H = 1$, $S = 1$, $Pt_p = 30.0$ s (from discussion above).

$t_v = 200/8 = 25$ s; $T = 3.5 + 3.5 + 30 = 37$ s $t_s = T - t_v = 37 - 25 = 12$ s

Using Equation (4.11):

$RTT = 2 \times 1 \times 25 + (1 + 1) \times 12 + 2 \times 30.0 = 134.0$ s

---

[1] Sears Tower, Chicago, USA.

[2] Petronas Towers, Kuala Lumpur, Malaysia.

Using Equation (8.1):

$$RTT = 2 \times 37 + 2 \times 30.0 = 134.0 \text{ s}$$

Such a shuttle lift would have a 5-minute handling capacity of nearly 100 persons. When a shuttle lift is serving a balanced two way traffic situation, then the lifts would fill fully at each lobby and the round trip time would increase by a further $2Pt_p$ seconds. If there were two sky lobbies then the first term of Equation (8.1) would be $3T$, etc.

The number of shuttle lifts that are installed world wide is not large. Their traffic design is relatively simple, but their application in a building requires expert consideration.

8.2.2 Double Decker Lifts

Double deck lifts comprise two passenger cars one above the other connected to one suspension/drive system. The upper and lower decks can thus serve two adjacent floors simultaneously. During peak periods the decks are arranged to serve "even" and "odd" floors respectively, with passengers guided into the appropriate deck for their destination. Special arrangements are made at the lobby for passengers to walk up/down a half flight of stairs/escalators to reach the lower or upper main lobby.

Double deck lifts, which are common in the USA and elsewhere, but unusual in Europe, are used in very tall buildings (see Section 8.3.2). Fortune (1996) indicated 465 double deck lifts in 34 buildings across the world.

Table 8.1 World wide location of double deck lifts

Location	Number	Buildings
North America	317	17
Singapore	55	5
Malaysia	29	1
Japan	17	3
Spain	15	3
Taiwan	12	1
Australia	11	1
England	4	1
Hong Kong	4	1
China	1	1
Total	465	34

There are many advantages and disadvantages to double deck operation (Fortune 1996) and special care has to be taken with the lobby arrangements (see Section 8.3.5). One advantage for double deck lifts is that the "hoistway" handling capacity is improved, as effectively there are two lifts in each shaft. A disadvantage for passengers during off peak periods is when one deck may stop for a call with no coincident landing, or car call, required in the other deck. Fortune (1996) describes special control systems that are available during off peak periods, such as skip/stop, trailing deck and restricted deck service.

Fortune (1996) expounds the advantages of double deck installations as:

1. Fewer lifts
2. Smaller car sizes
3. Lower rated speeds

4. Fewer stops
5. Increased zone size
6. Quicker passenger transit times
7. 30% less core space
8. Taller buildings on same footprint
9. Smaller lobbies
10. Fewer entrances
11. Faster installation
12. Reduced maintenance costs

and the disadvantages as:

1. One significant supplier
2. Passenger misuse
3. Zone populations must be large
4. Balanced demand from even and odd floors
5. Interfloor distance must be regular
6. Slightly larger hoistways
7. Increased pit and machine room loadings
8. Lobby exits need to be larger
9. Special facilities for disabled access to "other" floor.

Kavounas (1989) developed a very succinct analysis of double deck lifts following the direction of the uppeak analysis method described in Chapter 5, starting with Equation (4.11). He makes a number of assumptions:

- The double deck lift serves $2N$ floors above the main terminals.
- Both decks are the same size (CC) and carry identical passenger loads (P).[1]
- Both decks experience the same arrival process (pattern).

The expression for the high call reversal floor (H) is not changed and is given by Equation (5.12) as usual. The expected number of stops (S_d) the double deck lift will make is changed and can be derived by following the same arguments used in Section 5.3, but because the two decks together carry $2P$ passengers the expression becomes:

$$S_d = N\left[1 - \left[\frac{N-1}{N}\right]^{2P}\right] \tag{8.2}$$

The evaluation of this equation could be achieved by using the familiar probable stop table (Table 5.1). However, as $2P$ is likely to be larger (numerically) than 26.4 persons (equivalent to a 33 person rated load) then the evaluation often falls outside the range of the table. Fortunately Equation (8.2) can be simplified[2] to an easier expression:

$$S_d = 2S - \frac{S^2}{N} = S\left[2 - \frac{S}{N}\right] \tag{8.3}$$

---

[1] Kavounas counsels that a more accurate expression could be obtained if the variances of the number of passengers carried on each deck were to be considered. The improvements would only be secondary as most double deck cars will be designed to fill to capacity during uppeak, thus reducing the variances.

[2] To prove this, replace S in Equation (8.3) by the expression for S given by Equation (5.5), expand, combine and simplify to obtain Equation (8.2).

For interest Kavounas also derives expressions for coincident, non-coincident stops and a Figure of Merit.[1]

Definition 8.1: Coincident stops will occur when both decks load or unload at the same time during uppeak traffic.

Definition 8.2: Non-coincident stops will occur when one deck stops to unload without a simultaneous stop for the other deck during uppeak.

If each deck had been independent of the other and not connected together, each would have stopped S times. This leads to the determination of the how many stops are coincident (S_c):

$$S_c = 2S - S_d \tag{8.4}$$

The number of non-coincident stops will then be given by:

$$S_n = S_d - S_c \tag{8.5}$$

Definition 8.3: The Figure of Merit for double deck lifts is the number of coincident stops to all stops.

The Figure of Merit *FM* will be:

$$FM = S_c/S_d \tag{8.6}$$

This Figure of Merit should be as close to unity (100%) as possible. It can be derived from Equations (8.3) and (8.4):

$$FM = \frac{1}{2\dfrac{N}{S} - 1} \tag{8.7}$$

When S is close in value to N, as for example with large lifts serving a small number of floors, then *FM* approaches unity (100%). Siikonen (1998) has used this quality criterion in her analysis of double deck lifts during down peak traffic.

Kavounas also considers the efficiency of passenger transfers. For loading this is unchanged and is given by Pt_p as usual. For unloading with equal numbers of passengers exiting at each stop from both decks it will also be Pt_p. But in the worse cases, where there are no coincident stops it will twice this value at $2Pt_p$. In practice the total transfer time will lie between these two values. After some considerations Kavounas gives the unloading transfer time as:

$$t_u = P\left[2 - \frac{S}{N}\right]t_p \tag{8.8}$$

Using Equations (8.3) and (8.6) a modified expression for Equation (4.11) can be formed:

$$RTT = 2Ht_v + \left[S\left(2 - \frac{S}{N}\right) + 1\right]t_s + Pt_p + P\left(2 - \frac{S}{N}\right)t_p \tag{8.9}$$

---

[1] Fortune (1996) provides a comprehensive glossary of double deck lift terms.

Equation (8.9) has not been completely simplified, in order to indicate the changes clearly. If very large Rated Car Capacity double deck lifts serve a small number of floors then $S \rightarrow N$ making the modifying term $(2 - S/N)$ equal to unity and Equation (4.11) is regained. Strakosch (1983)[1] used this assumption for his analysis of double deck traffic by considering that the probable number of stops equal the number of floors served. For the fully loaded condition, the simplistic Strakosch procedure gives comparable results to Kavounas.

Peters (1995, 2000) using his "general analysis" technique has also analysed double deck calculations and has derived complex formulae for the general traffic case. For the special case of double deck lifts used during uppeak, Peters (1995) compares his approach with Kavounas and comments: "The results show a high degree of consistency for uppeak analysis". The simplicity of the Kavounas method is to be recommended for pure uppeak calculations.

The traffic design for double deck lifts is relatively simple, however when Equation (8.7) is evaluated the values of: t_v are for a double interfloor distance; $t_f(1)$ are for a double interfloor distance; and N is half of the $2N$ floors served. The handling capacity will be twice that calculated for one deck. The application of double deck lifts in large buildings requires expert attention.

Example 8.2

Consider a 16 floor building to be served by 6 double deck lifts with a 1250/1250 kg rated load. The interfloor distance is 3.9 m and the rated speed is 2.5 m/s. The flight time for a 3.9 m jump is 4.9 s and for a 7.8 m jump is 6.4 s. The door closing time is 2.7 s and the door opening time is 2.0 s. Assume the passenger transfer time is 1.0 s.

For double deck lifts, each deck will serve 8 floors from lower and upper lobbies.

From Table 5.1: taking the car occupancy as $P = 12.8$ persons then $H = 7.8$, $S = 6.6$. Also $t_v = 7.8/2.5 = 3.1$ s; $t_s = 2.7 + 2.0 + 6.4 - 3.1 = 8.0$ s

Using Equation (8.7):

$$RTT = 2 \times 7.8 \times 3.1 + [6.6(2 - 0.825) + 1]8.0 + 12.8 \times 1.0 + 12.8 \times 1.0(2.0 - 0.825)$$

$$= 48.4 + 70.0 + 27.8 = 146.2 \text{ s}$$

$INT = 24.4$ s, $HC = 157$ (one deck) $= 314$ (both decks) persons/5-minutes

In comparison, single deck lifts will serve all 16 floors from one lobby.

From Table 5.1: taking $P = 12.8$, then $H = 15.3$, $S = 9.0$ and $t_v = 3.9/2.5 = 1.6$ s; $t_s = 2.7 + 2.0 + 4.9 - 1.6 = 8.0$ s.

Using Equation (4.11):

$$RTT = 2 \times 15.3 \times 1.6 + [9.0 + 1]8.0 + 2 \times 12.8 \times 1.0$$

---

[1] Page 338.

$$= 49.0 + 80.0 + 25.6 = 154.6 \text{ s}$$

$INT = 25.8$ s, $HC = 149$ persons/5-minutes

The double deck group therefore delivers 2.1 times (314/149) the handling capacity of a single deck lift group, without much degradation in the interval. In general, a double deck installation provides over twice the single deck provision. Case Study CS14 illustrates a traffic calculation.

8.2.3 Firefighting Lifts

Firefighting lifts are discussed in Section 6 of the CIBSE Guide D: 2000, which provides an extensive list of further references. A wide range of regulations apply across the world, but the most comprehensive is prEN81-72, which repeats much of BS5588: Parts 5 & 8. All the regulations apply at the discretion of the local Fire Authorities. Generally firefighting lifts are required in buildings, which have occupiable space more than 18 m above and/or 9 m below the fire access level for every 900 m^2 of building footprint or part thereof. They must serve all occupiable levels, have at least a 630 kg rated load and be capable of reaching the highest (or lowest) level served in less that 60 s. They must not be used for the transport of goods and must be unobstructed at all times.

Firefighting lifts are often single lifts situated around the floor plate. Their size is often the lowest possible permitted (630 kg), and their speed is often the lowest possible to reach the highest floor served in 60 s. The handling capacity is therefore low and, as usually only a single lift is present at each location, the interval is equal to the round trip time. Firefighting lifts should not generally be considered as part of the vertical transportation provision, but they do provide a useful addition to the vertical transportation services of a building. For instance in a building with a large floor plate, occupants may be much nearer to a firefighting lift than the main group and may use it in preference, despite its poorer performance.

Sometimes a firefighting lift is part of a group and extra precautions are necessary to ensure its fire integrity, *viz*: protected stairways, additional doors, etc. (prEN81-72). These precautions may affect the traffic circulation to these lifts and should be taken into account when calculating the handling capacity of such a group. In particular the firefighting lift may be smaller than the other passenger lifts and may be equipped with additional doors and car operating panels. The speed, door dynamics, etc., may be different and the handling capacity of the group will need to be determined by calculating the round trip time for each lift, adding these together, and dividing the total by the number of lifts to obtain the average interval for the group.

No changes are required to Equation (4.11) in order to calculate the performance of a stand alone lift. The design procedure may need to be reviewed to take into account any different equipment specification, where the firefighting lift is part of a group.

8.2.4 Goods Lifts

The need for goods lifts has increased substantially in recent years. Despite the computer revolution the amount of paper into a building and scrap paper leaving a building has increased. Also it is quite common to find in any type of building one or more floors under refurbishment, with the requirement to bring in materials and equipment, and to remove rubbish and debris.

All buildings should be served by an adequate number of goods lifts of a suitable size. This will ensure that the passenger lifts are used for their designed purpose and are not "abused" as goods transporters to the detriment of passenger service and probable damage to the car interiors. Where passenger lifts are used as goods lifts, either generally, or in an emergency, the interiors should always be protected.

It is recommended that all office buildings contain at least one dedicated goods lift, particularly if the building is designed for single tenant occupancy. The following points should be noted:

■ For office buildings, provision should be made for one dedicated goods lift for floor areas up to 30,000 m², or part thereof.

■ For larger buildings, an additional goods lift should be provided for each additional 40,000 m² gross floor area.

■ Dedicated goods lifts should have a minimum capacity of 1600 to 2000 kg.

Dedicated goods lifts can be provided as single units or in small groups and should not be considered as part of the passenger vertical transportation provision. It is likely that they will be slower than the regular passenger lifts and, due to their condition, not attractive to most passengers.

Sometimes a goods lift is part of a passenger group and extra precautions should be made to protect it from damage. The goods lift may be larger and of a different rated speed than the adjacent passenger lifts and the handling capacity of the group will need to be determined by calculating the round trip time for each lift, adding these together, and dividing the total by the number of lifts to obtain the average interval for the group.

No changes are required to Equation (4.11) in order to calculate the performance of a stand alone goods lift. The design procedure may need to be reviewed to take into account any different equipment specification, where the firefighting lift is part of a group.

8.2.5 Observation Lifts

Observation, panoramic or scenic lifts contribute to the vertical transportation system of a building. They are often installed in offices, hotels and shopping complexes to provide a feature or visual impact and they can draw a large percentage of "pleasure" riders in the latter two locations. In offices, the whole or part of a group may be comprised of observation lifts. Shuttle lifts may be of this type.

In shopping centres particularly the lifts may be provided individually or in small groups. Here they will generally have lower rated speeds and hence longer flight times and the door times may also be longer, in order to operate the glass doors safely. There may also be long door dwell times to allow children, the elderly and pushchairs to cross the threshold safely. Also the car interiors are often shaped for aesthetic and viewing purposes rather than easy circulation in the car and can have a narrow entrance leading to a wider area at the back of the car. This has the effect of making passenger transfers inefficient.

All these factors can reduce the traffic handling performance to 50% of that obtained from a conventional lift. However, in shopping centres observation lifts are not the main vertical transportation facility, this is generally provided by escalators. They are invaluable, however, to the elderly, persons with mobility problems, persons carrying goods and parents with pushchairs.

In other locations, where observation lifts are part of a group, they may not have the same specification as the adjacent conventional lifts. The individual round trip times would need to be calculated separately, added together and divided by the number of lifts to obtain the average interval for the group.

No changes are required to Equation (4.11), but the design procedure may need to be reviewed to take into account different equipment specifications, inefficient passenger transfers and long door dwell times.

8.2.6 Platform Lifts and Lifts for the Disabled

Platform lifts comprise a platform, which travels a few metres in order to accommodate small level changes in a building. This permits persons with disabilities, who are ambulant or in wheelchairs to move between these levels. These lifts are to be used under supervision and do not contribute to the general vertical transportation facilities.

Lifts for the disabled may be regular lifts, which have been set aside for the use of persons with disabilities and which may not be available to able-bodied persons, ie: they may be card, key or code entry controlled. These lifts contribute little to the general vertical transportation facilities, as their usage is low.

Where a lift, which is part of a group, is designated for disabled use it may have slower door times, longer dwell times, etc., when used by a disabled person.

No changes are required to Equation (4.11), but the design procedure may need to be reviewed to take into account disabled usage.

8.3 CONSIDERATION BY BUILDING FORM

8.3.1 Tall Buildings

In modern high rise buildings each lift is not usually required to service every level, as this would imply a large number of stops during each trip and, consequently, long journey times for the passengers in the car.

Examination of Table 5.1 indicates that for a particular Rated Car Capacity of a lift, the number of stops (S) increases as the number of floors served (N) increases. As the round trip time in Equation (4.11) is dominated by the central term, which includes S, the effect is to increase the round trip time, which in turn increases the uppeak interval, the passenger waiting time and the passenger journey time. A similar deterioration of performance occurs for the other traffic conditions.

The solution is thus to limit the number of floors served by the lifts. A rule of thumb is to serve a maximum of $15-16$ floors with a lift, or a group of lifts. This introduces the concept of zoning. Zoning is where a building is divided so that a lift or group of lifts is constrained to serve a designated set of floors.

Definition 8.4: A zone is a number of floors, usually contiguous, in a building served by a group or groups of lifts.

Definition 8.5: A local zone is a zone adjacent to or close to the main terminal for that zone.

Definition 8.6: A high rise zone is a zone situated above any local zone in the middle or at the top of a building.

The use of the term "zone" is sometimes used to mean "sectors" in some traffic control systems. Sectors[1] are used in the implementation of a traffic controller algorithm and the term is also used to define a designated, but smaller, set of floors in a building. There may be as many sectors in a zone as there are cars to serve that zone.

There are two forms of zoning: stacked and interleaved, as illustrated in Figure 8.1.

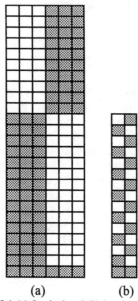

(a)   (b)

Figure 8.1 (a) Stacked and (b) interleaved zones

8.3.1.1 Stacked zones

A stacked zone building is where a tall building is divided into horizontal layers, in effect, stacking several buildings on top of each other, with a common "footprint" in order to save ground space, see Figure 8.1(a). It is a common and recommended practice for office and institutional buildings.

Each zone can be treated differently with regard to shared or separate lobby arrangements, grade of service, etc. The floors served are usually adjacent, although some buildings may have split subzones, where the occupants of each subzone are associated with each other and can be expected to generate some interfloor movements. The number of floors in a zone, the number of lifts serving a zone and the length of the express jump all affect the round trip time. The round trip time can be calculated by adding a time equal to the time taken to jump through the unserved floors in both directions. The changes required to Equation (4.11) are given by the last term in Equation (5.22), *viz*:

$$RTT = 2Ht_v + (S+1)t_v + 2Pt_p + 2t_e \tag{5.22}$$

This is illustrated in Example 8.3.

---

[1] See Definition 9.4.

Example 8.3

Consider a 36 floor building served by 16 person cars. All zones are to have an equivalent grade of service. Design one system with equal numbers of cars and another system with equal numbers of floors. Assume identical values for T, t_p, etc.

In order to accommodate the express jump to the first served floor for the zones not immediately above the main terminal, the calculation of the RTT will use Equation (5.22). This is a modification of Equation (4.11). Table 8.2 shows a design for an equal number of cars, with a decreasing number of floors in each zone. This is an attempt to equalise the RTTs at the main terminal (MT) to compensate for the express jumps (MT to 15; MT to 27) by reducing the number of floors to be served. Table 8.3 shows a system serving an equal number of floors, with an increasing number of cars to compensate for the time to transit the unserved express jumps. There are a wider range of values for *UPPINT*. It is not possible to fine tune the number of lifts, as there are less of them. It is usually easier to adjust the number of floors per zone as there are more of them, than the number of lifts per zone.

Table 8.2 Equal numbers of cars

Zone	No. of floors	Served floors	No. of cars	Speed (m/s)	UPPINT (s)
Low	14	1 – 14	6	2.50	24.0
Middle	12	15 – 26	6	3.15	24.3
High	10	27 – 36	6	4.00	23.5

Table 8.3 Equal numbers of floors

Zone	No. of floors	Served floors	No. of cars	Speed (m/s)	UPPINT (s)
Low	12	1 – 12	5	2.50	26.9
Middle	12	13 – 24	6	3.15	23.9
High	12	25 – 36	7	4.00	21.1

The introduction of groups of lifts to serve different zones of a building requires more space at the main terminal level. If the building footprint is small the occupiable space is also small with the lower levels of the building filled with hoistways. The positioning of the groups is important and adequate signs should be displayed to quickly and simply direct the passengers to the correct group. The arrangement of lifts within each group is given in Section 1.4.3.

Unless the machine rooms for all zone groups are placed at the top of the building, which is very unlikely, the machine rooms will be found just above the last served floors in each zone. Thus if a building has the same size of floor plate at every floor then the space occupied by the hoistways and machine rooms reduces at the higher levels making more rentable space available at the higher levels. As zone boundaries are changed the zone populations change not only by the number of floors, but also by the addition or removal of rentable space. This must be considered in the traffic calculation. Remember that hoistways and machine rooms will occupy at least two levels above the last served floor in a zone.

Traffic Case Study CS9 provides an example of a multi-zoned building.

8.3.1.2 Interleaved zones

An interleaved zone is where the whole building is served by lifts, which are arranged to serve either the even floors or the odd floors, see Figure 8.1(b). This has been a common practice in public housing and has been used in some office buildings (see traffic Case Study CS10). So, for example, in a 16 floor residential building one lift may serve: G,1,3,5,7,9,11,13,15, whilst another lift serves: G,2,4,6,8,10,12,14,16.

The effect is to reduce the number of stops a lift makes because there are fewer floors to be served. This also reduces the capital costs because there are fewer openings and landing doors to install. The service to passengers, however, is poorer than with a duplex serving all floors, because there is only one lift to take them to their dwelling. Tenants tend to solve this by calling both cars at the main terminal and if it is the "wrong" one, walking a flight of stairs to their floor (if they are able). Thus cars are unnecessarily brought to the main terminal.

Furthermore, it can also be shown (Barney, 1988) that the fewer landing doors perform 70% more operations than where the lifts stop at all floors. This is likely to cause an increased level of call backs to the most troublesome component of a lift system. Interleaved zoning is not recommended for the above reasons and is a "... proven disaster in the US" (Strakosch, 1988).[1]

When calculating the *RTT* for interleaved zoning remember that the interfloor distance is twice the standard spacing leading to an increased "single floor flight time" and the number of floors served is also halved (*N*/2).

There are no changes required to Equation (4.11), but care must be taken to use the correct values for the various variables.

Example 8.4

Consider a 16 floor residential building with two 10 person lifts. Each lift has the following characteristics: t_v = 2.0 s; t_s = 10.0 s; t_p = 1.5 s.

If the lifts operate as a duplex: P = 8; H = 14.7; S = 6.5 and

> $RTT = 2 \times 14.7 \times 2 + 7.5 \times 10 + 2 \times 8 \times 1.5 = 157.8$ s
>
> $INT = 78.9$ s
>
> $HC = 30.4$ persons/5-minutes

If the lifts operate as a simplex: P = 8; H = 7.5; S = 5.3 and

> $RTT = 2 \times 7.5 \times 2 + 6.3 \times 10 + 2 \times 8 \times 1.5 = 117$ s
>
> $INT = 117$ s
>
> $HC = 20.5$ persons/5-minutes (one lift) = 41 persons/5-minutes (both lifts)

The effect of providing a "skip-stop" installation is to increase the handling capacity by some 30%, but to increase the interval by some 48%. Not a satisfactory performance.

---

[1] Elevator World, page 102, November 1988.

8.3.1.3 Transfer floors

Most tall and very tall buildings (see next section) provide some means to travel between zones and stacks. This is sometimes achieved by overlapping zones (Petronas Towers), introducing extra stops (Sears Tower) or shuttle lifts (World Trade Center). A common served floor (other than the main terminal or sky lobby) is important where there are common facilities to be accessed, eg: restaurant, travel bureau, sports facilities, post room, reprographics, etc.

The effect on traffic handling can be disruptive. In general it is important to restrict access to such floors during uppeak and down peak, although the object of such a floor would be defeated at other times, ie: at the mid day break or during interfloor traffic. During up or down peak traffic, the transfer floor should be either included in the zone below or the zone above, whichever produces similar interval times at the main terminal floor. It is difficult to calculate the effect of a transfer floor and simulation techniques are often employed. Traffic Case Study CS11 offers a unique solution.

8.3.2 Very Tall Buildings

8.3.2.1 General

Very tall buildings might be defined as those buildings over 30 to 40 stories high. This height can be related to nature, as the tallest tree[1] ever measured was 132.6 m. Generally if service can be provided from the ground level main terminal floor to every floor in the building, this is a tall building. Once shuttle lifts to sky lobbies are required then the building could be called very tall. Fortune (1997) defines a tall building as a "skyscraper", ie: "A high rise building with more than one zone of elevators" and a very tall building as a "Mega High Rise building", ie: "A building with one or more sky lobbies and in excess of 75 floors". A monstrously tall building, which was never built, was Frank Lloyd Wright's mile high building (Fortune, 1992), see Section 8.3.2.7.

The Council on Tall Buildings and Urban Habitat survey of the 100 tallest buildings in 1999 (reported in Elevator World Source, 1999–2000) indicated 63 were in North America, 30 around the rest of the Pacific Rim, four in Europe (one in London) and three others. The heights of the top 50 ranged from 260–450 m tall, a 190 m range, whilst the bottom 50 ranged from 230–260 m tall, a 30 m range. There must be hundreds of buildings between 130 m (the highest tree) and 230 m.

Very tall buildings employ many techniques such as:

local zone lifts (all sections)
stacked zones (Section 8.3.1.1)
shuttle lifts (Section 8.2.1)
sky lobbies (Section 8.2.1)
transfer floors (Section 8.3.1.3)
double deck lifts (Section 8.2.2)
"top/down" service (Section 8.3.2.5).

---

[1] *Eucalyptus Regnans*, Watts River, Victoria, Australia, in 1872 was 132.6 m high (Guinness Book of records).

The previous sections indicated the modifications needed to the traffic analysis procedure. The techniques are best visualised by considering four of the tallest buildings in the world. A simple presentation is given only; no conclusions are drawn. Goods lifts, firefighting lifts and other lift services are not indicated.[1]

8.3.2.2 Double deck shuttles to double deck locals

Petronas Towers, Kuala Lumpur, Malaysia (1996).

The two Petronas Towers are 452 m high with 88 stories above ground. Uniquely the two towers are situated side by side and joined part the way up by a high level bridge.

The buildings are divided into two stacks (see Table 8.4).

Table 8.4 Lift and floor arrangements: Petronas Towers

Stack	Zone	Lifts	Serving Floors	Transfer Floors
1	Low rise	6DD	0/1 8−23	23
	High rise	6DD	0/1 23−37	
2	Low rise	6DD	41/42 44−61	61
	Mid rise	6DD	41/42 61−73	
	High rise	6DD	41/42 61/62 69−83	
	Shuttle	5DD	0/1 41/42	
X	Transfer	2SD	36−37 40−43	Stack 1 Stack 2

SD = single deck, DD = double deck

Stack 1 has 2 groups of 6 double deck lifts serving the main terminal and Floors 8−23 at 4 m/s and the main terminal and Floors 23−37 at 5 m/s. Floor 23 acts as a transfer floor.

Stack 2 is served by a group of 5 double deck shuttle lifts at 6 m/s to sky lobbies at Floors 41/42. From the sky lobbies there are 3 groups of 6 double deck lifts serving 3 zones: Floors 44−61 at 3.5 m/s; Floors 61−73 at 7 m/s; Floors 61/62 and 69−83 at 7 m/s. Floors 61 and 62 act as transfer floors.

There are 2 lifts linking Stack 1 to Stack 2 serving Floors 36−37 and 40−43 at 1.6 m/s.

There are escalators between upper and lower main terminal (0/1) levels and between Floors 41 and 42.

---

[1] The information presented has been derived from the Elevator World Source 1999-2000. The information there is more comprehensive and includes stack diagrams of lift and floor arrangements. The origin of the information was by Fortune (1997) again in Elevator World.

8.3.2.3 Double deck shuttles to single deck locals

The Sears Tower, Chicago, USA (1974).

The Sears Tower is 436 m high with 103 stories above ground. The building is divided into three stacks (see Table 8.5).

Table 8.5 Lift and floor arrangements: Sears Tower

Stack	Zone	Lifts	Serving Floors	Transfer Floors
1	Low rise	6SD	2 5–10	10
	Mid low rise	6SD	2 10–17	10/17
	Mid high rise	6SD	2 17–23	17/23
	High rise	5SD	2 23–28	23 /27 (1)
2	Low rise	6SD	33 35–42	42
	Mid low rise	6SD	33 27 (1) 42–49	42/49 /27 (1)
	Mid high rise	6SD	34 49–57	49
	High rise	5SD	34 58–63	63 (2)
	Shuttle	8DD	0/1 33/34	
3	Low rise	4SD	66 68–74	63 (2)
	Mid low rise	4SD	66 75–81	
	Mid high rise	4SD	67 82–87	
	High rise	5SD	67 88, 90–102	
	Shuttle	6DD	0/1 66/67	33/34 (3)

SD = single deck, DD = double deck
(1) Transfers Stack 1 and 2; (2) Transfers Stack 2 and 3, (3) Transfers Stack 2.

Stack 1 has 3 groups of 6 single deck lifts serving the main terminal and Floors 5−10 at 2.5 m/s; the main terminal and Floors 10−17 at 3.5 m/s; the main terminal and Floors 17−23 at 4 m/s and 1 group of 5 lifts serving the main terminal and Floors 23−28 at 5 m/s.

Stack 2 has sky lobbies at Floors 33/34 served by 8 double deck shuttle lifts at 7 m/s. From the sky lobbies there are 3 groups of 6 single deck lifts serving Floor 33 and Floors 35−42 at 2.5 m/s; Floor 33 and Floors 42−49 (2 lifts also serve Floor 27) at 3.5 m/s; Floor 34 and Floors 49−57 at 4 m/s and 1 group of 5 lifts serving Floor 34 and Floors 58−63 at 5 m/s.

Stack 3 has sky lobbies at Floors 66/67 served by 6 double deck shuttle lifts at 8 m/s, which also can stop at Floors 33/34 for service to Stack 2. From the upper sky lobby there are 3 groups of 4 single deck lifts serving Floor 66 and Floors 68−74 (2 lifts also serve Floor 63) at 2.5 m/s; Floor 66 and Floors 75−81 at 3.5 m/s; Floor 67 and Floors 82-87 at 4 m/s and 1 group of 5 single deck lifts serving Floor 67 and Floors 88−102 at 5 m/s. In addition, there are two observation lifts serving the main terminal and Floor 103 at 9 m/s.

8.3.2.4 Single deck shuttles to single deck locals

The World Trade Center, New York, USA (1972, destroyed by terrorists, 11 September 2001).

The World Trade Center comprised two towers was 416 m high with 110 floors. It was divided into three stacks (see Table 8.6).

Table 8.6 Lift and floor arrangements: World Trade Center

Stack	Zone	Lifts	Serving Floors	Transfer Floors
1	Low rise	6SD	MT 9−16	
	Mid low rise	6SD	MT 17−24	
	Mid high rise	6SD	MT 25−32	
	High rise	6SD	MT 33−40	
2	Low rise	6SD	44 46−54	
	Mid low rise	6SD	44 55−61	
	Mid high rise	6SD	44 62−67	
	High rise	6SD	44 68−74	
	Shuttle	8SD	MT 44	
3	Low rise	6SD	78 80−86	
	Mid low rise	6SD	78 87−93	
	Mid high rise	6SD	78 94−99	
	High rise	6SD	78 100−107	
	Shuttle	8SD	MT 78	
X	Transfer Shuttle	3SD	44 78	

SD = single deck

Stack 1 had 4 groups of 6 single deck lifts serving: the main terminal and Floors 9−16 at 4 m/s; the main terminal and Floors 17−24 at 5 m/s; the main terminal and Floors 25−32 at 6 m/s; and the main terminal and Floors 33−40 at 7 m/s.

Stack 2 had a sky lobby at Floor 44 served by 8 single deck shuttle lifts at 8 m/s. From the sky lobby there were 4 groups of 6 single deck lifts serving: Floors 46−54 at 2.5 m/s; and Floors 55−61 at 4 m/s; Floors 62−67 at 4 m/s; and Floors 68−74 at 5 m/s.

Stack 3 had a sky lobby at Floor 78 served by 8 single deck shuttle lifts at 8 m/s. From the sky lobby there were 4 groups of 6 single deck lifts serving: 78 and Floors 80−86 at 2.5 m/s; 78 and Floors 87−93 at 4 m/s; 78 and Floors 94−99 at 4 m/s; 78 and Floors 100-107 at 5 m/s.

In addition, there were a group of 3 single deck interzone shuttles between Floors 44 and 78 at 8 m/s.

8.3.2.5 Top/down service

A top/down lift installation is where a sky lobby is used to serve building zones or stacks both in the conventional up direction but also in the down direction (Schroeder, 1989a). This does mean that passengers may (psychologically) be concerned that they have travelled up a building only to be then required to travel down to their destination. This is a subtle defeat of Rule 3, and the technique has been applied in a few buildings.[1] The concept is illustrated in Figure 8.2.

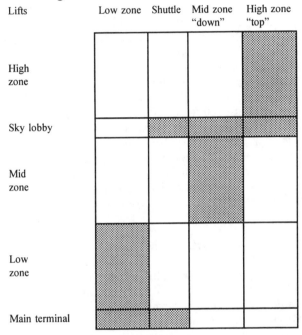

Figure 8.2 Illustration of the concept of a top/down sky lobby

Although more expensive in equipment terms (Fortune, 1990) the technique allows a small footprint building to provide a larger rentable space. Consider the UOB building as an example.

United Overseas Bank, Singapore (1992)

This is 280 m high with 66 storeys, divided into three stacks.

Stack 1 has a group of 6 single deck lifts serving the main terminal and Floors 7–20.

Stack 2 has a sky lobby at Floor 37 served by the lower deck of 6 double deck shuttles. From the sky lobby a group of 6 single deck lifts serve (down) Floors 20 and 23–36.

Stack 3 has a sky lobby at Floor 38 served by the upper deck of 6 double deck shuttles. From the sky lobby a group of 6 single deck lifts serve (up) Floors 41–59.

---

[1] Eg: UOB Plaza, Singapore; First Interstate Bank Plaza, Dallas, USA.

There is a transfer floor at Floor 20 for Stack 1 and 2 interchanges.

This building can be categorised as double deck shuttles to single deck locals.

8.3.2.6 Traffic design considerations for very tall buildings

Very tall buildings, sometimes described as "monumental" buildings, are few in number compared to the totality of buildings world wide and their traffic design requires expert consideration. It involves a mixture of zoning, shuttle lifts and double deck lifts. The formulae which have already been given deal with these traffic systems.

As a general rule a building of 60 floors can be served, where four groups of lifts (a practical limit) are used to serve four zones from a main terminal lobby. If double deck lifts are used, this permits up to 80 floors to be served from a main terminal lobby. Above an 80 floor building, sky lobbies must be used with shuttle lifts to serve them. This permits service to 120 to 160 floors with one sky lobby and 180 to 240 floors with two sky lobbies. Remember also that the maximum practical number of lifts that can be grouped together is eight cars with four facing four.

There are no fundamental changes required to Equation (4.11). However, care must be taken to apply the appropriate formulae, eg: Equation (8.1) for shuttle lifts; Equations (8.2), (8.3) and (8.7) for double deck lifts and Equation (5.22) to deal with express jumps. Care must be also taken to use the correct values for the various parameters.

8.3.2.7 Very Tall Buildings: a Postscript

Frank Lloyd Wright proposed a mile high building (1600 m) in 1956 (described by Fortune, 1992) to be built in Chicago, USA. This building would have 528 storeys and would have accommodated 130,000 occupants. Wright proposed to install 76 quintuple (5) deck lifts. Fortune estimates that more than twice that number would have been needed to obtain current day performance standards. This building would have been four times higher than the world's current tallest building, the Petronas Towers, described above. It was never built.

8.3.3 Basement Service

The provision of lift services to basements has been considered by convention to be either a costly nuisance, or a mere appendage to the normal lift system serving between the main entry terminal and the floors above. However, basements are provided in buildings for many purposes and at many levels of usage and cannot be ignored. Pearce (1995) provides a comprehensive analysis of basement service.

Passengers requiring lift service in the basements are generally those arriving by car or as a result of a facility located in the basement area, such as restaurants, leisure facility, health clubs, etc. The number of passengers arriving by car and requiring lift service from the basements is limited by a number of factors. These include: the number of vehicles which can be admitted to the basements during the uppeak period; the number of car spaces in the basements; and the expected number of persons in each car entering the basements. Planning authorities generally place limits on the number of car parking spaces to be provided in a building. Quite often this represents only 10% of the total building occupancy.

Service to floors above the main terminal is treated as the important part of the lift design, since it provides transportation for almost all of the building population. Basements are usually a very small addition to this, but can increase the passenger waiting time, and thus cause annoyance to waiting passengers either when a full lift bypasses the main terminal or the lift stops partly full, allowing only a small number of the passengers to enter. Strakosch (1967)[1] gave advice which is equally relevant today, but is still being ignored by building developers and designers, as an economy measure:

"All elevators in a group should serve the same floors. This is a common-sense rule that is often violated for false economy. If, for example, only one car out of a group of three serves the basement, people wishing to go to the basement from an upper floor have only one chance out of three that the next elevator that comes along will take them to the basement. Conversely, people in the basement must wait three times as long for elevator service than upper floor passengers. Ideally, all cars should serve the basement, but if not, a special shuttle elevator must be considered, to run only between the main floor and the basement."

There are three basic service arrangements for basements. The first arrangement is that less than all lifts serve the basements, which is not satisfactory as Strakosch (1967) comments:

"The expedient of providing a separate call button at an upper floor to call the single car that serves the basement has often been tried and never proved satisfactory. The average person will operate both the normal call button and the basement call button, take the first car that comes along, and cause the basement car to make a false stop. Such false stops will add up in lost elevator efficiency over the years to more than pay for the cost of the extra entrances on all elevators."

The second arrangement is where all lifts serve all the basement levels. It is the most expensive arrangement, but the most versatile arrangement with one or two basements but would involve a time penalty on the overall system. The availability of lifts to serve the basements can be limited during peak traffic conditions by modern control systems.

Finally, separate basement service is appropriate in a building with multiple basements, and where there are high rise groups, or high speed lifts serving high zones, as it would prove to be less expensive than the alternative of all lifts serving the basements. An illustration of these arrangements is provided in Figure 8.3.

Peters (1997a) has analysed a system of four lifts, two of which serve the basement. He found the two lifts not serving the basement had an interval of 40 s, whereas the two lifts serving the basement had an interval of 45 s. He concludes the interval at the lobby would be 21 s and at a basement 45 s. See Example 12.7.

In the uppeak traffic condition, the presence of a served basement will introduce at least one extra stop (if the main terminal is bypassed on the way down) and probably two extra stops (if the lift stops at the main terminal on the way down and again on the way back up). The second circumstance will arise if passengers press both buttons at the main terminal ("in order to make the lift come quicker") as they invariably do. The outcome is passenger loading delays at the main terminal as passengers try to decide whether to

---

[1] Page 35.

enter a down going lift or wait for an up going lift. Another effect of service to the basement area during uppeak is that cars arrive at the main terminal already partly full, thus causing more confusion.

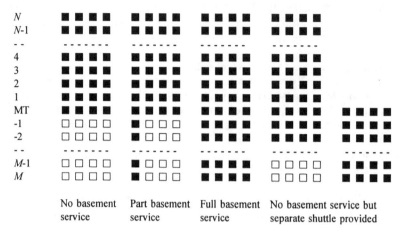

Figure 8.3 Basement service arrangements
For this example a four lift group is shown as an illustration.
M refers to the number of floors served below the main terminal floor.

Barney (1993) in the first edition of CIBSE Guide D wrote:

"The time penalty for the extra stops can be between 10 s and 20 s and say between 5 s to 10 s for the increased passenger loading times i.e. some 15 s to 30 s to be added to the *RTT*. In the case of a 16 person lift serving 16 floors the *RTT* could be 150 s and the extra times would add 10% to 20% to the *RTT*."

During down peak any trip down below the main terminal to, say, leisure facilities will add one extra stop and the extra time to transit the extra interfloor distance; say 10 s to 12 s. This will reduce the handling capacity during down peak.

Designers will need to take account of these factors when sizing an installation with served levels below the main terminal. An important factor is the ratio of arrivals from the basements to the arrivals at the main terminal, ie: is the ratio 10:90 or 40:60, etc. The ratio will depend on the building size. Three possible ways to account for this in traffic design are:

(a) Add an estimate of the time penalty for the extra stops to the *RTT*, as above.
(b) Calculate the probable stops and reversal floor in a similar way to the upward service calculation (Nahon, 1990). The resulting additional time can then be added to the normal *RTT*.
(c) Increase the value of *N* in the use of Equations (4.11) or (5.22) by the number of basement floors to obtain a value for the *RTT*.

Barney and Pearce (2000, 2001) found that the method suggested by Nahon gave the most accurate results. Nahon assumed that passengers boarding at the basement floors can be modelled as if they were alighting. This is a reasonable assumption. He did not give details of his method. The Nahon method provides a separate calculation for any number of basement levels based upon the uppeak model, which is then added to the above ground round trip time. This method also allows for the normally different basement interfloor heights.

To understand the method it is necessary to modify Equation (4.11) to incorporate Nahon's method. Consider the spatial movement of a lift serving above and below the main terminal, as shown in Figure 8.4.

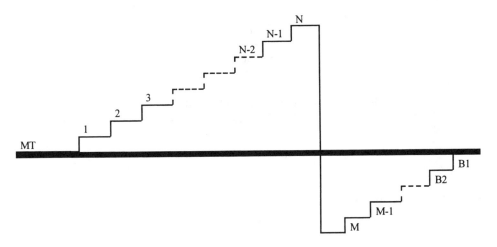

Figure 8.4 Spatial movements of a lift serving above and below ground

The round trip consists of a number of elements resulting from passengers travelling from the basement floors or from passengers joining a lift at the main terminal:

(a) passenger loading time t_l
(b) passenger unloading time t_u
(c) door closing and opening times t_c and t_o
(d) interfloor flight time (for a single floor assuming rated speed is reached) $t_f(1)$
(e) time to travel in the upward direction at rated speed for jumps greater than a single floor t_a
(f) time to travel from the highest floor to the main terminal t_e
(g) passenger loading time for xP passengers
(h) passenger unloading time for xP passengers
(i) door closing and opening times
(j) interfloor flight time (for a single basement interfloor distance assuming rated speed is reached)
(k) time to travel in the upward direction at rated speed for jumps greater than a single floor
(l) time to travel from the main terminal to the lowest floor

where:
    x is the percentage of passengers boarding at the basement floors and present in the lift on its arrival at the main terminal.

The terms (a) to (f) have been considered in Section 4.5.3 and give rise to Equation (4.11). Terms (g) to (l) are additional terms as a result of basement service. The addition of items (a) to (l) translates into the equation:

$$RTT = 2Ht_p + (S+1)t_s + 2Pt_p + 2H_M t_{vm} + S_M t_{sm} \tag{8.10}$$

where:

H_M is the average lowest reversal floor
S_M is the average number of stops below the main terminal
t_{vm} is the interfloor transit time below the main terminal (s)
t_{sm} is the stopping time below the main terminal (s)

It will noted that the first three terms are the same as Equation (4.11). The additional times from the last two terms represent the time penalty to serve the basements.

Example 8.5

Five lift systems are presented in Table 8.7. They represent installations in the "four corners of the (lift) world" and its "centre of gravity" (see Section 5.12.3), ie:

$N = 10$	$N = 10$	$N = 16$	$N = 20$	$N = 20$
$CC = 10$	$CC = 24$	$CC = 16$	$CC = 10$	$CC = 24$

Data assumed:

$L = 1$, $T = 10$ s, $t_p = 1.0$ s.
$d_f = 3.3$ m (above MT), 2.5 m (below MT).
$v = 1.6$ m/s (for $N = 10$), 2.5 m/s (for $N = 16$), 3.15 m/s (for $N = 20$).
$M = 1, 3, 5$, $x = 10\%, 20\%$.

Table 8.7 indicates the extra round trip time that will be incurred to serve one, three or five basement floors with two levels of basement traffic demands at 10% and 20%, ie: cars are 10% or 20% occupied on arrival at the main terminal. The extra time incurred will increase the service interval at the main terminal and reduce the overall handling capacity. For the systems considered in Table 8.7 the range is 7% to 18%. The extra time to serve one basement is some 8 s to 9 s, which increases to some 30 s for a large installation. What is important is the ratio of the above ground round trip time with the round trip time obtained with basement service. This can result in up to an 18% loss of handling capacity for the examples considered. The figures obtained are close to the estimates Barney (1993) suggests in Guide D: 1993.

The effect of serving floors below the main terminal has an effect on the main traffic patterns. During interfloor traffic there will be no appreciable deterioration in service. But during uppeak and down peak, the loss of cars below the main terminal will affect service. The Nahon calculation method is very plausible, as it does mathematically calculate the effect of basement service. It does, however, depend on the theory that a "reversed" down peak boarding policy is reasonable.

Equation (8.8) can be used to calculate an incoming "uppeak" traffic pattern, where most of the passengers arriving at the main terminal travel to floors above the main terminal in the conventional way and others travel to floors below the main terminal.

There are changes required to Equation (4.11), as indicated and care must be taken to select the correct values for basement demand (see Traffic Case Study CS12).

Table 8.7 Results of the five examples of basement service

Basement demand	N/CC	RTT(N)	RTT(N+1)	RTT(N+3)	RTT(N+5)
	10/10	108			
Nahon @ 10%			115	118	122
Nahon @ 20%			115	122	127
	10/24	156			
Nahon @ 10%			164	172	177
Nahon @ 20%			164	176	184
	16/16	153			
Nahon @ 10%			161	166	170
Nahon @ 20%			161	170	177
	20/10	123			
Nahon @ 10%			131	134	137
Nahon @ 20%			131	138	143
	20/24	200			
Nahon @ 10%			208	216	222
Nahon @ 20%			208	220	229

8.3.4 Multiple Entry Levels and Entrance Bias

Some buildings have more than one main entrance and each entrance may be served by its own group of lifts. Other buildings may be designed with the main entry points at more than one level (Pearce, 1995). The effect of more than one main terminal is disruptive and, in the interests of efficient circulation, buildings should not be designed in this way. If there is more than one entrance, means should be provided to bring the two routes together at a single lift lobby. If this is not possible then the lift system sizing should take into account the extra times incurred stopping and loading at multiple entry floors. This could be achieved by using Equation (8.8).

Another difficulty is deciding whether the building population will use each entrance (and their associated group of lifts) equally. In the absence of any guidance, the solution is to assume an entrance bias of 60% and size the lift groups to meet this demand at an extra cost.

There are no changes required to the round trip time equation. Any separate lift systems may not be equally loaded and some extra handling capacity may be needed. Extra time may be needed to take care of multiple passenger loading and unloading times. Traffic Case Study CS13 illustrates some aspects of multiple entrances.

8.3.5 Lobby Design

The design of lift lobbies, especially at the main terminal, where there is most activity is a neglected area. Pearce (1996) suggested that many lobbies considered "a trivial element in lift design". Very often the design of a lift lobby is considered against the requirements of aesthetic design, security, noise limitation, smoke penetration and fire precautions, rather than the movement of people. If a lift lobby becomes too crowded, especially during uppeak, then intending passengers often reach a lift and cannot board it because it is too full, or cannot reach it and it leaves partially empty. The latter event reduces the handling capacity and the former may delay the despatch of the lift as passengers may need to leave as it has become not only crowded, but also overloaded.

Lobby arrangements have been discussed in Section 1.4.3, where guidance regarding lobby size is indicated, and in circulation Case Study 1, where the number of passengers who can be accommodated are calculated. Pearce (1996) suggests that sometimes this guidance is breached for the convenience of the machine room layout or structural efficiency. The shape of lobbies is also important. For example, a group of lifts whose doors are fitted to a convex shaped lobby would be most inefficient, whereas if the doors were fitted to a concave shaped lobby the lobby design would be nearly perfect.

What effect does lobby design have on the round trip equation ? Section 1.4.3 suggested if lifts are served from large lobbies that the lobby door dwell time may need to be increased. The Australian Computer Aided Design Service (Pearce, 1996) suggests:

> "Should a lobby be narrow or the lift cars a shape that restricts passenger movement or the lift doorways are fitted with very thick walls it would be reasonable to allow an increase on loading and unloading times, typically 10%."

It would appear, therefore, that the effect of lobby design cannot be completely quantified and that the best guess, where a problem is anticipated, is to increase the passenger transfer time (t_p) used in Equation (4.11) by 10% to account for the inefficiency. In severe situations some designers add 10% to the round trip time.

The lobby design for double deck lifts presents problems as there are two lobbies, one above the other. Fortune (1995) indicates several solutions and the care to be taken in the lobby design. One arrangement is by using split level lift lobbies reached by stairs, pedestrian ramp or escalators from an entrance lobby, positioned mid way between the lift lobby levels. Another arrangement is where the entrance lobby is also either the upper, or the lower lift lobby, and the other lift lobby is by stairs, pedestrian ramp or escalators.

8.3.6 Effect of Door Width and Car Shape

This effect is not specifically due to building form, but could be and so is discussed here. For example, the available core shape may influence the well shape, or an observation lift of an eccentric design may be installed. The main effect will be on the efficiency of passenger transfers. Phillips (1973) has a graph for evaluating passenger transfer times (made into equations by Jones, 1971) and Strakosch (1967) has tables giving suggested times, which in later editions he terms transfer inefficiencies. The estimation of passenger transfer time is often simply common sense (see Section 5.9).

Table 8.8 shows the car platform sizes and door widths, which are commonly available and recommended in ISO4190: 1999. The table shows four classifications: residential, health care, general purpose and intensive traffic. Taking each type in turn.

8.3.6.1 Residential lifts

These lifts are generally small, ie: up to rated loads of 1000 kg and fitted with narrow doors in the range 700 mm to 900 mm. The car width is never larger than 1100 mm and therefore loading/unloading will generally be single file. This will influence the value used for t_p, which will generally be at the high end of the range at 2 s per passenger.

Table 8.8 ISO4190 Car platform and door sizes

	Residential				Heath care			
Rated load	320	450	630	1000	1275	1600	2000	2500
Platform width	900	1000	1100	1100	1200	1400	1500	1800
Platform depth	1000	1250	1400	2100	2300	2400	2700	2700
Door width	700	800/900	800/900	800/900	1100	1300	1300	1300/1400
	General purpose				Intensive traffic			
Rated load	630	800	1000	1275	1275	1600	1800	2000
Platform width	1100	1350	1600	2000	2000	2100	2350	2350
Platform depth	1400	1400	1400	1400	1400	1600	1600	1700
Door width	800/900	800/900	900/1100	1100	1100	1100	1200	1200

8.3.6.2 Health care lifts

These lifts are generally large and spacious, ie: from 1275 kg to 2500 kg, in order to accommodate wheelchairs, beds and trolleys. The doors are wider, in the range 1100 mm to 1400 mm, to accommodate movement of the equipment. Calculations of round trip times are not particularly relevant in this situation.

8.3.6.3 General purpose lifts

These are generally at the low end of the rated load range. Lifts with rated loads of up to 800 kg have doors of 700–800 mm size, thus restricting loading/unloading to single file. The car platform shape allows circulation in the car. Lifts with rated loads above 800 kg generally have doors of 1100 mm or larger and this allows double file entry after (say) the first six passengers have transferred. The platform shape allows circulation in the car.

8.3.6.4 Intensive traffic lifts

These lifts are generally larger than 1275 kg and are fitted with doors of 1100–1200 mm width. This allows double file passenger transfers and the car platform shape allows easy circulation around the car.

8.3.6.5 Unusual car shapes

Some lifts may be very deep, but narrow. This will hinder circulation in the car and slow passenger transfers as passengers jostle to move in/out. This type of platform shape occurs frequently with observation/scenic lifts, which tend to provide as much rear surface in the car, in order that passengers obtain a good view. Other lifts may be wide but shallow. This means that the doorway will be obstructed by passengers at the sides of the car. All these factors should be considered when carrying out round trip calculations.

8.3.7 Effect of Large Floor Plates

Some floor plates are the size of football fields. Here, the discussion indicated in Section 1.4.1 should be considered. No special changes to Equation (4.11) are necessary with respect to the calculation method.

8.3.8 Effect of Building Facilities

There will be facilities in buildings which will distort traffic movements. Examples are restaurants (positioned at the top of the building, in the basement, and even half way up the building); drinks and sandwich machines; leisure club facilities (swimming pools, gymnasia), toilet facilities; post rooms; trading floors, etc. These floors will provide a powerful attraction at different times of the day and must be considered in the traffic design.

No special changes are necessary to Equation (4.11) with respect to the calculation method. Traffic Case Study CS9 illustrates the effect.

8.4 CONSIDERATION BY BUILDING FUNCTION

8.4.1 Shopping Centres

Chapter 2 dealt with the circulation in shopping centres and indicated that they are built on two or three levels of retail with several levels of car parking. Lifts do not play a major part in the transportation arrangements, which are usually centred on escalators. Often scenic, or observation lifts are provided not only for transportation, but as an enjoyable experience. These lifts are usually hydraulic with slow flight times and slow door times. The traffic handling in a shopping centre is eased by the enjoyment aspects and the many modes of movement available.

Attention must be given to the transportation of the disabled and pushchairs. Often "pram" lifts are provided. The shape of these lifts is important as is their positioning on the concourse.

The use of car park lifts is determined mainly by the maximum rate of entry of vehicles and the average occupancy of each vehicle. These figures are usually determined from an associated (road) traffic study.

As a rough guide to lift provision, assume one person of lift capacity is required for each 100 m² of gross lettable retail area. So a 4000 m² store would require two 20 person cars. Lifts should always be located in pairs and not singly in order to provide a reasonable interval of 40−60 s and security of service during breakdowns and maintenance. It is unlikely that the lifts will fill to more than 50% and if trolleys are available this could be an optimistic value.

There are no changes required to Equation (4.11), but care will need to be taken in the assumptions made regarding vehicle occupancy levels.

A major problem in multi-level shopping complexes is the movement of shopping trolleys from one level to another. As indicated in Section 2.4.4, escalators can be designed to accept a trolley but are often dangerous in use as goods on or in them can dislodge and injure other escalator users. A commonly applied solution to this problem is to install moving ramps and this greatly improves circulation.

8.4.2 Airports

Most airports are arranged on two main levels with the arrival level below the departure level. There may then be other levels above and below providing various services (eg: baggage handling, catering) and facilities (offices). Another common characteristic is an adjacent, underground or elevated railway station.

Passengers at airports with baggage will use the trolleys provided. Most airports have sufficiently large halls and corridors and no problems should arise when the trolleys are used on one level. However, when the passenger requires to move from one level to another, difficulties can arise. Escalators should not be used, although moving ramps are ideal. Lifts are the main means of vertical movement.

Generally each baggage trolley will be attended by two persons plus their baggage. The weight of the baggage is generally restrained by the 20 kg allowance most (economy) passengers are allowed plus some 5 kg of hand luggage. Thus a trolley will weigh (including its own weight) some 75 kg, ie: equivalent in weight to one person. However, it will occupy the space taken by three or four persons. Thus the total weight of two passengers and their trolley will be some 225 kg and occupy the space of some 5 people. This occupancy and loading requirement must be taken into account.

Consider a nominal 50 person rated car capacity (rated load 3750 kg) lift. According to EN81-1: Table 1.1, the maximum available car area for an electric traction lift must then be 7.0 m^2. According to the body template (Figure 2.1), only 33.3 passengers can be accommodated. They will weigh only 2500 kg, some 67% of the rated load. If each pair of passengers and their baggage trolley occupy 1.05 m^2 (5 human spaces) then the nominal 50 person lift can accommodate 13.4 passengers and 6.7 trolleys. The total load will be 1500 kg. This is 40% of the rated load. Thus it can be seen that in these circumstances lifts are very unlikely to be overloaded.

If a hydraulic lift were to be installed with a nominal 50 person rated car capacity (3750 kg rated load) then the platform area from EN81-2: Table 1.1 would still be 7.0 m^2. However if EN81-2: Table 1.1A were used, a maximum area of 13.64 m^2 is allowed.[1] It is then possible to accommodate 26 passengers and 13 trolleys. This is a load of 2925 kg, which is some 80% of rated load. This would indicate that hydraulic lifts are most suitable for this environment and their poorer dynamic performance would not be a significant disadvantage. The number of up starts required by hydraulic lifts may be a limiting factor rather than the traffic calculation.

There are no changes required to Equation (4.11), but care will need to be taken in the assumptions of car occupancy levels.

As with shopping centres (Section 8.4.1), the movement of baggage trolleys from one level to another is a problem. A solution to this problem is to install moving ramps and this greatly improves circulation.

8.4.3 Car Parks

Car parks can be attached to shopping centres, offices, airports, railway stations, etc. They are often multistorey, although those at out of town shopping centres and railway stations may be a single level. The pedestrian demand is more likely to be restrained by the vehicle entry and exit ramp handling capacities. Another factor is the vehicle occupancy, which is likely to be two persons per car.

---

[1] The 1987 version of EN81-2 stated "when there is a low probability of the car being overloaded".

For offices the peak demand is often in the evening, when building occupants are attempting to reach their cars. The office lifts, which may not serve the car parking levels, will bring large numbers of people to the lobby. Those persons with cars will then make a significant demand on any lifts serving the car park levels. The demand on the car park lifts is similar to that experienced by the main lifts during the morning peak period, but this demand is downwards.

Once the occupants have reached their vehicles they may then spend significant time reaching the vehicle exit. Another factor is the vehicle occupancy, which is likely to be low at about 1.2 persons per car, unless car pools are in operation. However, there will still be a demand for an efficient lift service. The car park lifts must therefore be designed to meet this demand.

The traffic design should use Equation (4.11) if the car park lifts are separate to the main lifts, and Equation (8.8) if the lifts are part of a basement service.

8.4.4 Hospitals

The building form is important, ie: whether the building has a small footprint and is tall (US practice) or has a large footprint and is low (UK practice). In the former case where lifts are used as a primary circulation element their proper operation is vital, particularly when dealing with operating theatre emergencies. In Britain most hospitals are designed on a 2−3 storey low rise principle, although many city hospitals have high rise elements.

The principal corridors are sized to accommodate bed and trolley movements and therefore present no difficulties when handling pedestrian movements. Lifts are provided mainly as a means of moving bed bound patients from floor to floor.

The traffic designer will need to understand the *modus operandi* of the hospital before finalising a design. Factors to be considered include: numbers of staff and shift patterns, numbers of visitors and visiting hours, location of theatres, X-ray, etc., distribution and deliveries of food, beverages, supplies, waste disposal, patient emergency evacuation, porterage, etc. It is important that patient bed lifts are separate from the visitor and staff lifts, to avoid cross infection. The Health Technical Memorandum (HTM: 1995) gives some general guidance for UK hospitals.

There are no changes required to Equation (4.11).

8.4.5 Hotels

Lifts play an important part in the circulation of guests and service staff in a hotel. Escalators should be employed for short range movements, eg: to connect function levels with the lobby.

The traffic patterns in hotels are complex, and are not comparable to the morning and afternoon peaks in an office. The most demanding times are at check-out (08:00 to 10:00) and check-in (17:00 to 19:00). At these times heavy two way traffic occurs with guests going to and from rooms, restaurants and in and out of the hotel. Calculations should therefore assume equal numbers of up and down stops at these times.

At most times lifts are unlikely to load to more than 50%. The lift sizes should be at least 16 persons, in order to accommodate luggage and provide guests with uncrowded travel conditions. As a rule of thumb assume one lift for every 90−100 keys. This rule must be used with care as it would not be suitable for a low rise hotel with 30% of its rooms at the entrance level. Neither would it be suitable for a high rise hotel with a small footprint. There are also differences between the operational needs of a "transit" hotel near to airports, etc. where guests stay one night and hotels used by longer term and holiday guests.

It is recommended that the service traffic (baggage, goods, room service, messengers, etc.) be served by a secondary vertical transportation system, leaving the main lifts for the guests. As a rule of thumb there should be one passenger/goods lift for every two passenger lifts.

There are no changes required to Equation (4.11).

8.4.6 Railway Stations

Railway stations may be served mainly by stairs and pedestrian ramps, although some, particularly the deeper underground stations, will use escalators. Generally railway stations, whether above or below ground, have poor provision of lifts. This will change as the requirements to assist persons with limited mobility are applied.

When passengers require to move from one level to another with hand baggage difficulties arise. When baggage trolleys are used these difficulties increase. **Escalators should not be used**. As with shopping centres (Section 8.4.1) a solution to this problem is to install moving ramps and this also greatly improves circulation.

Lifts should be considered to be the main means of vertical movement in this case. There are no changes required to Equation (4.11).

8.4.7 Department Stores

This category applies to large departmental and chain stores. These stores will have many entrances, some of which may open to a main street whilst others open into shopping mall areas. The opportunity therefore exits for "leakage" into and out of shopping centres. Many stores will own lifts and escalators inside their occupancies. These facilities may be used by shopping centre shoppers to move between mall levels. Thus store facilities enhance those provided by a shopping centre or mall to the mutual advantage of both. Many of the considerations concerning shopping centres discussed in Section 8.4.1 apply to department stores.

8.4.8 Universities and Other Education Buildings

Universities buildings can be classified as institutional buildings, where the occupants receive a service. Where universities occupy city sites many have tall buildings (10−20 stories) and even those on out of town sites follow suit in order to reduce land use and keep a compact campus. Most buildings are mixed function: lecture rooms, laboratory and offices, although some buildings may specialise as lecture blocks. There are hourly cycles of 10 minutes of demand before and after each 50 minute lecture, tutorial or seminar session. In between the peaks the activity levels are low.

Often a university campus will have a mixed collection of office type buildings, halls of residence, catering services and factory like units containing teaching and research equipments (laboratory reactors and the like). The office type buildings can be treated in the same way as detailed in Chapters 4−7. The halls of residence can be treated in a similar way to hotels, although perhaps at lower levels of demand and performance. The catering services can be attached to either the office type buildings or halls of residence and should be treated similarly to those provided in office facilities or hotel facilities. The research buildings will probably be low rise and be subject to special movement provisions associated with the equipment installed, ie: barriers to radioactive areas.

The dominant feature of a university campus is the lecture change-over periods. To install lifts in a tall building to suit this demand is not cost effective (and universities do not have large capital budgets). Somehow the demands made for 10 minutes every hour must be reduced. A solution is to try to re-arrange the activities in the building to reduce the load on the lifts. An example in Tutt and Adler (1979)[1] illustrates the relationships in a small firm. A set of relationships can be formed for a university building. Some suggestions are:

- Place lecture facilities in the lower levels, say basement, ground and three to four floors above the entrance level. The general public will generally ascend one to two floors up and descend two to three floors down. Students (who are mainly young and fit) will probably go one more floor. For this activity to occur it is essential to provide wide, well lit and visible stairs.

- Laboratory, bulk service facilities (computer clusters, libraries) and student administration (Registrars, Bursars, Careers Advisory, etc. can be placed from the fourth floor upwards. These are either used for periods longer than one hour (laboratories), shorter than one hour (administration) or randomly (libraries).

- Offices should be placed at the top of the building. Their occupants will generally use the lifts on a more random basis.

There are no changes required to Equation (4.11), but care must be exercised in its application.

8.4.9 Residential Buildings

The customary basis for the traffic design of a residential building (apartments/flats) is to determine the number, car capacity and speed of passenger lifts necessary to adequately handle a 5-minute, two way traffic period based on the type of occupancy. The estimation of the population in a residential building is usually based on the number of bedrooms and the occupancy per bedroom. Suitable rules of thumb for the number of persons occupying a flat are given in Table 8.9.

Table 8.9 Occupancy factors for residential buildings

Type	Luxury	Normal	Low income
Studio	1.0	1.5	2.0
1 Bedroom	1.5	1.8	2.0
2 Bedroom	2.0	3.0	4.0
3 Bedroom	3.0	4.0	6.0

The commonly used design period for a residential building is the afternoon 5-minute, two way traffic condition, which is considered the most demanding traffic period. During this period of time, people are both entering and leaving the building. The lifts are loading passengers at the main lobby, distributing these passengers to various upper floors, reversing at the uppermost hall call, stopping in the down direction for additional passengers and transporting them to the main lobby. In low income housing,

---

[1] Page 121.

where many children and adults are leaving for school and work at the same time, the morning down peak may also be very heavy. Table 8.10 gives guidance.

Table 8.10 Design criteria: Residential buildings (5−minute, two way)

Type	Interval (s)	Handling capacity
Low income	≤50−70	≥5−7%
Normal	≤50−60	≥6−8%
Luxury	≤45−50	≥8%

Often in residential and low income (Local Authority) flats one passenger lift is generally arranged to allow furniture movement, coffins and stretchers and to handle other service needs. Luxury flats may include a separate goods lift for furniture, tradespeople and domestic help. These goods lifts are usually "hospital shaped" with capacities of around 2000 kg.

There are also requirements that each flat shall have access to an alternative lift during maintenance or out of service conditions. This has often been achieved in the past for low income blocks of flats, by using two simplex lifts operating on an interleaved (skip-stop) basis. This solution is not recommended today. Another solution for low income residential blocks is high level walkways to an alternative lift.

There are no changes required to Equation (4.11), but care must be exercised in its application.

8.4.10 Residential Care Homes and Nursing Homes

Homes generally have a low traffic requirement, which can be catered for by a single lift. Larger homes might acquire a second lift giving security of service in the event of break down or maintenance.

There is little opportunity to use the round trip time equation.

8.4.11 Entertainment Centres, Cinemas, Theatres, Sports Centres and Concert Halls

Buildings providing these functions can specialise in one of the activities or many of them. Many sports centres are low rise and do not require lifts. Town centre buildings such as cinema complexes, concert halls and theatres will be of higher rise. Such complexes generally use escalators as the main vertical transportation element. Lifts provided in these circumstances do not have to meet a large demand and may only have to satisfy the requirements for the handicapped and firefighting.

There are no changes required to Equation (4.11), but care must be exercised in its application.

8.5 SUMMARY

Table 8.11 provides a summary of the changes which may need to be made to the round trip time Equation (4.11) when designing for special situations.

Table 8.11 Summary of special situations

Situation		Comments	Equation
8.2.1	Shuttle lifts & sky lobbies	Long travel distances affect t_v. Increased passenger transfer times with two way traffic.	4.11, 8.1
8.2.2	Double decker lifts	Double "values" for t_v, $t_f(1)$; "half" values for N. Handling capacity twice the value for one deck.	8.7
8.2.3	Firefighting lifts	When stand alone use Equation (4.11). When part of a group may have different specification.	4.11
8.2.4	Goods lifts	When stand alone use Equation (4.11). When part of a group may have different specification.	4.11
8.2.5	Observation lifts	Low performance.	4.11
8.2.6	Platform lifts and lifts for the disabled	When stand alone not considered as part of circulation. When part of a group may have different specification.	4.11
8.3.1.1	Stacked zones	Express jumps to be considered.	5.22
8.3.1.2	Interleaved zones	Double "values" for t_v, $t_f(1)$; "half" values for N.	4.11
8.3.1.3	Transfer floors	Decide which zone a transfer floor belongs to during peak traffic.	4.11, 5.22
8.3.2	Very tall buildings	Complex situation. Apply general, sky lobby and double deck equations.	4.11, 5.22, 8.7
8.3.3	Basement service	Use for service from basement and to basement. Select percentage demand carefully.	8.8
8.3.4	Multiple entry and entrance bias	Apply basement equations. Care in deciding entrance demand split.	4.11, 8.8
8.3.5	Lobby design	Adjust t_p to account for inefficiencies.	4.11
8.3.6	Door width & car shape	Adjust t_p to account for inefficiencies.	4.11
8.4.1	Shopping centres	Lifts have lower performance. Car occupancy levels affected by prams and shopping.	4.11
8.4.2	Airports	Lifts have lower performance. Car occupancy levels affected by baggage trolleys.	4.11
8.4.3	Car parks	Demand affected by vehicle ramp capacity. Critical demand during evening for outgoing passengers.	4.11, 8.8
8.4.4	Hospitals	Lifts critical in tall hospitals.	4.11
8.4.5	Hotels	Highest demands at check-out and check-in. Low car occupancy.	4.11
8.4.6	Railway stations	Poor provision of lifts.	4.11
8.4.7	Department stores	Lifts have lower performance. Car occupancy levels affected by prams and shopping.	4.11
8.4.8	Educational buildings	Highest demand when teaching periods begin/end.	4.11
8.4.9	Residential buildings	Highest demand late afternoon.	4.11
8.4.10	Care and nursing homes	Low usage.	n/a
8.4.11	Entertainment buildings	Difficult application of Equation (4.11).	4.11

CHAPTER NINE

General Philosophy of Lift Traffic Design by Calculation

9.1 INTRODUCTION

This chapter deals with traffic calculations only, computer simulations are dealt with in Chapter 16. Generally, a large number of traffic designs, maybe as many as 95%, can be carried out by calculation, hence the importance of understanding this method.

The traffic design procedure should always start by applying the classical calculation methods discussed in Chapters 4 – 8. It is often possible to modify these methods to take into account some unusual features in a building, or an unusual passenger demand. This may be all that is needed to produce a satisfactory traffic design.

It is important to remember that traffic design is not an exact science and another designer will make different decisions and assumptions, which will affect the outcome. Each design will have validity.

Remember also that defining a distance to the nearest millimetre or a time to one hundredth of one second is not necessary. Any traffic design that is implemented will be used by human beings, who tend to modify their behaviour to suit the environment in which they exist. Thus errors of as much as ten or so per cent are often not significant. Gross errors, however, of say one lift too few (underlifted) will be noticed, as is evident during peak periods, when a lift is out of service ! In contrast, too many lifts (overlifted) are a waste of capital and recurrent expenditure.

Where the design requirements are more complicated, computer simulations may need to be carried out. Simulation models allow all traffic conditions to be examined more exactly. It is important to remember, however, that all calculations are based on a mathematical model and deliver an average answer to a design. The designs obtained from a mathematical model and a simulation model may be quite different as a result.

This chapter describes a methodical (philosophical) process for the traffic design of lifts. It takes the form of a case study. The building used to illustrate the process is based on a real project, but has been modified to emphasise certain points. The discussion should therefore not be related to any real building.

Calculation programs will be used in this chapter to automate the underlying equations. These should only be used when a traffic designer is familiar with the principles on which they are based as described in Chapters 4 – 8. Traffic Case Studies 9 – 14 should also be consulted. See Appendix 2 for a description of the programs used.

9.2 PERFORMANCE CRITERIA

The performance criteria that will be used are those discussed in Section 6.6 and Table 6.5. In Table 6.5 the performance of a system would be graded for one hour of activity, eg: balanced interfloor. The numerical values would reduce by one grade for 15 minutes of unbalanced activity, eg: lunch time, ie: an "excellent" system would use the "good" values. For 5 minutes of high demand activity the numerical values to use would be two grades down, eg: uppeak, ie: an "excellent" system would use the "fair" values.

9.3 THE BUILDING TO BE CONSIDERED

The building to be considered is a 31 storey building containing trader floors, offices associated with the traders, at the lowest level, a prestigious mid zone used as offices and a less prestigious upper zone used as a call centre. The owner is able (unusually) to provide exact zonal boundaries (see Figure 9.1) and occupancy figures and floor activity (see Table 9.1). Thus one design freedom has been lost where the traffic designer decides where the zonal boundaries might be placed but there will be no debate on the occupancy levels. In this design suitable equipment is to be selected to service the requirements of each zone.

Table 9.1 Building data

Floor	Activity	Population
1−3	Traders	600/floor
4−6	Office	160/floor
7	Facilities	80/floor
8−9	Office	160/floor
10−20	Office	153/floor
21−30	Call Centre	160/floor

Elevation. Floor 1: 11.7 m, Floor 10: 49.5 m, Floor 21: 98.8 m.

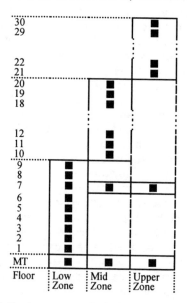

Figure 9.1 Zonal arrangement in the building to be considered

A Facilities floor is situated at Floor 7 in the Low Zone and provides restaurant, snack bar, reprographics, etc. for the whole building. Stops are possible to this floor by Mid Zone and Upper Zone lifts. It will be assumed that about 5% of the passengers arriving during the 5 minute uppeak period will stop at the Facilities floor. There is a restricted hours access to and from this level to a high level Promenade connecting to other buildings and transportation facilities. The use of this entrance will be ignored.

There are no transfer floors between the mid and upper zones except at Floor 7 (or ground). This allows the building to be let as three separate zones or to one tenant with appropriate security control of the access to each zone.

9.4 CLASSICAL DESIGN OF LOW ZONE BY CALCULATION METHOD

The lower zone comprises nine floors (Floors 1–9) above the ground with an escalator service provided to the lower three floors (Floors 1–3). The peak hour for arrivals is from 09:00 to 10:00 and the traffic design calls for a 15% uppeak demand with a 25 s interval. This zone contains the Facilities floor at Floor 7.

As the first three floors are served by escalators, the lifts might deal with 60 persons travelling to Floor 1, 150 persons travelling to Floor 2 and 300 persons travelling to Floor 3, if the guidance of Table 1.12 is followed.

Thus there are a total of 510 persons travelling to Floors 1–3. Floors 4–6 and 8–9 have populations of 160 persons per floor, ie: 800 persons. Some passengers will have their workplace on the Facilities Floor 7 (say 80) and other persons in the Low Zone may call there during the uppeak period (say 80). It will be assumed that 160 persons travel to Floor 7. Thus the total number of passengers to be transported by the Low Zone lifts is 1470 persons, which for a 15% peak arrival demand is 220 persons.

As the number of lifts, their rated speed and load, dynamics, etc. are not known, the iterative balance procedure (Section 6.10) will be used to examine the options. Table 9.2 shows the calculation using the IBM program. The 5-car group would not be suitable[1] as the interval is too long (>25 s) and the cars would be very large (>2500 kg), in order to accommodate 23.8 persons. The 6-car group provides an interval of 24.3 s and a car occupancy of 17.8 persons.

A 6-car group will be satisfactory if lifts with a Rated Car Load of 2500 kg (Table 7.2) are installed. They permit a design occupancy of 19 persons (Rated Car Capacity of 26 persons). It is likely that the number of stops will be slightly lower than the number (7.9) stated in Table 9.2, as the demand for Floor 1 will be lower and the demand for Floor 3 will be higher, than the demand for the other floors. Probably in practice a lift of 2000 kg rated load will be suitable.

Table 9.2 Low Zone calculation using the IBM program

```
*****************************************************************
Lift traffic calculations                    Iterative Balance Method
*****************************************************************
Design name: Low Zone            Run No.: 1          Date: 2002
.................................................................
Desired {I}nterval (s)    = 25     {AR}rival rate (5-min) =  220
{N}umber of floors        =  9     Building {POP}ulation  = 1470
Interfloor {D}istance (m) = 4.2    {E}xpress {J}ump (m)    = 11.7
{P}ass. {T}ransfer time (s) = 1.0  Speed (m/s) {v}         = 3.15
{AC}eleration (m/s2)  = 0.9   {J}erk (m/s3) = 1.3 {S}tart {D}elay (s) = 0.3
Door times (s)-{C}lose = 2.6  {O}pen       = 1.8 {AD}vance        = 0.5
.................................................................
Flight time (s) = 5.07              Performance time (s) = 9.27
.................................................................
Average high reversal floor =  8.9    Average number of stops =  8.5
Passengers =   23.8  Interval = 32.5  Number of lifts =  5 %POP = 15.0
Average travel time =  87.5
.................................................................
Average high reversal floor =  8.9    Average number of stops =  7.9
Passengers =   17.8  Interval = 24.3  Number of lifts =  6 %POP = 15.0
Average travel time =  75.0
*****************************************************************
```

---

[1] Generally, odd numbers of lifts in a group are not desirable. Three lifts arranged in line is acceptable, but groups with five or seven lifts often result in illogical lobby arrangements.

9.5 CLASSICAL DESIGN OF MID ZONE BY CALCULATION METHOD

The mid zone comprises 11 floors (Floors 10−20), the peak hour is from 08:30 to 09:30 and the traffic design calls for a 13% arrival rate with a 25 s interval.

The population of each floor is 153 persons, ie: a total of 1683 persons. The demand to reach Floor 7 might be some 77 persons during the uppeak period, making the total Mid Zone population 1760 persons and the 13% uppeak demand will be 230 persons/5-minutes.

As the number of lifts, their rated speed and load, dynamics, etc. are not known the iterative balance procedure, using the IBM program (Section 6.11), will be used to examine the options.

To accommodate the Facilities floor at Floor 7, it is necessary to modify the calculation data. Floor 7 is two floors below the first stopping floor (Floor 10) in the Mid Zone. To accommodate this and enable a calculation to be carried out, it will be assumed that Floor 7 is positioned as Floor 9. This will introduce an extra stop increasing the floors served from eleven to twelve floors ($N = 12$). The express jump must also be reduced by 4.2 m (= 45.3 m), as shown in Table 9.3.

Table 9.3 shows that a 7-car group will provide the handling capacity required at a suitable interval, but the cars would again be very large (2500 kg). The 8-car group provides the handling capacity and also a much lower interval of 21.2 s with an average load of 16.3 passengers in the cars.

The occupancy level in the car indicates a lift with a Rated Car Load of 2000 kg (see Table 7.2). The interval with 8 lifts is much better than specified.

Table 9.3 Mid Zone calculation using the IBM program

```
************************************************************************
Lift traffic calculations                         Iterative Balance Method
************************************************************************
Design name: Mid Zone           Run No.: 2           Date: 2002
.........................................................................
Desired {I}nterval (s)     = 25    {AR}rival rate (5-min) =  230
{N}umber of floors         = 12    Building {POP}ulation  = 1760
Interfloor {D}istance (m)  = 4.2   {E}xpress {J}ump (m)    = 45.3
{P}ass. {T}ransfer time (s) = 1.0  Speed (m/s) {v}        =    4
{AC}eleration (m/s2)  = 1.0  {J}erk (m/s3) = 1.5 {S}tart {D}elay (s) = 0.3
Door times (s)-{C}lose = 2.6  {O}pen       = 1.8 {AD}vance          = 0.5
.........................................................................
Flight time (s) = 4.82             Performance time (s) = 9.02
.........................................................................
Average high reversal floor = 11.8   Average number of stops =  9.9
Passengers =  20.1  Interval =  26.3  Number of lifts =  7 %POP =  13.1
Average travel time =  95.0
.........................................................................
Average high reversal floor = 11.7   Average number of stops =  9.1
Passengers =  16.3  Interval =  21.2  Number of lifts =  8 %POP =  13.0
Average travel time =  85.2
************************************************************************
```

The traffic design for the Mid Zone is complicated by the presence of the Facilities Floor just below the lowest floor in the zone. There will be an error caused by the amendment of the design data to accommodate the Facilities Floor traffic. However, the error should not be significant. This is a good case for study by simulation. These studies can look at all traffic patterns of uppeak, down peak, interfloor and mid day activity.

9.6 CLASSICAL DESIGN OF UPPER ZONE BY CALCULATION METHOD

The upper zone comprises 10 floors (Floors 21−30) and the occupants are required to be at their desks by 09:00 to provide a call centre service. A 15% peak arrival rate should be assumed. A 30 s interval would be acceptable for this zone.

The floor population of 160 persons per floor, ie: a total of 1600 persons. The demand to reach Floor 7 might be some 80 persons during the uppeak period, making the total Upper Zone population 1680 persons.

As the number of lifts, their rated speed and load, dynamics, etc. are not known the iterative balance procedure, using the IBM program (Section 6.10), will be used to examine the options.

Floor 7 is 14 floors below the first stopping floor (Floor 21) in the Upper Zone. To accommodate this and enable a calculation to be carried out, it will be assumed that Floor 7 is positioned as Floor 20. This will again introduce an extra stop ($N = 11$) and again the express jump must be reduced by 4.2 m (= 94.6 m), as shown in Table 9.4.

Table 9.4 shows that an 8-car group provides the handling capacity of 15% and a suitable interval of 23.0 s with 19.3 passengers in the cars. This occupancy level is higher than the design occupancy level of 16 persons for lifts with a rated load of 2000 kg (see Table 7.2). It is not sensible to install nine lifts, so eight, 2500 kg should be provided although they will be large lifts. The designer might assume that, as these floors are heavily populated with call centre staff, there would be absences owing to illness, etc., which would reduce the demand. In which case lifts with a rated load of 2000 kg might be considered.

Table 9.4 Upper Zone calculation using the IBM program

```
********************************************************************
Lift traffic calculations                    Iterative Balance Method
********************************************************************
Design name: Upper Zone                   Run No.: 3        Date: 2002
. . . . . . . . . . . . . . . . . . . . . . . . . . . . . . . . . . .
Desired {I}nterval (s)     = 25    {AR}rival rate (5-min) =  252
{N}umber of floors         = 11    Building {POP}ulation  = 1680
Interfloor {D}istance (m)  = 4.2   {E}xpress {J}ump (m)   = 94.6
{P}ass. {T}ransfer time (s) = 1.0  Speed (m/s) {v}        = 6
{AC}eleration (m/s2)  = 1.0  {J}erk (m/s3) = 1.5 {S}tart {D}elay (s) = 0.3
Door times (s)-{C}lose = 2.6  {O}pen     = 1.8 {AD}vance       = 0.5
. . . . . . . . . . . . . . . . . . . . . . . . . . . . . . . . . . .
Flight time (s) = 4.82             Performance time (s) = 9.02
. . . . . . . . . . . . . . . . . . . . . . . . . . . . . . . . . . .
Average high reversal floor = 10.9    Average number of stops =  9.9
Passengers = 23.8  Interval = 28.3  Number of lifts =  7 %POP = 15.0
Average travel time = 105.2
. . . . . . . . . . . . . . . . . . . . . . . . . . . . . . . . . . .
Average high reversal floor = 10.8    Average number of stops =  9.3
Passengers = 19.3  Interval = 23.0  Number of lifts =  8 %POP = 15.0
Average travel time =  95.6
********************************************************************
```

The traffic design for the Upper Zone is complicated by the presence of the Facilities Floor so far below the lowest floor in the zone. Again there will be errors owing to the method of accommodation used to enable a calculation to be carried out. This again is a good case for study by simulation. The traffic design for the Upper Zone is considered further in the Simulation Case Studies CS18−21. These studies look at uppeak, down peak, interfloor and mid day activity.

9.7 DISCUSSION

The equipment chosen for the lift systems considered in this chapter will provide better than average performance with reasonable door timings and modest drive dynamics.

The passenger transfer time chosen of 1.0 s is pessimistic and allows 16 seconds of transfer time, when 16 persons (see Section 5.9) transfer into a lift. In practice if the first six passengers enter single file and the remaining 10 enter in double file this is 11 transfers [6 + 5(2)]. This implies a transfer time per passenger of 1.5 s each (16/11).

The values obtained for the intervals (see Table 9.5) are often lower than those specified, but are low in order to achieve the specified handling capacities. No intervals are lower than 15 s when passenger loading inefficiencies might occur, where more than one lift is loading at the same time at the main terminal.

Table 9.5 Comparison of results by calculation

Zone	N	L	Load	%POP	Interval	P
Low	9	6	2000 kg	15.0	24.3 s	17.8
Mid	11	8	2000 kg	13.0	21.2 s	16.3
Upper	10	8	2500 kg	15.0	23.0 s	19.3

The results indicate very good interval times and that the handling capacities required are met. The zonal split was defined by the owner and probably cannot be improved. The Upper Zone has eight lifts and serves 10 floors. A possibility would be to reduce the number of lifts to six (an equal number to ease the lobby design) and reduce the number of floors served in the Upper Zone, with an increased express jump. The number of floors served in the Low and Mid Zones would then need to be adjusted and eight lifts would be required for the Low Zone. No core space would be saved in the lower parts of the building and only a little in the upper parts. Readers might like to consider these zonal changes.

9.8 ESCALATOR SERVICE

Finally, the design is not complete until the usage of the escalators at the lower three floors is considered. Can the escalators adequately serve these lower three (trader) floors? The escalator usage was assumed in the calculation of the lift system to be as given by Table 1.12, ie:

- 90% of passengers travelling to Floor 1
- 75% of passengers travelling to Floor 2
- 50% of passengers travelling to Floor 3.

Then the lowest escalator must be able to transport all the persons not using the lifts up to Floor 1. That is 540 persons to Floor 1, 450 persons to Floor 2 and 300 persons to Floor 3, making a total of 1290 persons. Assuming a 15% peak arrival, similar to that experienced by the lifts, this is 193.5 persons in 5 minutes. A 1000 mm wide escalator with a rated speed of 0.5 m/s has a handling capacity of 75 persons per minute, ie: 375 persons/5-minutes. Thus the escalators will be able to meet the demand and will be relatively lightly loaded, especially those connecting Floors 1/2 and 2/3.

TRAFFIC CASE STUDIES

Contents

CASE STUDY NINE

Tall Building with Separated Lift Lobbies

CS9.1 INTRODUCTION

This case study considers a slender 32 floor building, built in the 1970s with more consideration to its shape and form rather than the practicality of its building operations. The ideal shape which allows efficient vertical transportation is "compact". The use of a design, which places the load bearing structure at the ends of the building, means the centre of the building is too weak to support the lift systems. Thus the lifts are placed at the ends of the building, contrary to good circulation practice. The main entrance is central to the building. Thus we have an unconventional building and of which can be said:

> "Beauty stands
> In admiration only of weak minds
> Led captive"

[Milton, Paradise Regained, I. 497]

There were two groups of three lifts situated at each end of the building with one group serving a zone of odd floors and the other group serving a zone of even floors. The lift dynamics were poor and two panel side opening doors were fitted to the cars. The owner wanted to take the 30-year-old building "upmarket" and thus needed to improve the vertical transportation system.

CS9.2 THE ORIGINAL SYSTEM

The original skip/stop system was designed circa1970 probably to design criteria for high prestige buildings, ie: 15% uppeak handling capacity; a 30 s or less uppeak interval; a density of occupation of one person per 14 m²; and a 100% daily attendance. The rentable size of each floor is 420 m². As the building is tall and slender, ie: an inefficient design (Section 1.1), the usable area (Section 6.4.4) is 80% of this, ie: 336 m². This implies 24 persons per floor, ie: 384 persons in each zone.

The data for the original system are:

Odd zone:	serving 16 floors: MT, 1−31
Even zone:	serving 16 floors: MT, 2−32
Number of lifts per zone:	3
Rated load:	950 kg/12 persons
Rated speed:	3.0 m/s
Interfloor distance:	6.0 m (between served floors)
Doors:	1100 mm wide, two panel side opening
Door opening time:	2.0 s
Door closing time:	4.2 s
Single floor flight time:	6.5 s (6.0 m)

A lift of rated capacity 12 persons will accommodate a maximum of 10 persons (see Table 7.2), therefore assume $P = 8$ persons. From the data it is possible to calculate the original performance. Both zones will perform similarly. The output from a simple computer program[1] is shown in Table CS9.1[2] and Table CS9.2 summarises the results.[3]

Table CS9.1 Calculation of original performance using "Lift Traffic Design" (LTD) program

```
****************************************************************************
Lift traffic calculations                              Lift Traffic Design
****************************************************************************
Design name: Case Study Nine           Run No.: 1          Date: 2002
............................................................................
{N}umber of floors         = 16     Number of {P}assengers  =  8
Number of {L}ifts          = 3      Building {POP}ulation   =  384
Interfloor {D}istance (m)  = 6      {E}xpress {J}ump (m)    =  0
{P}ass. {T}ransfer time (s) = 1.0   Speed (m/s) {v}         =  3.0
{AC}eleration (m/s2)   = 0.7 {J}erk (m/s3) = 1.2 {S}tart {D}elay (s) = 0.5
Door times (s): {C}lose = 3.7 {O}pen   = 2.0 {AD}vance         =  0
............................................................................
Flight time (s) = 6.47              Performance time (s) = 12.67
............................................................................
Average high reversal floor = 14.7    Average number of stops =  6.5
RTT = 154.2  Interval =  51.4  Handling capacity = 46.7  %POP = 12.2
Average journey time =  86.8
****************************************************************************
```

Table CS9.2 Summary of original performance

Parameter	Odd zone	Even zone
Number of floors (N)	16	16
Number of passengers (P)	8	8
Highest reversal floor (H)	29.7	30.7
RTT (s)	154	154
INT (s)	51	51
HC (Persons per 5-minutes)	47	47
Zone population	384	384
%POP (%)	12	12

The lift system cannot serve the original design requirements either in respect of the handling capacity required and the interval is 51 s. The only way that the lift system could cope would be for: either the building to have a lower population than indicated; or the demand to be less than 15%; or the daily attendance to be less than 100%; or a combination of all possibilities. For example, if the attendance was 85% and the peak arrival rate was 13% then the 5-minute peak arrivals would be 42 persons,[4] lower than the system capability of 47 persons/5-minute. This is may be what happened, ie: the handling capacity was sufficient, but the interval was unacceptable.

Eventually the population of the building increased as it aged and the lift system was unable to cope. Thus 30 years later a new design was required.

---

[1] In these case studies the underlying mathematics described in earlier chapters is not presented, instead a simple computer program is used. Readers can write such a program, either in a computer programming language, or by using a spreadsheet technique.

[2] Explanation to Table CS9.1 for LTD calculations. The table has three main fields. The first is the banner and design information. The second indicates all the design data provided for the calculation. The third field indicates the calculated data, which includes: flight and cycle times; values for *H, S, RTT, INT, HC, %POP* and *ATT*. There is an editing fourth field at the bottom of the table. See Appendix 2 for details.

[3] Some figures are rounded.

[4] $42 = 384 \times 0.13 \times 0.85$

CS9.3 A NEW DESIGN

A slender building of 32 floors will always be difficult to "elevate" as (a) it is not quite tall enough for three zones and (b) up to five lifts could be required in each group, in order to achieve the necessary Quality of Service, ie: with respect to the interval and the average passenger waiting time. In 1972 the British Code of Practice CP407 indicated: "For excellent quality of service, 1 lift is required for every 3 floors ... interconnected in one group". Strakosch stated in 1967[1] that a skip/stop arrangement was not to be recommended.

The building was provided with two lobbies so this gives the opportunity to divide the building into a stacked configuration served from the two lobbies. The installation of additional lifts would assist, but this is impossible owing to the building structure. Thus two groups of only three lifts arranged to serve two zones can be proposed. The lift and door dynamics (rated speed, door configuration, door times) and passenger handling facilities (passenger detection) should also be improved. This was achieved by increasing the rated speeds, improving the motion profile and by the fitting centre opening doors.

A transfer floor needs to be provided at the zone interface. The transfer floor should be placed as high as possible in the building in order to compensate for the express jump to the first served floor in the high zone, without compromising the Quality of Service to either zone. Calculations show the best position for the transfer floor to be Floor 19. This will therefore be the highest floor in the low zone and the lowest floor in the high zone. During uppeak traffic, only the high zone lifts would serve Floor 19, thus the last stopping floor in the low zone would be Floor 18. The data for the new system are:

Low zone:	3 lifts serving 18 floors: MT, 1−19 (1−18 uppeak)
High zone:	3 lifts serving 14 floors: MT, 19−32 (19−32 uppeak)
Rated load:	950 kg/12 persons
Rated speed:	3.15 m/s (low zone) 4.0 m/s (high zone)
Interfloor distance:	3.0 m (between served floors)
Express jump:	59 m (high zone)
Doors:	1100 mm wide, two panel centre opening
Door opening time:	1.8 s
Advance opening time:	0.4 s
Door closing time:	2.3 s
Single floor flight time:	4.2 s (3.0 m interfloor distance)

From this data it is possible to calculate the new performance using the LTD program. Both zones will perform differently. Tables CS9.3 and CS9.4 show the results of the computer program and Table CS9.5 summarises the results (figures rounded).

Table CS9.5 Summary of improved performance

Parameter	Low zone	High zone
Number of floors (N)	18	14
Number of passengers (P)	8	8
Highest reversal floor (H)	16.6	30.9
RTT (s)	102	128
INT (s)	34	43
HC (Persons per 5-minutes)	70	56
Zone population	432	336
%POP (%)	16	17

---

[1] Page 38.

Table CS9.3 Calculation of improved performance using the LTD program: low zone
```
***********************************************************************
Lift traffic calculations                             Lift Traffic Design
***********************************************************************
Design name: Case Study Nine         Run No.: 2          Date: 2002
.......................................................................
{N}umber of floors           = 18    Number of {P}assengers  =  8
Number of {L}ifts            = 3     Building {POP}ulation   =  432
Interfloor {D}istance (m)    = 3     {E}xpress {J}ump (m)    =  0
{P}ass. {T}ransfer time (s)  = 1.0   Speed (m/s) {v}         =  3.15
{AC}eleration (m/s2)   = 1.0  {J}erk (m/s3) = 1.5  {S}tart {D}elay (s) = 0.3
Door times (s): {C}lose = 2.3  {O}pen   = 1.8  {AD}vance       =  0.4
.......................................................................
Flight time (s) = 4.19              Performance time (s) = 8.19
.......................................................................
Average high reversal floor = 16.5   Average number of stops =  6.6
RTT =  102.4  Interval =  34.1  Handling capacity =  70.3  %POP =  16.3
Average journey time =  60.8
***********************************************************************
```

Table CS9.4 Calculation of improved performance using the LTD program: high zone
```
***********************************************************************
Lift traffic calculations                             Lift Traffic Design
***********************************************************************
Design name: Case Study Nine         Run No.: 3          Date: 2002
.......................................................................
{N}umber of floors           = 14    Number of {P}assengers =  8
Number of {L}ifts            = 3     Building {POP}ulation  =  336
Interfloor {D}istance (m)    = 3     {E}xpress {J}ump (m)   =  59
{P}ass. {T}ransfer time (s)  = 1.0   Speed (m/s) {v}        =  4.0
{AC}eleration (m/s2)   = 1.0  {J}erk (m/s3) = 1.5  {S}tart {D}elay (s) = 0.3
Door times (s): {C}lose = 2.3  {O}pen   = 1.8  {AD}vance      =  0.4
.......................................................................
Flight time (s) = 4.19              Performance time (s) = 8.19
.......................................................................
Average high reversal floor = 12.9   Average number of stops =  6.3
RTT =  128.2  Interval =  42.7  Handling capacity =  56.1  %POP =  16.7
Average journey time =  76.9
***********************************************************************
```

Table CS9.5 shows that the Quality of Service represented by the interval at the main terminal is better for both zones. The round trip and interval times are still too long for this type of building. But are the underlying handling capacities really required?

The floor populations are still set at 24 persons per floor, which returns percentage population values of 16% and 17% respectively. Suppose now that the density of occupation had changed to one person/10 m², then the floor populations would be 34 persons and the zone populations would then be 612 and 476 persons respectively. If the daily attendance was only 85% and the arrival rate were now to be 12% (which is more typical in 2000), then handling capacities of 62 and 49 persons/5-minutes would be required. These values are less than the underlying capacities of 70 and 56 persons/5-minutes. An iterative balance calculation would need to be carried out for these arrival rates in order to determine how the lift systems would then perform. Both zones will perform differently. Tables CS9.6 and CS9.7 show the results of the computer program "Iterative Balance Method" (IBM) and Table CS9.8 summarises the results.

Table CS9.8 Matching the performance by the demand

Parameter	Low zone	High zone
Number of floors (N)	18	14
Number of passengers (P)	6.1	6.4
Highest reversal floor (H)	15.9	30.6
RTT (s)	88	117
INT (s)	29	39
HC (Persons per 5-minutes)	62	49

Table CS9.6 Calculation of improved performance using the IBM program: low zone[1]
```
*****************************************************************************
Lift traffic calculations                          Iterative Balance Method
*****************************************************************************
Design name: Case Study Nine          Run No.: 4           Date: 2002
...........................................................................
Desired {I}nterval (s)      = 30      {AR}rival rate (5-min) = 62
{N}umber of floors          = 18      Building {POP}ulation  = 432
Interfloor {D}istance (m)   = 3       {E}xpress {J}ump (m)   = 0
{P}ass. {T}ransfer time (s) = 1.0     Speed (m/s) {v}        = 3.15
{AC}eleration (m/s2)   = 1.0  {J}erk (m/s3) = 1.5  {S}tart {D}elay (s) = 0.3
Door times (s): {C}lose = 2.3  {O}pen       = 1.8  {AD}vance           = 0.4
...........................................................................
Flight time (s) = 4.19                Performance time (s) = 8.19
...........................................................................
Average high reversal floor = 17.3    Average number of stops =  10.4
Passengers =  15.0 Interval =  72.6   Number of lifts =  2 %POP =  14.3
Average journey time =  104.6
...........................................................................
Average high reversal floor = 15.9    Average number of stops =   5.3
Passengers =   6.1  Interval =  29.3  Number of lifts =  3 %POP =  14.4
Average journey time =  50.3
*****************************************************************************
```

Table CS9.7 Calculation of improved performance using the IBM program: high zone
```
*****************************************************************************
Lift traffic calculations                          Iterative Balance Method
*****************************************************************************
Design name: Case Study Nine          Run No.: 5           Date: 2002
...........................................................................
Desired {I}nterval (s)      = 30      {AR}rival rate (5-min) = 49
{N}umber of floors          = 14      Building {POP}ulation  = 336
Interfloor {D}istance (m)   = 3       {E}xpress {J}ump (m)   = 59
{P}ass. {T}ransfer time (s) = 1.0     Speed (m/s) {v}        = 4.0
{AC}eleration (m/s2)   = 1.0  {J}erk (m/s3) = 1.5  {S}tart {D}elay (s) = 0.3
Door times (s): {C}lose = 2.3  {O}pen       = 1.8  {AD}vance           = 0.4
...........................................................................
Flight time (s) = 4.19                Performance time (s) = 8.19
...........................................................................
Average high reversal floor = 12.6    Average number of stops =   5.3
Passengers =   6.4  Interval =  39.1  Number of lifts =  3 %POP =  14.6
Average journey time =  69.7
...........................................................................
Average high reversal floor = 11.7    Average number of stops =   3.6
Passengers =   4.0  Interval =  24.7  Number of lifts =  4 %POP =  14.6
Average journey time =  53.6
*****************************************************************************
```

CS9.4 COMMENTARY

The performance of the low zone is now within the desired specification of handling capacity and interval. The high zone interval is improved, but fails to meet the desired interval. The number of passengers travelling in the cars is much lower (6.1 and 6.4) than the available capacity, giving a more enjoyable journey for the passengers. Little more could be done to improve the installation, without structural changes, for example to add another lift to each group.

---

[1] Explanation to Table CS9.6 for IBM calculations. The table has three fields. The first is the banner and design information. The second indicates all the design data provided for the calculation. The third field indicates the calculated data, which includes: flight and cycle times; values for *H, S, P, INT, L, %POP* and *ATT*. This is given for two possible options, changing the number of lifts. See Appendix 2 for details.

Basement Service in a Low Rise Building

CS10.1 INTRODUCTION

Basements were considered in Section 8.3.3 and Table 8.6 indicated a range of results. Now consider a practical example of a nine storey building where there are three basement parking areas. The relevant data provided are:

LB	43 car parking spaces
B	59 car parking spaces
LG	74 car parking spaces + 40 cycle spaces
G	1160 m² [83]
1	1170 m² [84]
2−4	1270 m² [91]

The areas given are net internal and are all usable. The density of occupation is taken as one person per 14 m² to give the population figures shown as [nn] above. What lift system should be installed ?

CS10.2 DISCUSSION

In this case the basement demand is very high and will distort service from the main terminal floor (Floor G) to Floors 1−4, particularly as the cars will be, on average, 56% full on arrival at the Ground floor.

The total building population is 440 persons. The population on Floors 1−4 will be 357 persons. They will all reach their work destinations using the lifts, whether they arrive via the basements, or via Floor G. The population on Floor G is 83, some of whom will arrive via the basement car parks, thus using the lifts and some will arrive directly to Floor G via the main entrance and not use the lifts at all. There are 176 car parking spaces. Assuming 1.2 passengers per car, this would indicate a daily arrival via the car parks of 211 persons. Add to this the 40 cyclists, gives a total daily basement arrival of 251 persons. This implies that 189 persons (440−251) arrive via the G floor entrance.

The effect of some persons arriving directly at the G floor and not using the lifts can be considered to balance other persons travelling to the G floor from the basements and therefore alighting at the G floor. Therefore it is reasonable to take the population of Floors G−4 of 440 persons when calculating the percentage population figures.

The arrival of cars into a car park is generally limited by the number of entry ramps. The vehicle entry rate is given as one vehicle every 10 s, which is the time to negotiate the entry barrier control. Thus in the basement there could be a maximum 5-minute arrival rate of 36 persons. Some of these arrivals may work on the Ground floor, but will use the lifts to get there.

If a 15% peak is assumed then some 28 persons (15% × 189) arrive at the G floor over the peak 5-minute period for travel to Floors 1−4. The peak arrivals at the Ground floor can occur in a worst case scenario at the same time as the peak arrivals from the basements. The arrivals at the three basement floors (36 persons) are larger than the numbers arriving at the Ground floor (28 persons). Thus lifts arriving at the Ground floor will be 56% occupied. This is calculated as 36/64, where 64 represents the total peak 5-minute arrivals at the Ground floor.

A computer program "Lift Traffic Design with Basements" (LTDB)[1] designed to deal with basement service and using Equation (8.8) was used to select a suitable lift system. Table CS10.1 indicates the first design with two lifts. The percentage basement passenger occupancy at the arrival at Floor G was taken as 56%. The rated speed was 1.6 m/s, which is appropriate to the height of the building. The rated capacity was 13 persons with a design occupancy of 9.1 persons. The relevant other parameters are shown in the table.

Table CS10.1 First calculation of performance using the LTDB program

```
*****************************************************************************
Lift traffic calculations                          Basement traffic design
*****************************************************************************
Design name: Case Study Ten              Run No.: 1          Date: 2002
.......................................................................
{N}umber of floors              = 4      Number of {P}assengers      =  9.1
Base{M}ent floors               = 3      %{B}asement {P}as'gers       = 56
Number of {L}ifts               = 2      Building {POP}ulation       = 440
Interfloor {D}istance (m)    = 3.8      {B}asement {D}istance (m)   =  2.7
{P}ass. {T}ransfer time (s) = 1.0      Speed (m/s) {v}             =  1.6
{AC}eleration (m/s2)     = 0.8      {J}erk (m/s3) = 1.2  {S}tart {D}elay (s) = 0.5
Door times (s): {C}lose = 2.6  {O}pen        = 2.0  {AD}vance            = 0.4
.......................................................................
Flight time (s) = 5.08                   Performance time (s) = 9.78
.......................................................................
Average high reversal floor = 3.9    Average number of stops = 3.7
HM =  2.9    SM =      2.6    tfB =     4.40
RTT =  95.9  Interval = 47.9  Handling capacity = 57.0  %POP = 12.9
*****************************************************************************
```

Table CS10.2 Final calculation of performance using the LTDB program

```
*****************************************************************************
Lift traffic calculations                          Basement traffic design
*****************************************************************************
Design name: Case Study Ten              Run No.: 2          Date: 2002
.......................................................................
{N}umber of floors              = 4      Number of {P}assengers      =  6.2
Base{M}ent floors               = 3      %{B}asement {P}as'gers       = 56
Number of {L}ifts               = 3      Building {POP}ulation       = 440
Interfloor {D}istance (m)    = 3.8      {B}asement {D}istance (m)   =  2.7
{P}ass. {T}ransfer time (s) = 1.0      Speed (m/s) {v}             =  1.6
{AC}eleration (m/s2)     = 0.8      {J}erk (m/s3) = 1.2  {S}tart {D}elay (s) = 0.5
Door times (s): {C}lose = 2.6  {O}pen        = 2    {AD}vance            = 0.4
.......................................................................
Flight time (s) = 5.08                   Performance time (s) = 9.78
.......................................................................
Average high reversal floor = 3.8    Average number of stops = 3.3
HM =  2.7    SM =      2.3    tfB =     4.40
RTT =  84.6  Interval = 28.2  Handling capacity = 66.0  %POP = 15.0
*****************************************************************************
```

CS10.3 COMMENTARY

The first design shows a very long interval of 47.9 s and a handling capacity of 12.9%. This indicates that three lifts are needed. Designs with three lifts and a car occupancy of 9.1 persons gave too high a handling capacity. Once this was reduced to 6.2 persons a suitable handling capacity of 15% was achieved at an interval of 28.2 s.

---

[1] Explanation to Table CS10.1 for LTDB calculations. The table has three fields. The first is the banner and design information. The second indicates all the design data provided for the calculation. The third field indicates the calculated data, which includes: flight and cycle times for the above ground data; the flight time for the basement data; values for *H, S, HM, SM, RTT, INT, HC* and *%POP*. See Appendix 2 for details.

Tall Building with Total Connectivity

CS11.1 INTRODUCTION

The building to be considered is a 54 level triangular shaped tower with 51 office floors from Level 4 to Level 54 for a single prestige tenant (Godwin, 1993). The levels are arranged in "villages" of three floors with a garden at the lowest floor. Three villages share a central atrium of nine floors. As the villages progress up the building they rotate by 120° producing a cycle of three villages before the original orientation is regained. Three cores are positioned at the apexes of the triangle and include the lift hoistways and toilet pods. Each core was designed to accommodate six lifts.

CS11.2 THE BRIEF

(1) To minimise or eliminate the necessity for transfer levels.

(2) Users to alight at a core furthest from the garden.

(3) Male and female toilet pods to be accommodated in two of the three cores.

(4) Levels 4 and 47 are common facilities accessible to all occupants.

(5) Levels 48 – 54 to be a secure area at lower levels of occupancy (total 300 persons).

(6) Each floor is to accommodate a nominal 60 persons.

(7) Lifts to be visible from the outside of the building.

CS11.3 THE DESIGN

Designs do not appear spontaneously, but in a progressive way, as this did (Godwin, 1993). The actual core each passenger must use is defined by design brief point (2), which requires a passenger to alight at a core furthest from the garden. This means that every third village is in the correct orientation. This leads to the idea of a non-contiguous 15 floor building zone comprising three floors, skip six floors progressing up the building. As this is not a conventional zone of contiguous floors, it would be important that occupants of each village group have common relationships (Tutt & Adler, 1990), ie: sales with marketing and training with human resources, in order to reduce interzonal traffic.

To meet the requirement of design brief point (1) the idea of secondary lift lobbies emerges. There are two secondary lobbies on each floor. They have to accommodate the toilet pods, one male and one female per floor. However, space can be made available to enable access to three of the six lifts in each of the secondary cores, one for a floor orientated 120° "ahead" and one for a floor orientated 120° "behind". This secondary circulation route would accommodate interfloor movements only and car calls would not be possible. The secondary core lifts would only respond to landing calls, car calls would be inhibited.

The service to the secure Floors 48−54 is ignored, as it would be served by two separate lifts via a screening desk in the Blue core at level 47. Thus it was possible to envisage each group of lifts serving some 15 floors in each village group, having substantially the same population. The groups of floors are identified as "red", "yellow" and "blue" village groups. The floors served are shown in Table CS11.1.

Table CS11.1 Floor service arrangements

Village Group	Floors served							Total floors
Red	4	5−6	13−15	22−24	31−33	40−42	47	16
Yellow	4	7−9	16−18	25−27	34−36	43−45	47	17
Blue	4	10−12	19−21	28−30	37−39	46	47	15

The associated interfloor service is not shown in Table CS11.1 but is illustrated in Figure CS11.1. Note that primary interchange floors are provided at Levels 4 and 47.

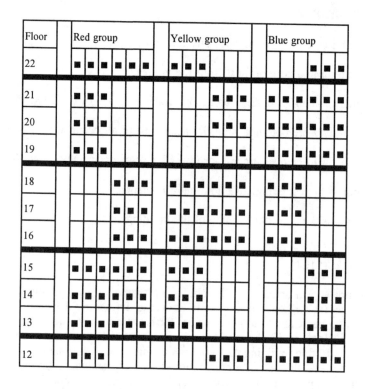

Figure CS11.1 Segment of design strategy: Floors 12−22.

All design briefs were realised. However, passengers need some initial orientation for interfloor travel. Consider a passenger at Floor 14 (red group). If the passenger wished to travel to: another red group floor they will go to a red lobby (with six lifts); to a yellow group floor they will go to a yellow lobby (with three lifts); and to go to a blue group floor they will go to a blue lobby (with three lifts).

CS11.4 TRAFFIC PERFORMANCE

Consider the Red group. The analysis for the other two groups will be similar.

CS11.4.1 Uppeak Analysis

Design requirements

Uppeak arrival rate:	15%
Uppeak interval:	<30 s
Uppeak car load:	80% of actual capacity
Average waiting time (AWT):	<25 s
Average passenger time to destination (ATT):	<90 s

Design data

Rated speed:	7.0 m/s
Rated acceleration:	1.2 m/s^2
Rated jerk:	2.0 m/s^3
Average interfloor distance:	3.7 m
Single floor flight time:	4.2 s
Door size:	1100 mm
Door opening time:	1.6 s
Door advance opening:	0.9 s
Door closing time:	2.5 s
Rated Car Capacity:	1600 kg
Nominal capacity:	21.0 persons
Actual capacity:	16.9 persons
Design capacity:	13.5 persons
Passenger transfer time:	1.0 s

Design assumptions

Lifts always stop at Floor 47 carrying one third of the Floor 47 population.
Lifts do not stop at Floor 4 during uppeak.
Passengers for Floors 48−54 use the Blue group.

Calculation

Population:	$14 \times 60 + 1 \times 20 = 860$ persons
Handling capacity required:	$860 \times 0.15 = 129$ persons/5-minutes
Calculation data:	$N = 16, P = 13.5, H = 47, S = 9.7$
	$t_v = 3.7/7.0 = 0.53, t_p = 1.0$ s
	$t_s = 2.5 + 1.6 - 0.9 + 4.2 - 0.53 = 7.47$ s.

Using Equation (4.11):

$$RTT = 2 \times 47 \times 0.53 + 10.7 \times 7.47 + 2 \times 13.5 \times 1.0 = 156.8 \text{ s}$$

Correction for a larger average jump:

Average jump $H/S = 47/9.7 = 4.9$ floors

Time (from dynamics program[1]) to jump 4.9 floors is 8.5 s

Assumed time to jump 4.9 floors = $4.2 + 3.9 \times 0.53 = 6.3$ s

Error is $8.5 - 6.3 = 2.3$ s

Total RTT error = $10.7 \times 2.3 = 24.6$ s

Corrected $RTT = 156.8 + 24.6 = 181.4$ s

Interval for 6 lifts = 30.2 s

Handling capacity = $300 \times 13.5/30.2 = 134.1$ persons/5-minutes

The average waiting time will be $0.85 \times 30.2 = 25.7$ s

The average journey time using Equation (6.6) will be:

$0.5 \times 47 \times 0.53 + 0.5 \times 10.7 \times 7.47 + 1.5 \times 13.5 \times 1.0 + 25.7 = 98.4$ s

The interval is slightly longer, the handling capacity slightly larger and the passenger average waiting time is slightly longer than specified, but are acceptable.

CS11.4.2 Down Peak Analysis

A conventional lift system will handle 50% more traffic during down peak than during uppeak. If the design provides 15% uppeak handling capacity then the system will provide 22.5% handling capacity during down peak. It is likely that as the villages are arranged in three floor sectors that a lift calling at the top floor of every village will fill at that floor and the lower two floors and be completely full before leaving the village. Any spare capacity can of course be used as it reaches the next village. Also as there are five villages of three floors and six lifts for each of the Red, Yellow and Blue village groups, this allows one lift to always be travelling to the next village to be served.

CS11.4.3 Interfloor Analysis

Design assumptions

Six lifts serve each group of villages.

At any given floor there are three lifts from each of the other two groups serving the other village groups.

---

[1] See Appendix 1 and Appendix 2 for details.

All activity groups are clustered together.

A busy system is where a third of the building occupants use the lifts every hour.

Assume 50% of these passengers travel within their group and the other 50% are split equally between the other two village groups.

Consider the whole building. There are 44 floors, Floors 4−47 plus the population of Floors 48−52.

Population using all lifts every hour: $= (44 \times 60 + 300)/3 = 2940/3 = 980$ persons

Population using all lifts every 5 minutes: $= 980/12$ $= 82$ persons

Population using lifts in one zone every 5 minutes: $= 27$ persons

In any 5-minute period this is 27 landing stops or approximately one landing call per floor served.

Assume the worst case of each landing call causing a new car call, ie: no coincident calls.

Assume the worst case of calls to both terminal floors during a cycle of activity.

Thus there are (say) 60 stops per 5-minute period shared by six lifts.

Therefore each lift will stop 10 times per 5-minute period.

Calculation: typical case

The standard calculations are not geared to deal with non-contiguous floor zones, so a first principles approach is necessary. Assume one person in a lift, which travels 47 floors. It stops twice and one person travels down 47 floors in a lift and stops twice. Thus:

$$RTT = 2 \times 47 \times 0.53 + 4 \times (8.0 - 0.53) + 4 \times 1 \times 1 = 83.7 \text{ s}$$

Number of round trips per lift per 5 minutes $= 300/83.7 = 4.2$

Number of stops per 5 minutes $= 4 \times 3.6 = 14$

This is more than the required activity (10 stops).

The design will be suitable.

CS11.5 COMMENTARY

This unique design permits travel within each zone using six lifts and travel to the other two zones with three lifts. The interfloor demand is thus balanced. The design requires some extra 45 entrances per lift group, ie: 135 in total. Some space is taken to create the secondary lobbies. The design brief is fulfilled.
    This design was not applied (Jappsen, 2000). The number of office floors was reduced from 51 to 46. The requirement for a secure service to the upper floors was removed. The original design had an empathy with three: triangular building, 3 sides, 3 apexes, 3 cores, 3 floors to a village, 3 villages to an atrium of 9 floors, 6 lifts to a

zone of 15 floors, 3 lifts for interfloor movements. The new design was a triangular building: 3 sides, 3 apexes, 3 cores, 4 floors to a village, 3 villages to an atrium of 12 floors, 5 or 6 lifts to a zone of 15/16/17 floors, 3 transfer floors for interfloor movements.

Godwin (1993) remarked "Never follow blindly conventional solutions to non conventional buildings". This case study illustrates a unique solution to the interfloor movement problem.

Medium Rise Trader Building

CS12.1 INTRODUCTION

Most buildings do not have the luxury of total interconnectivity as described in Case Study 11. This case study considers a regular (not prestige) trader building where the lower three floors are occupied by trading floors and the upper nine floors contain office accommodation occupied by multiple tenants. The building is designed with three zones: a lower zone (Floors 1−3) served by escalators; a middle zone (Floors 4−7) and upper zone (Floors 9−12) served by lifts with a common facilities floor between these two lifted zones.

The client asked two questions: for the original design to be confirmed for two population densities of one person per 10 m^2 and one person per 14 m^2 on uppeak traffic; and the effect on interfloor traffic of moving a common facilities floor from Floor 8 to Floor 6. The trading floors are populated at one person per 7 m^2. The client asked for the British Council of Offices (BCO) best practice criteria[1] to be applied, viz: an interval of 30 s and a peak arrival rate of 15%.

CS12.2 THE ORIGINAL SYSTEM

Table CS12.1 The basic building data

Floor	NIA (m^2)	Population @ 1/7	Population @ 1/10	Population @ 1/14	Elevation	LZ	MZ	UZ
12	3500		350	250	55.6			▓
11	3500		350	250	51.4			▓
10	3500		350	250	47.2			▓
9	3500		350	250	43.0			▓
8	3500		350	250	38.8		▓	▓
7	3500		350	250	34.6		▓	
6	3500		350	250	30.4		▓	
5	3500		350	250	26.2		▓	
4	3500		350	250	22.0		▓	
3	3000	430			16.5	▓		
2	3000	430			11.0	▓		
1	3000	430			5.5	▓		
Ground	2000	n/a	n/a	n/a	0	▓	▓	▓
Legend:					Equipment	E	L1	L2

NIA: Net Internal Area. Shaded area indicates floor service.
Population: 1/7 = one person/7 m^2, 1/10 = one person/10 m^2, 1/14 = one person/14 m^2.
LZ: Lower Zone. MZ: Middle Zone. UZ: Upper Zone.
Equipment: E=escalators 3 pairs, 1000 mm, 0.5 m/s. L1=6, 1600 kg, 2.5 m/s. L2=6, 1600 kg, 3.15 m/s.

---

[1] BCO recommendations for lifts are very brief. They are intended for the lay person, ie: architect, owner, developer, etc., not a lift industry specialist, as a means towards initial equipment sizing. They do not represent the current practice contained in the CIBSE Guide D: 2000.

CS12.2.1 Escalator provision

The demand at the Ground floor for the first set of escalators will be to populate Floors 1−3 with 1290 persons. If the intending passengers arrive as a 15% demand over 5 minutes then this is 194 persons. The theoretical handling capacity of a 1000 mm wide escalator with a rated speed of 0.5 m/s is 150 persons/minute (Table 1.8). The actual handling capacity is half of this. So over any 5 minutes a 1000 mm escalator can handle 375 persons. This is nearly twice the likely demand and thus the escalators can easily accommodate a sudden 2:1 surge in demand.

CS12.2.2 Lift Provision

Floor 8 is the common facilities floor where restaurant, travel agency, reprographics, post room, etc. are located. In view of the small number of floors each lift group is serving and the large rated load of the lifts, it is reasonable to assume the occupants of Floor 8 are split equally between the two groups. Then the population to be served to Floors 4−7 and half of Floor 8 is 1575 persons at 1/10 and 1125 at 1/14. There will be an identical population demand in the Upper Zone. Detailed information was not provided regarding the lift equipment except the rated load, rated speed and number proposed (see Legend to Table CS12.1). The equipment values selected for operating times, etc. can be seen in the LTD calculation Tables CS12.2 and CS12.3.

Table CS12.2 Middle Zone: underlying performance at highest population

```
*********************************************************************************
Lift traffic calculations                                    Lift Traffic Design
*********************************************************************************
Design name: Case Study Twelve          Run No.: 1              Date: 2002
.................................................................................
{N}umber of floors          = 5      Number of {P}assengers = 13.5
Number of {L}ifts           = 6      Building {POP}ulation  = 1575
Interfloor {D}istance (m)   = 4.2    {E}xpress {J}ump (m)   = 22.0
{P}ass. {T}ransfer time (s) = 1.0    Speed (m/s) {v}        = 2.5
{AC}eleration (m/s2)  = 1.2  {J}erk  (m/s3) = 1.5  {S}tart {D}elay (s) = 0.3
Door times (s): {C}lose = 2.6  {O}pen    = 1.8  {AD}vance        = 0.5
.................................................................................
Flight time (s) = 4.63               Performance time (s) = 8.83
.................................................................................
Average high reversal floor = 4.9    Average number of stops = 4.8
RTT =  108.1  Interval = 18.0  Handling capacity = 224.8  %POP = 14.3
Average journey time = 55.3
*********************************************************************************
```

Table CS12.3 Upper Zone: underlying performance at highest population

```
*********************************************************************************
Lift traffic calculations                                    Lift Traffic Design
*********************************************************************************
Design name: Case Study Twelve          Run No.: 2              Date: 2002
.................................................................................
{N}umber of floors          = 5      Number of {P}assengers = 13.5
Number of {L}ifts           = 6      Building {POP}ulation  = 1575
Interfloor {D}istance (m)   = 4.2    {E}xpress {J}ump (m)   = 38.8
{P}ass. {T}ransfer time (s) = 1.0    Speed (m/s) {v}        = 4.0
{AC}eleration (m/s2)  = 1.2  {J}erk  (m/s3) = 1.5  {S}tart {D}elay (s) = 0.3
Door times (s): {C}lose = 2.6  {O}pen    = 1.8  {AD}vance        = 0.5
.................................................................................
Flight time (s) = 4.63               Performance time (s) = 8.83
.................................................................................
Average high reversal floor = 4.9    Average number of stops = 4.8
RTT = 109.8  Interval = 18.3  Handling capacity = 221.3  %POP = 14.1
Average journey time = 57.6
*********************************************************************************
```

For the worst case occupancy of 1575 persons, Tables CS12.2 and CS12.3 show that the BCO recommendation for a 15% handling capacity is nearly obtained at some 14%. The interval of some 18 s is most satisfactory. This means the client's original BCO requirements are met.

Today this type of building would expect (according to CIBSE Guide D and Tables 6.1 and 6.3) to meet a 25 s interval (the lower value in the range) and a 13% arrival demand (a mid value in the range). These requirements are more stringent with respect to the interval value and a less stringent with respect to the arrival demand.

Table CS12.4 Middle Zone: Balanced performance with 13% demand and highest population

```
*************************************************************************
Lift traffic calculations                        Iterative Balance Method
*************************************************************************
Design name: Case Study Twelve        Run No.: 3          Date: 2002
.......................................................................
Desired {I}nterval (s)      = 25      {AR}rival rate (5-min) = 205
{N}umber of floors          = 5       Building {POP}ulation  = 1575
Interfloor {D}istance (m)   = 4.2     {E}xpress {J}ump (m)   = 22.0
{P}ass. {T}ransfer time (s) = 1.0     Speed (m/s) {v}        = 2.5
{AC}eleration (m/s2)   = 1.2   {J}erk (m/s3) = 1.5  {S}tart {D}elay (s) = 0.3
Door times (s): {C}lose = 2.6  {O}pen        = 1.8  {AD}vance           = 0.5
.......................................................................
Flight time (s) = 4.63                Performance time (s) = 8.83
.......................................................................
Average high reversal floor =  5.0    Average number of stops =  5.0
Passengers =   21.5  Interval =   31.4  Number of lifts =   4 %POP =   13.0
Average journey time =   70.8
.......................................................................
Average high reversal floor =  5.0    Average number of stops =  4.8
Passengers =   15.4  Interval =   22.5  Number of lifts =   5 %POP =   13.0
Average journey time =   59.7
*************************************************************************
```

Table CS12.5 Upper Zone: Balanced performance with 13% demand and highest population

```
*************************************************************************
Lift traffic calculations                        Iterative Balance Method
*************************************************************************
Design name: Case Study Twelve        Run No.: 4          Date: 2002
.......................................................................
Desired {I}nterval (s)      = 25      {AR}rival rate (5-min) =  205
{N}umber of floors          = 5       Building {POP}ulation  =  1575
Interfloor {D}istance (m)   = 4.2     {E}xpress {J}ump (m)   =  38.8
{P}ass. {T}ransfer time (s) = 1.0     Speed (m/s) {v}        =  4.0
{AC}eleration (m/s2)   = 1.2   {J}erk (m/s3) = 1.5  {S}tart {D}elay (s) = 0.3
Door times (s)-{C}lose = 2.6   {O}pen        = 1.8  {AD}vance           = 0.5
.......................................................................
Flight time (s) = 4.63                Performance time (s) = 8.83
.......................................................................
Average high reversal floor =  5.0    Average number of stops =  5.0
Passengers =   22.0  Interval =   32.1  Number of lifts =   4 %POP =   13.0
Average journey time =   73.8
.......................................................................
Average high reversal floor =  5.0    Average number of stops =  4.8
Passengers =   15.7  Interval =   23.0  Number of lifts =   5 %POP =   13.0
Average journey time =   62.5
*************************************************************************
```

The IBM calculation Tables CS12.4 and CS12.5 show the handling capacity requirement has been met with five lifts and the intervals are some 23 s. The passenger occupancy of the cars is high at some 16 passengers, ie: higher than the design capacity of 13.5 persons for a 1600 kg lift. This over occupancy should be taken as significant and a five lift system rejected.

However, there are some other factors that could be considered. The floor areas used in the calculations above were the Net Internal Areas and these can be reduced by (say) 10% to obtain net usable area (Definition 6.5). Also 100% attendance of the building population is unlikely on any day and therefore the effective population (Section 6.5) can be reduced to (say) 90% of possible population. This means the effective population of each zone can be reduced to 81% (0.9 × 0.9), ie: to 1275 in the 1/10 persons/m² case and 911 persons in the 1/14 persons/m² case. At a 13% arrival rate this becomes a demand over 5 minutes of 166 and 118 persons, respectively.

Table CS12.6 Middle Zone: Balanced performance with 13% demand and effective population (1/10)

```
***********************************************************************
Lift traffic calculations                    Iterative Balance Method
***********************************************************************
Design name: Case Study Twelve       Run No.: 5          Date: 2002
.......................................................................
Desired {I}nterval (s)      =  25    {AR}rival rate (5-min) =  166
{N}umber of floors          =  5     Building {POP}ulation  =  1275
Interfloor {D}istance (m)   =  4.2   {E}xpress {J}ump (m)   =  22.0
{P}ass. {T}ransfer time (s) =  1.0   Speed (m/s) {v}        =  2.5
{AC}eleration (m/s2)   =  1.2   {J}erk (m/s3) =  1.5 {S}tart {D}elay (s) = 0.3
Door times (s):{C}lose =  2.6   {O}pen       =  1.8 {AD}vance          = 0.5
.......................................................................
Flight time (s) = 4.63               Performance time (s) = 8.83
.......................................................................
Average high reversal floor =  5.0    Average number of stops =  4.9
Passengers =  15.7  Interval =  28.3  Number of lifts =  4 %POP =  13.0
Average journey time =  63.0
.......................................................................
Average high reversal floor =  4.9    Average number of stops =  4.6
Passengers =  11.4  Interval =  20.5  Number of lifts =  5 %POP =  13.0
Average journey time =  53.9
***********************************************************************
```

Table CS12.7 Upper Zone: Balanced performance with 13% demand and effective population (1/10)

```
***********************************************************************
Lift traffic calculations                    Iterative Balance Method
***********************************************************************
Design name: Case Study Twelve       Run No.: 6          Date: 2002
.......................................................................
Desired {I}nterval (s)      =  25    {AR}rival rate (5-min) =  166
{N}umber of floors          =  5     Building {POP}ulation  =  1275
Interfloor {D}istance (m)   =  4.2   {E}xpress {J}ump (m)   =  38.8
{P}ass. {T}ransfer time (s) =  1.0   Speed (m/s) {v}        =  4.0
{AC}eleration (m/s2)   =  1.2   {J}erk (m/s3) =  1.5 {S}tart {D}elay (s) = 0.3
Door times (s):{C}lose =  2.6   {O}pen       =  1.8 {AD}vance          = 0.5
.......................................................................
Flight time (s) = 4.63               Performance time (s) = 8.83
.......................................................................
Average high reversal floor =  5.0    Average number of stops =  4.9
Passengers =  16.0  Interval =  28.9  Number of lifts =  4 %POP =  13.0
Average journey time =  65.8
.......................................................................
Average high reversal floor =  4.9    Average number of stops =  4.6
Passengers =  11.6  Interval =  21.0  Number of lifts =  5 %POP =  13.0
Average journey time =  56.5
***********************************************************************
```

The IBM calculation program, reported in Tables CS12.6 and CS12.7, indicates that an installation of four lifts would provide a performance of 13% handling capacity at the higher arrival rate. The middle and upper zone performances would be some 16 passengers in each car at an interval of 29 seconds. The performance of both zones has a raised interval and high car loading. The degradation is significant and a five lift system would be recommended. With five lifts the interval is some 21 seconds and the the car loading some 12 persons, which would be very comfotable in a 1600 kg lift.

Now what will happen at the lower occupancy level ? Calculations for the lower levels of occupation have not been carried out, but obviously the lifts will provide a better performance in the face of a lower demand. Readers might like to carry out these calculations.

CS12.3 COMMENTARY

■ The escalator provision is excellent.

■ Two groups of six lifts would meet the BCO recommendations.

■ Two groups of five lifts would meet the requirements of this building for a Guide D: 2000 recommendations

■ It is always important to consider what happens when one lift is not available, when being serviced or as a result of a breakdown. Two groups of six lifts would therefore provide comfort to the building management.

CS12.4 MOVING THE FACILITIES FLOOR

Assume that the characteristics of Floors 6 and 8 are completely interchangeable. The previous situation was:

Middle zone: G, 4−7, 8 (50%)
Upper zone: G, 8 (50%), 9−12

The new situation is:

Middle zone: G, 4−5, 6 (50%), 7−8
Upper zone: G, 6 (50%), 9−12

The number of stops in the middle and upper zones would remain unchanged. The highest reversal floor in the middle zone would be slightly higher as Floor 8 now has a bigger demand. The performance of the lifts would be little changed.

Persons from the upper zone wishing to reach floors in the middle zone would have to transfer at Floor 6 and move back up the building to reach Floors 7−8. Similarly, persons from Floors 7−8 wishing to travel to the upper zone would first have to travel down to Floor 6 and then back past their originating floors to enter the upper zone. This would increase the travel time of such passengers and is clearly illogical.

The main effect is to destroy the symmetry of the original arrangement. This is a subjective rather than an objective view.

CASE STUDY THIRTEEN

Multiple Entrance Modernisation

CS13.1 INTRODUCTION

A rent review between a long-term tenant and an owner is often contentious and generally carried out by agents, rather than the principal parties. The main bone of contention is where a tenant has improved or proposes to improve the building to the advantage of both parties and wishes this to be reflected in the ensuing rent. This case study considers such a building prior to such a modernisation. Factors to be considered in this building are: that it has two entrances; two cores serving different numbers of floors; and an escalator service to the lower four floors.

CS13.2 THE BUILDING AND ITS POPULATION

The building, built in the 1980s, is rectangular in shape with a main entrance in the middle of a long side. The two cores (A and B) serve 12 floors and are located to the left and right of the main entrance. The first four floors are served by escalators through an atrium. Four lifts in each core are arranged in line. The vertical transportation facilities in this building thus defy the principles of good circulation and lift location practice. However, as occupants must walk some distance to each group, this has the effect of smoothing any peak flows that occur and easing the demand on the lift system.

When built the likely density of occupation of the floors was one person per 14 m². The 1980s performance criteria, if the recommendations of BS CP407 (1972) were followed, were probably a 15% handling capacity at an interval of 30 s. Today the density is one person per 10 m². The lift system is performing badly and clearly was not designed to service the larger numbers of occupants. The tenant wishes to bring the building up to modern standards.

To perform a traffic analysis of this building three uppeak scenarios should be considered:

(1) Worst condition, where all the 830 possible occupants use the lifts to reach their destinations (Scene 1).

(2) Probable condition, where only 670 of the possible occupants use the lifts and some of the attendees use the escalators to reach Floors 1−4 (Scene 2).

(3) Likely condition, where only 603 of the possible occupants use the lifts, as there is a 90% attendance and some of the attendees use the escalators to reach Floors 1−4 (Scene 3).

A number of other assumptions must be made:

(1) The Net Internal (rentable) Areas will be reduced by 90% to calculate the Net Usable Area (see column 3 of Table CS13.1).

(2) The lifts in Cores A and B will serve Floors 1−12 equally, ie: at 50%.

(3) When calculating the performance with escalator usage, it will be assumed the escalators transport:

90% of passengers: ground−Floor 1
67% of passengers: ground−Floor 2
33% of passengers: ground−Floor 3
10% of passengers: ground−Floor 4.

Figures shown in square brackets [*nn*%] in Table CS13.1 are passengers using the lifts.

(4) Stair circulation will not be considered.

Table CS13.1 summarises the main data about this building.

Table CS13.1 Building data

Floor	Rentable area (m²)	Usable area (m²)	Population (1/10 m²)	Demand Cores A & B				
				Scene 1	Scene 2		Scene 3	
1	1777	1600	160	80	8	[10%]	7	[9%]
2	1777	1600	160	80	27	[33%]	24	[30%]
3	1777	1600	160	80	54	[67%]	48	[60%]
4	1777	1600	160	80	72	[90%]	65	[81%]
5	1777	1600	160	80	80		72	
6	1777	1600	160	80	80		72	
7	1333	1200	120	60	60		54	
8	1333	1200	120	60	60		54	
9	1333	1200	120	60	60		54	
10	1333	1200	120	60	60		54	
11	1333	1200	120	60	60		54	
12	1111	1000	100	50	50		45	
Totals				830	670		603	

CS13.3 THE SYSTEM TO BE IMPROVED

The equipment installed in the mid-1980s was not up-to-(that)-date, being a Ward−Leonard drive with poor door control and inadequate signalling fixtures. All twelve lifts are identical except for the number of floors served. Measurements made of the unmodified system showed reasonable door timings, very poor motion dynamics and excessive door dwell times. These latter deficiencies lead to a cycle time of 18.6 s, although the performance time is reasonable at 13.0 s.

The LTD calculation Table CS13.2 shows the equipment data[1] and performance for Cores A and B, and Table CS13.3 summarises the results. Table CS13.2 indicates that the underlying handling capacity as 75.9 persons/5-minutes at an interval of 51.4 s. The 75.9 persons/5-minute handling capacity represents a percentage handling capacity of 9.2% for Scene 1, 11.3% for Scene 2 and 12.6% for Scene 3. Only if the building population were to be 506 persons would the original criterion of 15% be meet. At this population the interval would still be 51.4 s, which means the other original criterion would not be met.

---

[1] Notice that in order to set the cycle time to 18.6 s the advance door opening time is set to a negative value of −7.28 s and the passenger transfer time (valid at the main terminal floor only) is set to 0.5 s.

Table CS13.2 Performance of original system

```
**********************************************************************
Lift traffic calculations                          Lift Traffic Design
**********************************************************************
Design name: Case Study Thirteen        Run No.: 1         Date: 2002
......................................................................
{N}umber of floors          = 12     Number of {P}assengers =  13.0
Number of {L}ifts           = 4      Building {POP}ulation  =  830
Interfloor {D}istance (m)   = 4.5    {E}xpress {J}ump (m)    = 0
{P}ass. {T}ransfer time (s) = 0.5    Speed (m/s) {v}         = 2.75
{AC}eleration (m/s2)    = 0.7 {J}erk (m/s3) = 1.0 {S}tart {D}elay (s) = 0.9
Door times (s): {C}lose = 2.7 {O}pen      = 1.9 {AD}vance          = -7.28
......................................................................
Flight time (s) = 5.82               Performance time (s) =18.60
......................................................................
Average high reversal floor = 11.6    Average number of stops =  8.1
RTT =  205.6  Interval =  51.4  Handling capacity =  75.9  %POP =  9.1
Average journey time = 113.8
**********************************************************************
```

CS13.4 THE IMPROVED SYSTEM

A modern system employing a variable voltage, variable frequency (VVVF) control can deliver excellent performance times of some 9.5 s over the 4.5 m interfloor distances. This combined with landing call dwell times set at 3 s and car call dwell times set at 2 s and with passenger detectors capable of initiating differential door timing will make the cycle time dependent on passenger movements rather than equipment timers. Improved signalling will enable passengers to quickly identify the next lift.

The LTD calculation Tables CS13.4−6 give the results and Table CS13.3 summarises the results.

Table CS13.3 Summary of performance for improved system

Scene	Scene 1	Scene 2	Scene 3
P	11.5	7.6	6.2
INT (s)	32	26	24
HC (P/5-min)	108	87	79
%POP (%)	13	13	13

All scenarios can provide handling capacities of 13% and with intervals of less than 32 s. In one of the three cases the intervals are lower than the modern performance criterion of 25 s. None of the cars fill to the design capacity of 13 persons.

CS13.5 COMMENTARY

The underlying equipment can be modernised to meet the modern performance criteria. Even for the worst case scenario the performance would be most acceptable. There is room for negotiation on rents.

Table CS13.4 Performance of improved system: Scene 1 — worst scenario

```
****************************************************************************
Lift traffic calculations                              Lift Traffic Design
****************************************************************************
Design name: Case Study Thirteen        Run No.: 2            Date: 2002
............................................................................
{N}umber of floors          = 12     Number of {P}assengers =  11.5
Number of {L}ifts           = 4      Building {POP}ulation  =  830
Interfloor {D}istance (m)   = 4.5    {E}xpress {J}ump (m)   =  0
{P}ass. {T}ransfer time (s) = 1.0    Speed (m/s) {v}        =  2.75
{AC}eleration (m/s2)    = 1.0 {J}erk (m/s3) = 1.5 {S}tart {D}elay (s) = 0.4
Door times (s): {C}lose = 2.7 {O}pen      = 1.9 {AD}vance          = 0.5
............................................................................
Flight time (s) = 4.96               Performance time (s) = 9.46
............................................................................
Average high reversal floor = 11.5    Average number of stops = 7.6
RTT =  127.7  Interval = 31.9  Handling capacity = 108.1  %POP =  13.0
Average journey time =  72.3
****************************************************************************
```

Table CS13.5 Performance of improved system: Scene 2 — probable scenario

```
****************************************************************************
Lift traffic calculations                              Lift Traffic Design
****************************************************************************
Design name: Case Study Thirteen        Run No.: 3            Date: 2002
............................................................................
{N}umber of floors          = 12     Number of {P}assengers =  7.6
Number of {L}ifts           = 4      Building {POP}ulation  =  670
Interfloor {D}istance (m)   = 4.5    {E}xpress {J}ump (m)   =  0
{P}ass. {T}ransfer time (s) = 1.0    Speed (m/s) {v}        =  2.75
{AC}eleration (m/s2)    = 1.0 {J}erk (m/s3) = 1.5 {S}tart {D}elay (s) = 0.4
Door times (s): {C}lose = 2.7 {O}pen      = 1.9 {AD}vance          = 0.5
............................................................................
Flight time (s) = 4.96               Performance time (s) = 9.46
............................................................................
Average high reversal floor = 11.1    Average number of stops = 5.8
RTT =  104.6  Interval = 26.2  Handling capacity =  87.2  %POP =  13.0
Average journey time =  56.2
****************************************************************************
```

Table CS13.6 Performance of improved system: Scene 3 — likely scenario

```
****************************************************************************
Lift traffic calculations                              Lift Traffic Design
****************************************************************************
Design name: Case Study Thirteen        Run No.: 4            Date: 2002
............................................................................
{N}umber of floors          = 12     Number of {P}assengers =  6.2
Number of {L}ifts           = 4      Building {POP}ulation  =  603
Interfloor {D}istance (m)   = 4.5    {E}xpress {J}ump (m)   =  0
{P}ass. {T}ransfer time (s) = 1.0    Speed (m/s) {v}        =  2.75
{AC}eleration (m/s2)    = 1.0 {J}erk (m/s3) = 1.5 {S}tart {D}elay (s) = 0.4
Door times (s): {C}lose = 2.7 {O}pen      = 1.9 {AD}vance          = 0.5
............................................................................
Flight time (s) = 4.96               Performance time (s) = 9.46
............................................................................
Average high reversal floor = 10.8    Average number of stops = 5.0
RTT =   94.7  Interval = 23.7  Handling capacity =  78.6  %POP =  13.0
Average journey time =  49.5
****************************************************************************
```

Double Deck Installation

CS14.1 INTRODUCTION

Double deck installations are few in number compared to the millions of single deck installations. This case study indicates how a double deck solution solved a problem.

The building under consideration is a 16 storey headquarters for an international bank. Before the Pacific Rim crisis it was intended to populate the building in a prestigious manner with one person per 14 m² of usable space. This resulted in an actual population of 1250 persons. The building was in the course of erection, and the core had been constructed, when the client decided to retrench and place more staff in the building at a density of occupation of one person per 10 m². At the same time the footprint of the building was increased by 50%. Thus the potential population rose to 2625 persons, without the opportunity to provide more hoistways in the core. What to do ?

CS14.2 THE ORIGINAL DESIGN

The original design was for a group of six, 1600 kg lifts with a rated speed of 4 m/s. The data are shown in the LTD calculation of Table CS14.1. This table indicates a handling capacity of 168.2 persons/5-minute, ie: 15%, is possible with a 24 s interval, when serving an effective population of 1125 persons. (The 1125 figure is 90% of the possible population of 1250 persons.) This is an appropriate performance (<25 s interval, 15% handling capacity).

If the possible population rises to 2625, ie: an effective population, at 90% of 2363 persons, even eight lifts could not cope. Eight lifts would have a handling capacity of 224 persons (8/6 × 168), ie: 9.5%. An answer is to install double decker lifts.

CS14.3 DOUBLE DECK DESIGN

Using Equation (8.7) embedded in the Lift Traffic Design for Double Decks (LTDD) calculation program, the performance of six double deck lifts serving eight floors each is given in Table CS14.2. The table shows each deck can serve 1181 passengers, with a handling capacity of 173.5 persons/5-minutes (14.7%) at an interval of 23 s. This is slightly more than twice, ie: $(2 \times 173.5)/168.2 = 2.06$, the uppeak performance, but with a small reduction in the percentage handling capacity. This is acceptable.

CS14.4 CONSEQUENCES OF USING DOUBLE DECK LIFTS

The double deck installation serving eight floors provides 2.06 times the handling capacity of a single deck serving 16 floors. Thus the lift installation will now be able to handle the uppeak traffic demand. There will be little effect on the down peak traffic, mainly due to the small number of floors served by each deck and the likelihood that the cars will fill at three or four floors and then travel to the main terminal. There will be difficulties for interfloor traffic, particularly during the unbalanced situation in the middle of the day, when there will smaller numbers of coincident stops for both decks.

Table CS14.1 The original design

```
********************************************************************
Lift traffic calculations                          Lift Traffic Design
********************************************************************
Design name: Case Study Fourteen      Run No.: 1         Date: 2002
....................................................................
{N}umber of floors        =  16    Number of {P}assengers =  13.5
Number of {L}ifts         =  6     Building {POP}ulation  =  1125
Interfloor {D}istance (m) =  4     {E}xpress {J}ump (m)   =  0
{P}ass. {T}ransfer time (s) = 1.0  Speed (m/s) {v}        =  4
{AC}eleration (m/s2)      =  1.0   {J}erk (m/s3) = 1.5  {S}tart {D}elay (s) = 0.4
Door times (s): {C}lose = 2.8 {O}pen       = 2     {AD}vance       = 0.5
....................................................................
Flight time (s) = 4.72              Performance time (s) = 9.42
....................................................................
Average high reversal floor = 15.3  Average number of stops = 9.3
RTT = 144.4  Interval = 24.1  Handling capacity = 168.2  %POP = 15.0
Average journey time = 79.1
********************************************************************
```

Table CS14.2 The double deck design

```
********************************************************************
Lift traffic calculations            Lift Traffic Design for Double Decks
********************************************************************
Design name: Case Study Fourteen      Run No.: 2         Date: 2002
....................................................................
{N}umber of floors        =  8     Number of {P}assengers =  13.5
Number of {L}ifts         =  6     Building {POP}ulation  =  1181
Interfloor {D}istance (m) =  8     {E}xpress {J}ump (m)   =  0
{P}ass. {T}ransfer time (s) = 1.0  Speed (m/s) {v}        =  4.0
{AC}eleration (m/s2)      =  1.0   {J}erk (m/s3) = 1.5  {S}tart {D}elay (s) = 0.4
Door times (s): {C}lose = 2.8 {O}pen       = 2     {AD}vance       = 0.5
....................................................................
Flight time (s) = 6.36              Performance time (s) =11.06
....................................................................
Average high reversal floor = 7.8   Average number of stops = 6.7
RTT = 140.1  Interval = 23.3  Handling capacity = 173.5  %POP = 14.7
Average journey time = 70.0
********************************************************************
```

Note the handling capacity of both decks is 2 × 173.5, ie: 347 persons per 5 minutes.

Traffic Design Data Tables

The tables in this Data Sheet have been extracted from Chapters 5 to 7. They are presented here to provide a quick reference to all the traffic data discussed. The text should be referred to for specific details of use.

Traffic Data Tables

Table 5.2 Total time required to travel between terminal floors in different types of building

Building type	Transit time (s)
Offices:	
large	17–20
small	20
Hotels:	
large	17–20
small	20
Hospitals	24
Nursing and residential homes	24
Residential buildings	20–30
Factories and warehouses	24–40
Shops	24–40

Table 5.3 Typical lift dynamics

Lift travel (m)	Rated speed (m/s)	Acceler-ation (m/s^2)	Single floor flight time (s)
<20	<1.00	0.4	10.0
20	1.00	0.4–0.7	7.0
32	1.60	0.7–0.8	6.0
50	2.50	0.8–0.9	5.5
63	3.15	1.0	5.0
100	5.00	1.2–1.5	4.5
120	6.00	1.5	4.3
>120	>6.00	1.5	4.3

Table 5.6 Typical door closing and opening times

Door operation	Opening (advanced)		Opening (normal)		Closing (normal)	
Door type	800	1100	800	1100	800	1100
Side	1.0	1.5	2.5	3.0	3.0	4.0
Centre	0.5	0.8	2.0	2.5	2.0	3.0

Table 6.1 Estimation of population

Building type	Population estimate
Hotel	1.5–1.9 persons/room
Flats	1.5–1.9 persons/bedroom
Hospital	3.0 persons/bedspace[*]
School	0.8–1.2 m^2 net area/pupil
Office (multiple tenancy):	
regular	10–12 m^2 net area/person
prestige	15–25 m^2 net area/person
Office (single tenancy):	
regular	8–10 m^2 net area/person
prestige	12–20 m^2 net area/person

[*] Patient plus three others (doctors, nurses, porters, etc.).

Table 6.2 Percentage arrival rates

Building type	Arrival rate
Hotel	10–15%
Flats	5–7%
Hospital	8–10%
School	15–25%
Office (multiple tenancy):	
regular	11–15%
prestige	17%
Office (single tenancy):	
regular	15%
prestige	17–25%

Table 6.3 Uppeak intervals

Building type	Interval (s)
Hotel	30–50
Flats	40–90
Hospital	30–50
School	30–50
Office (multiple tenancy):	
regular	25–30
prestige	20–25
Office (single tenancy):	
regular	25–30
prestige	20–25

Traffic Data Tables

Table 6.4 Uppeak performance – numerical values

Car load(%)	AWT/INT (%)	Car load(%)	AWT/INT (%)
30	0.32	75	0.74
40	0.35	80	0.85
50	0.40	85	1.01
60	0.50	90	1.30
70	0.65	95	1.65

Table 6.6 Summary of times

Time	Aim for:	Poor
AWT	<20 s	>25 s
ATT	<60 s	>70 s
AJT	<80 s	>90 s

Legend
AWT Average waiting time
ATT Average travel time
AJT Average journey time

Table 6.5 Office Building Average System Response Time Performance

Grade of Service	Percentage of calls answered in		Time to answer call (s)	
	30 s	60 s	50%	90%
Excellent	>75	>98	20	45
Good	>70	>95	22.5	50
Fair	>65	>92	25	55
Poor/unacceptable	<65	<92	>25	>55

Table 7.2 Car loading and car capacity

Rated load (kg) (RL)	Max area (m²) (CA)	Rated capacity (persons) (CC)	Actual capacity (persons) (AC)	Design capacity (persons) (DC)	Capacity factor (%) (CF)	Actual load (kg) (AL)
320	0.95	4	4.5	3.6	90	338
450	1.30	6	6.2	5.0	82	465
630	1.66	8	7.9	6.3	79	593
800	2.00	10	9.5	7.6	76	713
1000	2.40	13	11.4	9.1	70	855
1275	2.90	16	13.8	11.0	69	1035
1600	3.56	21	16.9	13.5	64	1268
1800	3.92	24	18.6	14.9	62	1395
2000	4.20	26	20.0	16.0	62	1500
2500	5.00	33	23.8	19.0	58	1785

Table 8.9 Occupancy factors for residential buildings

Type	Luxury	Normal	Low income
Studio	1.0	1.5	2.0
1 Bedroom	1.5	1.8	2.0
2 Bedroom	2.0	3.0	4.0
3 Bedroom	3.0	4.0	6.0

Table 8.10 Design criteria: Residential buildings (5-minute, two way)

Type	Interval (s)	Handling capacity
Low income	≤50–70	≥5–7%
Normal	≤50–60	≥6–8%
Luxury	≤45–50	≥8%

Table 5.1 Values of H and S for EN81 rated capacities *

Floors	6 (4.8)		8 (6.4)		10 (8.0)		13 (10.4)		16 (12.8)		21 (16.8)		26 (20.8)		33 (26.4)	
N	H	S	H	S	H	S	H	S	H	S	H	S	H	S	H	S
5	4.6	3.3	4.7	3.8	4.8	4.2	4.9	4.5	4.9	4.7	5.0	4.9	5.0	5.0	5.0	5.0
6	5.4	3.5	5.6	4.1	5.7	4.6	5.8	5.1	5.9	5.4	6.0	5.7	6.0	5.9	6.0	6.0
7	6.2	3.7	6.5	4.4	6.6	5.0	6.8	5.6	6.8	6.0	6.9	6.5	7.0	6.7	7.0	6.9
8	7.1	3.8	7.4	4.6	7.5	5.3	7.7	6.0	7.8	6.6	7.9	7.2	7.9	7.5	8.0	7.8
9	7.9	3.9	8.2	4.8	8.4	5.5	8.6	6.4	8.7	7.0	8.8	7.8	8.9	8.2	9.0	8.6
10	8.7	4.0	9.1	4.9	9.3	5.7	9.5	6.7	9.7	7.4	9.8	8.3	9.9	8.9	9.9	9.4
11	9.6	4.0	10.0	5.0	10.2	5.9	10.5	6.9	10.6	7.8	10.8	8.8	10.8	9.5	10.9	10.1
12	10.4	4.1	10.8	5.1	11.1	6.0	11.4	7.1	11.5	8.1	11.7	9.2	11.8	10.0	11.9	10.8
13	11.2	4.1	11.7	5.2	12.0	6.1	12.3	7.3	12.5	8.3	12.7	9.6	12.8	10.5	12.9	11.4
14	12.1	4.2	12.6	5.3	12.9	6.3	13.2	7.5	13.4	8.6	13.6	10.0	13.7	11.0	13.8	12.0
15	12.9	4.2	13.4	5.4	13.8	6.4	14.1	7.7	14.3	8.8	14.6	10.3	14.7	11.4	14.8	12.6
16	13.7	4.3	14.3	5.4	14.7	6.5	15.0	7.8	15.3	9.0	15.5	10.6	15.7	11.8	15.8	13.1
17	14.5	4.3	15.3	5.5	15.6	6.5	16.0	8.0	16.2	9.2	16.5	10.9	16.6	12.2	16.8	13.6
18	15.4	4.3	16.0	5.5	16.6	6.6	16.9	8.1	17.1	9.3	17.4	11.1	17.6	12.5	17.7	14.0
19	16.2	4.3	16.9	5.6	17.4	6.7	17.8	8.2	18.1	9.5	18.4	11.3	18.5	12.8	18.7	14.4
20	17.0	4.4	17.8	5.6	18.2	6.7	18.7	8.3	19.0	9.6	19.3	11.6	19.5	13.1	19.7	14.8
21	17.9	4.4	18.6	5.6	19.1	6.8	19.6	8.4	19.9	9.8	20.3	11.7	20.5	13.4	20.6	15.2
22	18.7	4.4	19.5	5.7	20.0	6.8	20.5	8.4	20.9	9.9	21.2	11.9	21.4	13.6	21.6	15.6
23	19.5	4.4	20.4	5.7	20.9	6.9	21.4	8.5	21.8	10.0	22.1	12.1	22.4	13.9	22.6	15.9
24	20.3	4.4	21.2	5.7	21.8	6.9	22.4	8.6	22.7	10.1	23.1	12.3	23.3	14.1	23.5	16.2

* 80% capacity shown in parentheses. N is the number of floors above the main terminal.

Table 6.8a Values of H and S for different values of P from 5 to 12 persons

Floors	5		6		7		8		9		10		11		12	
N	H	S	H	S	H	S	H	S	H	S	H	S	H	S	H	S
5	4.6	3.4	4.7	3.7	4.8	4.0	4.8	4.2	4.9	4.3	4.9	4.5	4.9	4.6	4.9	4.7
6	5.4	3.6	5.6	4.0	5.7	4.3	5.7	4.6	5.8	4.8	5.8	5.0	5.9	5.2	5.9	5.3
7	6.3	3.8	6.4	4.2	6.5	4.6	6.6	5.0	6.7	5.3	6.7	5.5	6.8	5.7	6.8	5.9
8	7.1	3.9	7.3	4.4	7.4	4.9	7.5	5.3	7.6	5.6	7.7	5.9	7.7	6.2	7.8	6.4
9	8.0	4.0	8.2	4.6	8.3	5.1	8.4	5.5	8.5	5.9	8.6	6.2	8.7	6.5	8.7	6.8
10	8.8	4.1	9.0	4.7	9.2	5.2	9.3	5.7	9.4	6.1	9.5	6.5	9.6	6.9	9.6	7.2
11	9.6	4.2	9.9	4.8	10.1	5.4	10.2	5.9	10.3	6.3	10.4	6.8	10.5	7.1	10.6	7.5
12	10.5	4.2	10.7	4.9	11.0	5.5	11.1	6.0	11.2	6.5	11.3	7.0	11.4	7.4	11.5	7.8
13	11.3	4.3	11.6	5.0	11.8	5.6	12.0	6.1	12.1	6.7	12.3	7.2	12.3	7.6	12.4	8.0
14	12.1	4.3	12.5	5.0	12.7	5.7	12.9	6.3	13.0	6.8	13.2	7.3	13.3	7.8	13.4	8.2
15	13.0	4.4	13.3	5.1	13.6	5.7	13.8	6.4	14.0	6.9	14.1	7.5	14.2	8.0	14.3	8.4
16	13.8	4.4	14.2	5.1	14.5	5.8	14.7	6.5	14.9	7.0	15.0	7.6	15.1	8.1	15.2	8.6
17	14.6	4.4	15.0	5.2	15.3	5.9	15.6	6.5	15.8	7.1	15.9	7.7	16.0	8.3	16.1	8.8
18	15.5	4.5	15.9	5.2	16.2	5.9	16.5	6.6	16.7	7.2	16.8	7.8	16.9	8.4	17.1	8.9
19	16.3	4.5	16.8	5.3	17.1	6.0	17.4	6.7	17.6	7.3	17.7	7.9	17.9	8.5	18.0	9.1
20	17.1	4.5	17.6	5.3	18.0	6.0	18.2	6.7	18.5	7.4	18.6	8.0	18.8	8.6	18.9	9.2
21	18.0	4.5	18.5	5.3	18.8	6.1	19.1	6.8	19.4	7.5	19.6	8.1	19.7	8.7	19.8	9.3
22	18.8	4.6	19.3	5.4	19.7	6.1	20.0	6.8	20.3	7.5	20.5	8.2	20.6	8.8	20.8	9.4
23	19.6	4.6	20.2	5.4	20.6	6.2	20.9	6.9	21.2	7.6	21.4	8.3	21.5	8.9	21.7	9.5
24	20.5	4.6	21.1	5.4	21.5	6.2	21.8	6.9	22.1	7.6	22.3	8.3	22.5	9.0	22.6	9.6

N is the number of floors above the main terminal.

Table 6.8b Values of H and S for different values of P from 13 to 20 persons

Floors	13		14		15		16		17		18		19		20	
N	H	S	H	S	H	S	H	S	H	S	H	S	H	S	H	S
5	4.9	4.7	5.0	4.8	5.0	4.8	5.0	4.9	5.0	4.9	5.0	4.9	5.0	4.9	5.0	4.9
6	5.9	5.4	5.9	5.5	5.9	5.6	5.9	5.7	6.0	5.7	6.0	5.8	6.0	5.8	6.0	5.8
7	6.9	6.1	6.9	6.2	6.9	6.3	6.9	6.4	6.9	6.5	6.9	6.6	6.9	6.6	7.0	6.7
8	7.8	6.6	7.8	6.8	7.9	6.9	7.9	7.1	7.9	7.2	7.9	7.3	7.9	7.4	7.9	7.4
9	8.7	7.1	8.8	7.3	8.8	7.5	8.8	7.6	8.8	7.8	8.9	7.9	8.9	8.0	8.9	8.1
10	9.7	7.5	9.7	7.7	9.8	7.9	9.8	8.1	9.8	8.3	9.8	8.5	9.8	8.6	9.9	8.8
11	10.6	7.8	10.7	8.1	10.7	8.4	10.7	8.6	10.8	8.8	10.8	9.0	10.8	9.2	10.8	9.4
12	11.6	8.1	11.6	8.5	11.6	8.7	11.7	9.0	11.7	9.3	11.7	9.5	11.8	9.7	11.8	9.9
13	12.5	8.4	12.5	8.8	12.6	9.1	12.6	9.4	12.7	9.7	12.7	9.9	12.7	10.2	12.8	10.4
14	13.4	8.7	13.5	9.0	13.5	9.4	13.6	9.7	13.6	10.0	13.7	10.3	13.7	10.6	13.7	10.8
15	14.4	8.9	14.4	9.3	14.5	9.7	14.5	10.0	14.6	10.4	14.6	10.7	14.6	11.0	14.7	11.2
16	15.3	9.1	15.4	9.5	15.4	9.9	15.5	10.3	15.5	10.7	15.6	11.0	15.6	11.3	15.6	11.6
17	16.2	9.3	16.3	9.7	16.4	10.2	16.4	10.6	16.5	10.9	16.5	11.3	16.6	11.6	16.6	11.9
18	17.2	9.4	17.2	9.9	17.3	10.4	17.4	10.8	17.4	11.2	17.5	11.6	17.5	11.9	17.6	12.3
19	18.1	9.6	18.2	10.1	18.2	10.6	18.3	11.0	18.4	11.4	18.4	11.8	18.5	12.2	18.5	12.6
20	19.0	9.7	19.1	10.2	19.2	10.7	19.3	11.2	19.3	11.6	19.4	12.1	19.4	12.5	19.5	12.8
21	19.9	9.9	20.0	10.4	20.1	10.9	20.2	11.4	20.3	11.8	20.3	12.3	20.4	12.7	20.4	13.1
22	20.9	10.0	21.0	10.5	21.1	11.1	21.1	11.5	21.2	12.0	21.3	12.5	21.3	12.9	21.4	13.3
23	21.8	10.1	21.9	10.7	22.0	11.2	22.1	11.7	22.2	12.2	22.2	12.7	22.3	13.1	22.3	13.5
24	22.7	10.2	22.9	10.8	22.9	11.3	23.0	11.9	23.1	12.4	23.2	12.8	23.2	13.3	23.3	13.8

N is the number of floors above the main terminal.

CHAPTER TEN

Classical Traffic Control

10.1 BACKGROUND

It will be shown in this chapter that appropriate automatic traffic control systems can enable a single lift, or a group of lifts, to operate at very high efficiency, provided this is matched by high performance, reliable, electromechanical equipment. This chapter will deal with the traffic control of single and multiple lifts, by presenting a classical view. The ideas developed before the era of the digital computer were ingenious and were implemented by relays, without the benefit of Boolean algebra.[1] Many of the ideas are still valid today and have formed the basis of modern digital computer based traffic control systems. Their study here is illuminating.

10.1.1 History

The overall control of lift systems raises two different engineering problems:

- First, some means of commanding a car to move in both up and down directions and to stop at a specified landing must be provided.

- Second, in a group of lifts working together, it is necessary to coordinate the operation of the individual lifts in order to make efficient use of the lift group.

The former is concerned with drive systems and drive control and will not be discussed here, whereas the latter is concerned with (passenger) traffic control which will be discussed here.

Individual lift control is a basic necessity and, as such, was present from the very beginnings of lift usage. Early hydraulic and steam driven lifts were operated on "hand-rope" control (Strakosch, 1967): the operating device was a rope that ran the length of the shaft and actuated a valve in the basement. As shafts were not fully enclosed, a passenger requiring lift service at a particular landing could reach in and operate the rope to summon the lift. Although practical because it needed no attendant, this type of operation was very unsafe.

The advent of electrical lifts brought the electrical car switch operation, where a lift attendant had complete control of the car by moving a handle-operated switch to drive the car in the up or down directions. The attendant (manual) single car control is the simplest form of control that can be used. It was the attendant who decided the floors where stops would be made. Initially, by relying on direct observation of the landing halls through the entrance gates to know whether service was required. Later, in order to increase the efficiency of service and to allow for closed hoistways, signalling systems, able to give the attendant the information required about the traffic demand, were

---

[1] A system of mathematical logic used to represent digital computer operations, where true and false is represented by a "1" or a "0".

introduced in the car. At the same time other features concerning passenger safety and comfort were introduced and required by law, eg: landing door interlocks[1] and car gates.

Attendant control has almost disappeared today. In some buildings "It may be desirable to have the attendance of a commissionaire or lift operator as part of the atmosphere of service associated with the building" (Fletcher, 1954). An attendant is sometimes employed today in special buildings for security reasons.

Aware of the limitations of this type of control, lift engineers developed automatic electrically operated drive control and signalling systems for single lifts. Automatic drive and signalling control of lifts made rapid progress. It allowed better acceleration, slowing and levelling of cars, higher operating speeds and superior traffic handling capacities. The automatic control not only saved much labour for the attendant, but also brought better results than control dependent on human decisions.

The advent of automatic doors allowed the development of fully automatic pushbutton (FAPB) lift systems and completely eliminated the need for an attendant. Automation is particularly indispensable when a group of lifts are to be controlled because, in addition to the individual lift control, it is necessary to interconnect the lifts and provide a group control system capable of operating the lifts efficiently under various traffic patterns.

In broad terms it is possible to identify five generations in the history of the control of groups of lifts, as shown in Table 10.1.

Table 10.1 Generations of control

Era	Dates	Traffic control type
I	1850–1890	Simple mechanical control
II	1890–1920	Attendant and electrical car switch control
III	1920–1950	Attendant/dispatcher and pushbutton control
IV	1950–1975	Group control: IVa scheduled traffic control to 1960 IVb demand traffic control from 1960
V	1975–	Computer group control

Notice these generations are approximately equivalent to human generations. Also notice the change of expertise: mechanical (I), electro-mechanical (II & III), electrical (IV), and electronic (V).

10.1.2 Scheduled versus Demand or "on-call" Traffic Control

In Table 10.1, under Generation IV, the terms "scheduled" and "demand" traffic control are mentioned and require some explanation.

Lift traffic control systems need to respond to the necessity of providing efficient control for a group of lifts, in order to service a common set of landing calls. The traffic control system is required to present an improved performance, compared with the lifts working independently. A major problem that the traffic control system has to solve is to keep the lifts in the group equally distributed along the shaft height, especially during heavy demand periods. Under busy traffic conditions, the lifts have a tendency to bunch

---

[1] A door interlock prevents a car from moving if the doors are not properly closed and locked.

together and to "leap frog" each other, that is to stop at alternate floors with one lift frequently overtaking the other.

The first solution to this problem was provided by the scheduling systems, where the lifts are dispatched from the terminal floors at convenient time intervals and made to run like buses between these terminal floors. Scheduling systems do not work "on call", in the sense that the lifts are dispatched in response to a demand, but rather they are dispatched from the main terminal floors according to a schedule without the necessity of a call being registered. Improvements were introduced in the scheduling system in order to provide a number of different operating programs for the dispatch of lifts. Such programs are designed to deal with the various patterns of traffic flow and intensity of traffic encountered over the working day. However, scheduling systems suffer from a major disadvantage: the lifts spend a considerable amount of time at the top and bottom terminals waiting for a dispatch interval to expire, hence reducing the potential handling capacity. The stop at the top terminal is often pointless, and the lifts do no useful work while waiting for dispatch. Also, lifts frequently make fruitless journeys between terminal floors and thus run unnecessarily, when traffic is light.

Because of these disadvantages, lift traffic control engineers have developed non-scheduling control systems, which only respond when landing calls are made; the "on-call" system. There are several of these systems available on the market. They may behave quite similarly for certain traffic patterns, for example under uppeak, but they can also present dissimilar performance under other traffic conditions, such as down peak or local traffic demands. The main features available in the most representative conventional implementations of lift group control systems are analysed and discussed in this chapter.

10.2 SINGLE LIFT TRAFFIC CONTROL

10.2.1 Single Call Automatic Control

The simplest form of automatic lift control is single call automatic control. Single call pushbuttons are provided on the landings. This form of control is also termed non-collective or automatic pushbutton (APB) control. The passengers directly operate the lift by pressing landing and car buttons, so no attendant is necessary. Car calls are given absolute preference over landing calls, which are only answered if the lift is available. If a passenger in the car presses a call pushbutton corresponding to the required destination floor, the lift moves direct to this floor bypassing any intermediate floors at which landing call buttons have been pressed. When a landing call pushbutton is pressed and the lift is free, the call is immediately answered. If the lift is in use, a landing signal indicates a "lift busy", thus hopefully reducing the intending passenger's frustration.

This type of control is only suitable for short travel passenger lifts serving up to four floors in, for example, small residential buildings with a light traffic demand. It provides a very low carrying capability, as most of the time the lift carries a single passenger. It can also produce large passenger waiting times, owing to the many trips bypassing other passenger requests. The automatic pushbutton control is, however, suitable for goods lifts, particularly when a single item of goods can fit in the lift at one time.

10.2.2 Collective Control

Passengers today do not expect an attendant to be present in a lift. They were removed long ago for economic reasons, except in special circumstances. Passenger operated collective control provides better performance than attendant operation. Fletcher (1954) refers to a survey made in one installation, where a group of four identical lifts were arranged so that two of them operated as duplex passenger control and the remaining two operated with attendants. The duplex system provided a handling capacity 20% greater than the attendant controlled lifts.

The most common form of automatic control used is collective control. This is a generic designation for those types of control where all landing and car calls made by pressing pushbuttons are registered and answered in strict floor sequence. The lift automatically stops at landings for which calls have been registered, following the floor order rather than the order in which the pushbuttons were pressed. Collective control can either be of the single button, or of the two pushbutton types.

10.2.2.1 Non-directional collective

Single pushbutton collective control provides a single pushbutton at each landing. This pushbutton is pressed by passengers to register a hall call irrespective of the desired direction of travel. Thus, a lift travelling upwards, for example, and detecting a landing call in its path will stop to answer the call, although it may happen that the person waiting at the landing wishes to go down. The person is then left with the options to either step into the car and travel upwards before going down to the required floor; or to let the lift depart and re-registers the landing call. Owing to this inconvenience, this type of control is only acceptable for short travel lifts.

10.2.2.2 Down collective (up-distributive, down-collective)

Single pushbutton call registration systems may, however, be adequate in buildings where there is traffic between the ground floor and the upper floors only and no interfloor traffic is expected. A suitable control system is the down collective control (sometimes called up-distributive, down-collective) where all landing calls are understood to be down calls. A lift moving upwards will only stop in response to car calls. When no further car calls are registered, the lift travels up to the highest landing call registered, reverses its direction and travels downwards, answering both car and landing calls in floor sequence.

10.2.2.3 Full collective (directional collective)

The two pushbutton full collective control (also designated directional collective control) provides each landing with one UP and one DOWN pushbutton and passengers are requested to press only the pushbutton for the intended direction of travel. The lift stops to answer both car calls and landing calls in the lift direction of travel, in floor sequence. When no more calls are registered in the lift direction ahead of the lift, the lift moves to the furthest landing call in the opposite direction, if any, reverses its direction of travel and answers the calls in the new direction. This control system is suitable for single lifts or duplexes (two lifts) serving a few floors with some interfloor traffic. Typical examples are small office buildings, small hotels and blocks of flats.

Directional collective control applied to a single lift car is also known as simplex control. Duplex (two lifts), triplex (three lifts) or quadruplex (four lifts) control systems are available for groups of directional collective controlled lifts interconnected to work as a team. This is the simplest form of group control (see Section 10.3).

10.2.3 Important Rules for Passenger Movements

Two important rules emerge from the discussion of the movements of a single lift when serving passenger demands. They are:

Rule 1 Car calls *always* take precedence over landing calls.

Rule 2 A lift must *not reverse* its direction of travel with passengers in the car.

Rule 1 ensures that passengers already in the lift reach their destinations, whatever the demand on the landings may be. Rule 2 means that a lift must continue to serve the last car call in the direction of travel before reversing direction. These rules must also be obeyed by all group traffic control systems.

It might be thought that a more optimal solution might be to ignore Rule 2, in order to give a more equal service to all landing and car calls. This might involve a down travelling lift collecting an up travelling passenger at a floor, where a down travelling passenger alights, travelling down to a lower floor to allow another passenger to alight and then to travel up to the desired destination of the up travelling passenger. Closs (1970, 1972) showed this to be untrue and that it is always best (but not necessarily optimum) to always collect calls ahead and only reverse when the last passenger has exited the lift.

10.3 GROUP TRAFFIC CONTROL

10.3.1 Some Definitions

A single lift will not always be able to cope with all the passenger traffic in a building. Thus, a number of lifts may be installed, often side by side, and the problem arises of interconnecting the lifts. Many buildings have lifts installed where two lifts operate independently, although they are placed together. Such lifts do not operate efficiently, as people will tend to register landing calls at both lifts, thus motivating the two lifts to answer the same call. This causes extra trips and false stops and under heavy demands it produces bunching of lifts.

Where a number of lifts are installed together, the individual lift control mechanisms should be interconnected and also there should be some form of automatic supervisory control provided. In such a system the landing call buttons are common to all the lifts which are interconnected, and the traffic supervisory controller decides which landing calls are to be answered by each of the lifts in the group.

Definition 10.1: A group of lifts is a number of lifts placed physically together, using a common signalling system and under the command of a group traffic control system.

The lifts then serve a common set of landing calls, but are allocated calls according to a set of rules. The purpose of the group traffic control system is to coordinate the operation of the individual lifts. This is essentially intended to maximise the transport capability with the given facilities, to improve passenger service in terms of shorter waiting times and to make available a number of features to deal with specific traffic situations. With the development of high rise buildings and other large scale structures, group supervisory control of lifts is most important as a facility providing a central function of modern architecture.

Definition 10.2: A group traffic control system is a control mechanism to command a group of interconnected lift lifts with the aim of improving the lift system performance.

Modern group control systems are expected to provide more than one program or control algorithm in the traffic control system to allocate lifts to landing calls. The appropriate operating program is determined by the pattern and intensity of the traffic flow encountered by the lift system. The selection of the proper control algorithm can be done manually, when an attendant operates a key switch on a control panel. However, in the more complex systems the operating program is automatically selected by a traffic analyser which assesses the prevailing traffic conditions. Such a detection mechanism is based on either the measurement of car loads by means of weighing devices installed in the car floor, or on the counting of landing and/or car calls or on timing devices. Once a particular traffic condition applies, the control system may introduce only a few changes or adaptations in the control policy to cater for the specific circumstances inherent to the new traffic situation, or it may switch to a substantially different control algorithm.

Definition 10.3: A group traffic control algorithm is a set of rules defining the traffic control policy, which is to be obeyed by the lift system, when a particular traffic condition applies.

The complexity of the group traffic control system is related to the number of lifts in the group and the features that are required from the lift system. It varies from very simple two lift control systems to very sophisticated schemes, with up to eight lifts. The control systems can vary from very simple, single program systems to complex multiprogram systems, where a number of control algorithms are available to cover such conditions as uppeak, down peak, heavy sector demand, heavy floor demand, balanced traffic, off peak and night service.

The traffic control system complexity depends not only on the number of available control programs, but also on the complexity of the algorithms themselves. A lift system with a large variety of control algorithms is not necessarily the best system, as some problems may arise in the transfer of control from one algorithm to another, as an effective redistribution of lifts takes some time, making response to transient changes in traffic requirements very difficult to achieve consistently.

10.3.2 Call Allocation

The primary function of a group of lifts is to answer the car and landing calls belonging to the group in the most appropriate way. To make this possible it is obviously necessary that any car or landing calls once registered are memorised, until they are answered. The control system must know, at any instant, the demand placed on the group of lifts, in order to take the most suitable action to deal with the particular traffic pattern. Thus, a car call station is necessary for each lift in the group. The landing call system is common to all lifts in the group and must include one up and one down pushbutton at each floor. An obvious exception is made for the terminal floors, when one pushbutton is sufficient. For architectural balance or for ease of call registration, it is frequent to find the landing call buttons replicated two or more times at each floor. However, they are all interconnected in parallel.

A traffic control system must distribute the lifts equally around the building zone, in order to provide an even service at all floors. Also, it is important that only one lift be dispatched to deal with each landing call. Thus, an allocation policy is necessary to determine which lift answers each particular landing call.

Definition 10.4: Landing call allocation is the procedure by which a lift is assigned to service a particular landing call and prevents other lifts from starting to move, or continuing their travel, in response to that landing call.

The simplest method used to allocate landing calls to specific lifts is by sending the nearest lift. This works particularly well for the duplex, triplex or quadruplex control systems operating under interconnected directional collective control system, described in Section 10.2.2(c). A common method used to provide call allocation is by grouping the landing calls into sectors within each building zone and allocating lifts to each sector. If the supervisory control system parks any idle lifts in vacant sectors, then a good distribution of lifts is achieved and better performances may be obtained.

Definition 10.5: A sector is a group of landings, or of landing calls, considered together for lift allocation or parking purposes.

There are several classical ways of grouping landings or landing calls into sectors. The number of sectors in a building zone is generally dependent on the number of lifts in the group of lifts serving the zone. Two main methods can be used to group landing calls into sectors: static and dynamic sectoring.

10.3.2.1 Static sectoring

Where sectors are defined statically, a fixed number of landings are grouped together to constitute a sector. One of the existing schemes considers a number of levels in each sector and includes both up and down landing calls within the common sector limits.

Definition 10.6: A common sector is a fixed sector that is defined for both up and down landing calls originating from a number of (usually) contiguous landings.

A common design with this type of sectoring, where there are three to four lifts, is to define as many sectors as there are lifts. This allows a lift to be assigned to each sector under light to medium traffic conditions and therefore provide equalised service to all the sectors. A good service to the main terminal floor is also considered to be an important feature of the lift system. Where there are five or more lifts there may be one more lift than there are sectors, thus providing a "floating" lift to cover sudden demands.

As an example, consider a group of five lifts serving a local zone of 16 floors above the main terminal floor. Suppose that an equal floor demand is expected. A suitable partition might consist of defining four equal sectors of four floors each, with the main terminal considered as a single floor sector. This ensures preferential service to the main terminal floor and an even service to the other sectors.

An alternative method of static sectoring arranges the (usually) contiguous up-landings calls into a number of up-demand sectors, and contiguous down-landing calls into an independent number of down-demand sectors, thus defining directional sectors.

Definition 10.7: A directional sector is a fixed sector that includes a number of (usually) contiguous landing calls defined for one direction only.

Consider the same local zone discussed above, again served by a group of five lifts. For this type of scheme it might be appropriate to define as many down sectors as there are lifts, and one less up sector than the number of lifts. A solution is where the main terminal floor is again given an up sector to itself, in order to maintain a good service for the incoming passengers. The boundaries of the down sectors do not need to be the same as the boundaries of the up sectors.

In the above discussion, the main terminal floor is the bottom floor of the building. If one or more basement floors exist, they can be grouped into a low priority sector or, alternatively, the main terminal could be taken as the highest floor of the lowest sector and some restrictions to the basement service would be included.

The limits of static sectors are often dictated by the type of tenancy, the existence of unequal traffic demands, within the different floors in the zone, or the need for preferential service at predetermined floors. Thus, the sectors may contain a different number of floors. A sensible policy to adopt when defining the sectors is to try to equalise the amount of traffic originating at each floor, biased by the relative importance of the quality of service provided to the sector. For example, if a single floor sector is defined for an executive floor, the number of calls registered in this sector is certainly very small; however, a very good service is required at such a floor and may justify the procedure.

If equal traffic intensities are expected at each landing, then sectors of the same size should be provided. It often happens that this is not possible, because there is not an integer relationship between the number of floors and the number of sectors. A possible solution for common sectors and down directional sectors is either to make the lowest sector (main terminal not included) larger or the highest sector smaller. This is based on the assumption that more lifts pass through the lowest sector than any other, owing to the domination of traffic to and from the building entrance, and hence any spare car capacity will be taken up by stopping in this sector. In addition, passengers will frequently walk up one floor and down one or two floors rather than wait for a lift. In the case of up directional sectors, it may be appropriate to make the highest up sector larger, as under normal conditions less up traffic originates in this sector in comparison with lower up sectors.

10.3.2.2 Dynamic sectoring

In a dynamic sectoring scheme, the number of sectors and the position and limits of each sector depend on the instantaneous status, position and direction of travel of the individual lifts. Thus, dynamic sectors are not defined at the design stage, but are defined during normal lift operation and are continuously changing. The sector of each lift extends from the lift to the next lift ahead, which is either idle or travelling in the same direction. When a sector reaches a terminal floor, it continues in the opposite direction. A stationery idle lift (parked lift) is the boundary for one up and one down sector. Fully loaded lifts and idle lifts travelling for parking or any other special reason are not considered in the sector definitions. It may happen that two or more lifts are located at the same floor with identical direction of travel. In this case, one single floor is defined for the lifts and it is allocated to the lift with the highest reference symbol.

Definition 10.8: A dynamic sector is a sector that includes a variable number of floors defined by the position of moving and idle cars.

10.3.3 Assignment of Lifts to Sectors

There are several ways of assigning lifts to demand sectors. The dynamic sector concept assigns a lift in the sector definition itself, alleviating the necessity for a separate allocation algorithm. Thus, incoming landing calls in a sector are automatically allocated to the corresponding lift. Static sector systems, however, require a separate method of allocation.

In a static sector scheme a simple method of assignment consists in allocating a lift to a sector if the lift is present in that sector and the sector is not allocated to another lift. Special rules can then be developed for de-assignment, eg: a lift leaving a sector or by being fully loaded, etc. Special rules can be developed to assign a lift to a sector, eg: to an adjacent unoccupied sector, by a timer, by priority, etc. Some of these possibilities will be discussed in the next sections.

10.3.4 Information for Passengers: Signalling

For an efficient boarding of a lift it is important that passengers are given information. This would include:

■ Landing and car call registration pushbuttons should illuminate when operated to indicate the call has been accepted.

■ The illumination of the landing and car call registration pushbuttons should be extinguished when the lift arrives at the requested landing.

■ The arrival of a lift at a landing should be signalled by a visible directional (arrow shaped) lantern on the landing at least 4 s before the lift starts to open it doors. It would be convenient to colour code the up and down directions.

■ The arrival of a lift at a landing should be signalled by an audible gong on the landing at least 4 s before the lift starts to open it doors. It would be convenient to code the up and down directions by the number of gong strokes.

■ Arrow shaped direction indicators should be positioned in the rear of all lifts visible from the landing.

■ Direction and floor position indicators should be provided in the car above the doors.

■ To assist the persons with disabilities, Braille and raised characters should be provided on tactile fixtures, speech announcements may be provided on the landings and in the car, colour contrasting legends and numerals should be provided.

BS5655: Part 6: 2002 suggests "that the purpose of every pushbutton and indicator should be clearly understood by all passengers". BS5655: Part 7: 1983/A1 also gives some guidance on fixtures.

10.4 EXAMPLES OF CLASSICAL GROUP TRAFFIC CONTROL

The previous sections considered the features required for lift group traffic control. A number of examples of possible methods of lift group traffic control are now considered in some detail, ranging in complexity from a simple fixed logic system to classical traffic control schemes. Many of these traffic control systems discussed here have been commercially installed by such companies as Otis (VIP 260), Westinghouse (Mark 4) and Schindler (Aconic) and are still operating today. The descriptions here are deliberately generic rather than proprietary and may be at variance to actual implementations.

10.4.1 Simple Control — "Nearest Car" (NC)

The simplest type of group control is the directional collective control described in Section 10.2.2(c). It is suitable for a group of two, or three lifts, each operating on the directional collective principles and serving seven or so floor levels. The assignment of lifts to landing calls is achieved by the "nearest car" (NC) control policy.

A single landing call system with one UP and one DOWN pushbutton at each landing, except for the terminal landings is required. The NC control is expected to space the lifts effectively around the building, in order to provide even service, and also to park one or more lifts at a specified parking floor, usually the entrance lobby floor (main terminal). Other features, which might be included, are the bypassing of landing calls when a lift is fully loaded and the possibility of taking a lift out of the group for a special trip under independent car control for inspection, maintenance or because the car is faulty. The remaining lifts will continue to provide service to all floors.

Car calls are dealt with according to the directional distributive control principles. Landing calls are dealt with in the normal way by reversal at highest down and lowest up calls. Thus the lift answers its car and landing calls in floor sequence from its current position and in the direction of travel to which it is committed.

The only group traffic control feature contained in this simple algorithm is the allocation of each landing call to the lift that is considered to be the best placed to answer this particular call. The search for the "nearest car" is continuously performed until the call is cancelled after being serviced. The distance (d) between a particular landing call and a lift is "measured" in levels:

$$d = |\text{car floor} - \text{landing floor}| \text{ levels}^{1} \tag{10.1}$$

and can vary from zero to N in a building of $N + 1$ floors.

From the distance between landing call and car, and taking into account the landing call and the car directions, a figure of suitability (FS) is evaluated according to four rules:

Rule (a)   If the lift is committed to move towards the landing, in the same direction as that required by the landing call, then a position bias is given to the lift, such that the lift appears to be one floor nearer to the landing call. It is:

$$FS = (N + 1) - (d - 1) = (N + 2) - d \tag{10.2}$$

Equation (10.2) applies in the special case of a lift and a call being at extreme ends of the shaft, even though the lift is not technically moving in the same direction. Then:

$$FS = (N + 2) - N = 2 \tag{10.3}$$

Rule (b)   If the lift is committed to move towards the landing, in the opposite direction as that required by the landing call, then:

$$FS = (N + 1) - d \tag{10.4}$$

---

[1] The bars (|) indicate d is to be the modulus of the expression, ie: independent of sign.

The figure of suitability is therefore its maximum value equal to $N + 1$, when the lift is at the calling floor.

Rule (c)   If the lift is committed to move away from the landing call, it is considered to be "in service", and the figure of suitability is arbitrarily fixed as one, irrespective of the value of distance.

Rule (d)   A lift which is idle, ie: it has no car or landing call commitments, has a figure of suitability given by Equation (10.4).

The allocation procedure assigns each landing call to the lift presenting the highest figure of suitability for the particular call. If the figures of suitability are the same for two or more lifts, the call is assigned to the nearest of these lifts or, in the case of equality of distance, to the first lift that reached that figure of suitability some time earlier. Lifts that are fully loaded are not considered in the allocation procedure. However, if all the lifts are fully loaded, the landing calls are temporarily allocated to the "best" placed lift. A lift in these circumstances will not stop at such landing calls unless the lift becomes less than fully loaded as the result of a passenger alighting before reaching such a landing call. When a lift is not assigned to any landing call and no car calls are registered, it is made to stop at its current floor, with its doors closed. If no parking algorithm is provided, the distribution of lifts in the building will be poor, as stationary lifts are ready to move as soon as they become assigned to a landing call.

Consider a triplex control system serving a building of seven floors (main terminal and six floors). At a particular time instant, the lift positions and status and the registered car and landing calls are as illustrated in Figure 10.1.

To determine which lift is allocated to each landing call, the figures of suitability for each pair of landing call and lift are indicated in Table 10.2. To see how these figures of suitability are derived, consider the evaluation for the down call at Floor 4.

Car 1   - is moving towards the call
         - opposite directions for call and travel (Rule (b))
         - distance = 2 floors; therefore FS = 7 − 2 = 5
Car 2   - moving away from the call, FS = 1 (Rule (c))
Car 3   - not committed to direction (Rule (d))
         - distance = 4 floors; therefore FS = 7 − 4 = 3

The maximum figures of suitability in Table 10.2 are included for each landing call. They show that all the landing calls in the example are allocated to Car 1.

Table 10.2 Figures of suitability for the example in Figure 10.1

Call	Car 1 − UP @ floor 2	Car 2 − DOWN @ floor 1	Car 3 − EMPTY @ floor MT
4-UP	6	1	3
6-DN	3	1	1
4-DN	5	1	3

No other special features are included in the NC traffic control system. It provides a single program control algorithm designed specifically to cater for interfloor traffic in low rise buildings. The allocation procedure has some similarities with dynamic sectoring, however it does not take great advantage of the direction of the landing call. For instance, in the example the allocation of the down call at Floor 4 to Car 1 is highly inconvenient, as this lift is going to stop at least at two floors (4-UP, 6-DN) and maybe at Floor 5 if the 4-UP caller requests it, before answering the down call at Floor 4.

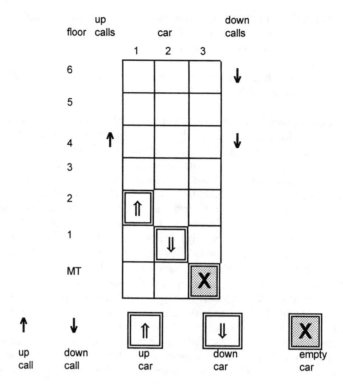

Figure 10.1 Example operation of a triplex traffic control system

A much better policy would be to allow the idle lift stationed at the ground floor (Car 3) to move up and help the first lift in this particularly unbalanced traffic situation. Actually (see Table 10.3), the down call at Floor 4 will eventually be served by Car 3, as it will be allocated to the call as soon as Car 1 leaves Floor 4, travelling upwards (assuming that no more landing calls have been registered). Meanwhile the call has been waiting for an unnecessary time.

Table 10.3 Revised figures of suitability

Call	Car 1 4−UP @ floor 4	Car 2 −EMPTY @ floor MT	Car 3 −EMPTY @ floor MT
6-DN	5	1	1
4-DN	1	3	3*

* Reached this value first.

Owing to these types of difficulties, this simple algorithm is suitable only for buildings with a small number of levels where the distribution of lifts is not very critical. Under such conditions, this algorithm provides reasonable interfloor traffic performances. This algorithm is not appropriate to serve down peak traffic. The allocation procedure will cause the lifts to travel to calls too high in the building, neglecting calls lower in the building, and thus giving better service to the higher floors. Again, in a low rise building and assuming that the occupants at the lower floors use the stairs rather than the lifts, the system may be acceptable.

It is under uppeak conditions that this algorithm presents its greatest disadvantages. Most control systems express the lifts down to the main terminal as soon as they discharge their last passenger whilst the uppeak condition applies. Unlike such systems, this simple algorithm uses the normal allocation procedure, assigning a lift to the ground floor only when a call is registered there. Lifts will park at upper floors for some time, thus providing a poor uppeak service.

10.4.2 A Fixed Sectoring Common Sector System (FSO)

A suitable fixed sectoring common sector control system can be devised for dealing with off peak traffic and can be complemented with special features to cater for heavy unbalanced traffic. The FSO system divides the building into a number of static demand sectors (Figure 10.2) equal to the number of lifts. A sector includes both the up and down landing calls at the floors within its limits. A lift is allocated to a sector if it is present in that sector and the sector is not committed to another lift. Fully loaded lifts are not considered for allocation.

An assigned lift operates on the directional collective principle within the limits of its range of activity. The lift responds to the landing calls registered inside its sector and to landing calls in adjacent vacant sectors above. It also responds to landing calls below the car sector, if the lift is assigned to the lowest occupied sector. If landing calls are detected behind a lift that is serving its sector and within the sector, then this sector is considered to be vacant, in order to allow the lift located below to answer these calls.

The de-assignment of a lift from its sector takes place when the lift leaves the sector. A lift picks up calls ahead when travelling in either direction, even if it is not assigned to the sector. So, the lift is allowed to start or continue its movement in response to:

(a) registered car calls
(b) landing calls registered inside its sector
(c) landing calls in contiguous unoccupied sectors above and adjacent to the sector to which the lift is assigned
(d) landing calls registered behind the lift assigned to the contiguous sector above, if this lift is travelling upwards
(e) all the landing calls registered below the car sector if the lift is assigned to the lowest occupied sector.

The de-assignment of a lift from its sector in this scheme implies that the lift leaves the sector or becomes fully loaded within the sector. A lift is allowed to pick up landing calls on its way even if it is not assigned. Fully loaded lifts do not, however, stop in response to landing calls.

The FSO system, by distributing the lifts equally around the building, presents a good performance under balanced interfloor traffic. It also performs well for uppeak and unbalanced interfloor traffic conditions. It lacks a proper procedure to cater for sudden heavy demands at a particular floor. Under heavy down-peak traffic conditions, a poor service may be provided to the lower floors of the building owing to problems in recycling the lifts to unoccupied sectors.

To consider how an assignment of floors to sectors might be made, consider a 16 floor building (including the ground floor) served by four lifts as shown in Figure 10.2.

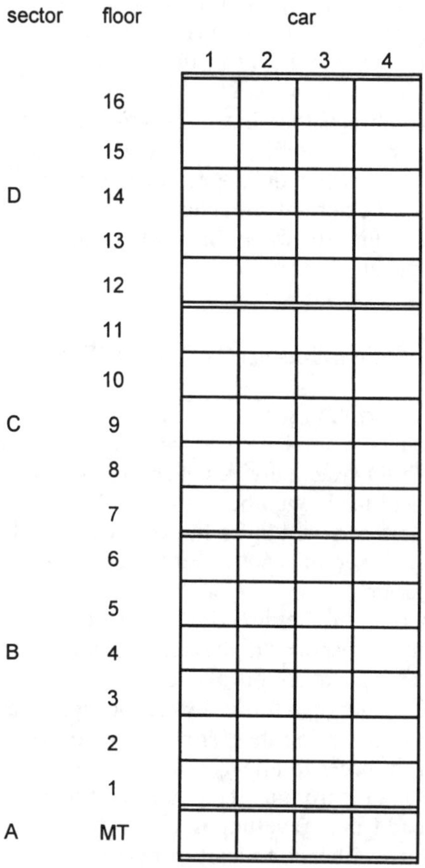

Figure 10.2 Illustration of the assignment of floors to fixed common sectors

There will be four sectors:

Sector A   Ground
Sector B   Floors 1 − 6
Sector C   Floors 7 − 11
Sector D   Floors 12 − 16

The ground floor is allocated solely to Sector A to emphasise its importance. Sector B above is made the larger of the remaining three sectors, on the basis that people may walk to/from Floor 1 and the travel distance to Sectors C and D is larger than to Sector B.

The FSO algorithm provides a good control algorithm for uppeak and balanced interfloor traffic.

10.4.3 A Fixed Sectoring Priority Timed System (FS4)

Static directional sectoring systems can also allocate the lifts on a priority timed basis. The landings in the building zone served by the group of lifts are grouped into up and down sectors. Each sector is timed as soon as a landing call is registered within its limits. The timing is measured in predefined periods of time, designated the priority levels. The time to reach a specific priority level may be different for each sector, in order to give some selected sectors an overriding priority. Also, the six priority levels may not comprise equal time steps. A common situation is to have six priority levels, which, for example, may follow the sequence 10, 15, 20, 30, 40, 60 seconds for a standard floor, but the sequence 30, 50, 70, 90, 105, 120 seconds for a low priority floor. Such arrangements are particularly useful to cater for executive floors, which require high priority service, and for basement floors, which can be given a low priority sector.

The assignment of lifts to the sectors takes into account the number and positions of the available lifts and the sector priority levels. A lift is available for allocation when it has completed its previous assignment and has dealt with all the car calls that have been registered. The sector with the highest priority is the first to be allocated a lift. If more than one lift is available, it is the nearest lift that is allocated to the sector. After the lift is assigned, it travels without stopping to the sector. However, after leaving the sector, the lift answers landing calls in its way, until the complete set of car calls is served. When the lift stops to answer its last car call, it becomes available and is allocated to the unattended sector with the highest priority. It may happen, however, that some passengers are waiting at the lift's last call floor. They are discouraged from entering the car, as no landing signal is given as the lift arrives and the lights in the car are dimmed. In addition, a flashing indicator reading "DO NOT ENTER" may be fitted in the rear of the car.

The FS4 system gives preferential service to the main terminal, when no lift is stationed there, and an available lift is despatched down, bypassing all but priority levels 5 and 6. The lift stationed at the main terminal floor, with open doors, has an indication that it is the next lift to depart. It is not available to answer demand on other sectors unless no other lift is available and the lift is stationed for more than 5 s at the main terminal.

Demands at Priority 6 call for an immediate response; the nearest lift travelling towards the top priority sector bypasses all landing calls before that sector. Use is made of this feature to cater (partially) for sudden traffic demands at particular sectors or floors. Should a lift fill up to capacity before completing its sector assignment, this sector is advanced to top priority level, thus applying for immediate service. The next lift to become available is despatched to the mid point of this sector. Fully loaded lifts do not stop in response to landing calls.

The FS4 control system provides a good uppeak performance. Its down peak performance is very good, especially under very heavy traffic conditions. Indeed, the system not only cycles the lifts to the sectors in a convenient way, but also provides extra service to sectors waiting too long to be assigned a lift. The interfloor traffic performance is fair, but not as good as can be obtained from dynamic sectoring. The main problem is that the system lacks a parking algorithm to redistribute the lifts in the building. It is true that the allocation procedure distributes the lifts to the sectors, but only after the landing calls are registered; thus the express jumps necessary for the redistribution are made when the passengers are already waiting for service. Uniform performance is obtained for all floors, but average waiting times tend to become extended.

The FS4 system is unique among the classical traffic control systems as it considers time when making an assignment, the other algorithms only consider position.

To understand this type of allocation algorithm, consider a group of five lifts serving a zone of 17 floors (including the main terminal). Assume that, at a particular time instant, the position and status of each lift are as indicated in Figure 10.3.

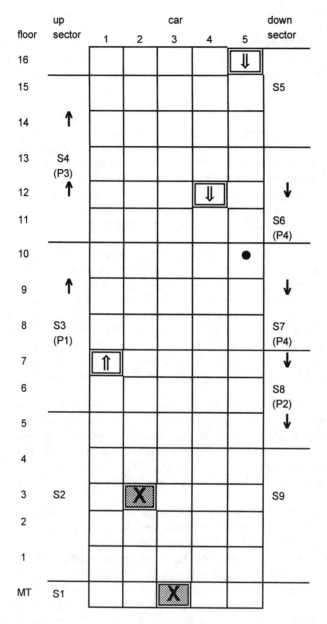

Figure 10.3 Illustration of the assignment of floors to fixed priority timed sectors
Car call shown ●

The lift activity can be summarised as:

■ Car 1 was allocated to Sector S4 and travels express to Floor 12, bypassing the up call at Floor 9, which is only at Priority 1 (P1).

■ Car 2 has just become available and two sectors are demanding service, the sector S3 with P1 and Sector S8 with P2. The lift is assigned to Sector S8, because this sector has the higher priority. It travels express to Floor 7, where it reverses its direction of travel and starts answering the traffic demand within its sector, ie: Floor 5.

■ Car 3 is the next lift to leave the ground floor, when required.

■ Car 4 has been allocated to Sector S7, where a landing call at Floor 9 has already reached the P4 stage. The lift travels express to Floor 9, unless a down call is registered at Floor 10 in the meanwhile.

■ Car 5 has serviced a down call at Floor 16 and is moving downwards to answer a car call at Floor 10. It has finished its assignment to Sector S5 and will answer the down call at Floor 12 on its way down in response to the car call.

10.4.4 A Dynamic Sectoring System (DS)

The dynamic sectoring group supervisory control system provides a basic algorithm and is suitable to deal with light to heavy balanced interfloor traffic. It is complemented by a number of other control algorithms to cater for unbalanced traffic conditions.

The basic DS algorithm groups landing calls into dynamic sectors, as described in Section 10.3.3. Each lift is allocated to a sector in the sector definition itself and answers the landing calls in the sector according to the directional collective procedure. Usually a lift defines a single sector, however, if a lift is located at the same floor as a higher reference lift, or if the lift is fully loaded, then the lift does not define a sector. Also, a lift that remains without a travel command for a specific length of time, typically 5 s, is declared as a free lift and given two sectors, one in each direction.

In parallel with the basic traffic algorithm, another DS algorithm is provided, the free lift algorithm. It operates in a similar way to the basic DS algorithm, but it considers only the free lifts and the heavy duty calls. This free lift control algorithm is intended to insert free lifts ahead of lifts allocated to a high demand sector, in order to provide service where the traffic is heavier. The algorithm continuously monitors the number of car and landing calls within the normal dynamic sectors, by counting the calls from the car position. After a predefined number of car and landing calls, "the optimum counter capacity", the remaining landing calls are registered as heavy duty calls in a free car zone memory. The optimum counter capacity is varied according to the number of free lifts, from a minimum where at least three lifts are free to a maximum when only one lift is free. As an example consider Table 10.4.

In order to express the free lifts to the heavy load centres, a free lift which is moving in response to a heavy duty call is taken out of the basic control algorithm.

To illustrate the working of this traffic control algorithm see Figure 10.4. A group of five lifts serves a local zone of 16 floors above the main terminal floor. The group operates under a dynamic sectoring control system, and the optimum counter capacity is as given in Table 10.4. Where will the sector boundaries be ?

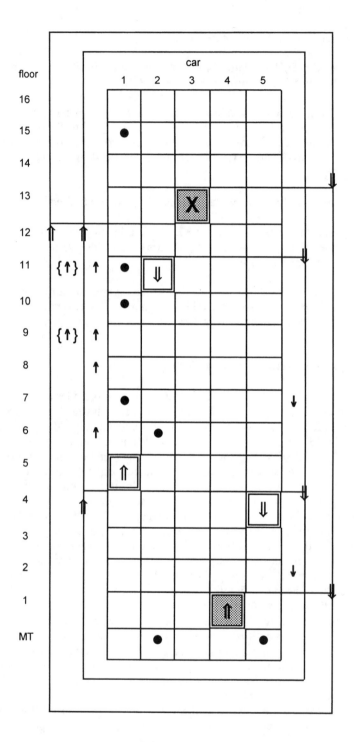

Figure 10.4 Illustration of the dynamic sectoring traffic control algorithm
Inner loop: basic dynamic sectors. Outer loop: free lift sectors.
New symbols: {↑} landing call assigned as heavy duty, hatched Car 4 at Floor 1 is an up travelling free lift.

Table 10.4 Optimum Counter Capacity

Number of free lifts	Optimum counter capacity
3	2
2	3
1	4

Consider first the normal dynamic sectoring procedure:

■ Car 1 is dealing with Floors 5 – 12 and the car calls at Floors 7, 10, 11, 15.

■ Car 2 is dealing with Floors 11 – 5 and the car calls at Floors 6, MT.

■ Car 3 is dealing with Floors 13 – 16 – 12 and is a parked free lift.

■ Car 4 is a moving free lift.

■ Car 5 is dealing with Floors 4 – MT – 4 and the car calls at MT.

In summary, the first, second and fifth lifts are moving in response to car and landing calls. The third and fourth lifts have no call demand placed on them and therefore are free. Note that the sector of each lift extends from the car position to the position of the next lift ahead which is free or moving in the same direction. When the sector reaches the terminal floor, as for Car 5, it continues in the opposite direction. Also note that Car 4 is taken out of the normal sectoring scheme, because it is a free lift committed to a direction of travel.

The traffic situation considered in the above example is severely unbalanced. The first car has four car calls and four landing calls to answer, which is disproportionate in comparison with the other lifts. The free car algorithm is in control of the situation and takes action to provide extra service to the sector assigned to Car 1. The optimum counter capacity is three, since two lifts are available, so the up landing calls at Floors 9 and 11 are declared as heavy duty calls. The two free lifts define two dynamic sectors in the free car zone memory and the two heavy duty calls are within the free car sector of Car 4. This is although Car 3 is nearer. The lift starts in response to the calls, moving express to the lowest heavy duty call at Floor 9. The insertion of a free lift into the heavy traffic section of the zone instead of attempting a redistribution of lifts, makes this control algorithm highly efficient.

The DS system provides a very good performance for uppeak and interfloor traffic conditions, but a poor performance for down peak.

10.5 OTHER FEATURES OF GROUP TRAFFIC CONTROL SYSTEMS

The control systems discussed in Section 10.4 illustrate possible methods of dealing with passenger demands (traffic). There are a number of other features which assist this process and lift group control systems must provide other specialised features, such as basement service, lobby floor preference service, car preference, director/VIP service, heavy demand floors, parking policy, automatic shut down, load bypass, maintenance, service, fire switches, passenger signals, etc.

10.5.1 Uppeak Service

Most lift group control systems detect and take special action for uppeak traffic conditions. Whilst the uppeak condition applies, the lifts answer car calls only, bypassing the landing calls. Thus, as soon as a lift discharges its last passenger on its way upwards, it reverses direction and travels nonstop to the main terminal. If, however, some astute passengers are still in the car and register car calls, the lift will answer the car calls before travelling to the main terminal (Rule 10.1).

There are several uppeak detection mechanisms. A common method is based on weighing devices installed in the car floor or by measuring the motor load current as an indicator of loading. When heavily or fully loaded lifts leaving the main terminal floor are detected, the uppeak control algorithm is selected for a specific time period. A variation of this method, which is able to cater for slight uppeak situations, detects a lift load at the main terminal in excess of a predefined level, say 50% or 60%. For a certain period of time a dummy call is set up at the main terminal to ensure that a lift is available there as soon as possible. Another method counts the car calls registered and when a predetermined number are registered, initiates the uppeak algorithm.

A more reliable and sophisticated approach employs an up/down logic counter which increments when loads are above a predefined level, and decrements for loads below this level. Additionally, the counter decrements on a timed basis, every 60 s or so, to ensure that the uppeak algorithm is switched off quickly as the uppeak traffic diminishes. To prevent instability, hysteresis is built into the mechanism by arranging that the uppeak turn on level is two or more counts above the turn off level. Also, there is a clock inhibit level for the decrementing time pulses, which is at least one unit above the turn off level. Two important counter levels are defined: the turn on level, which selects the uppeak state, and the turn off level, two or three units below the previous level, which switches the uppeak state off.

The automatic uppeak detectors may be substituted or complemented by a time clock or a manual switch in the supervisory controller. These may be a useful complement, but if they constitute the only available means to select the uppeak algorithm, they are poor alternatives. Strakosch (1967) states that in earlier systems, where an attendant was required to change the control program, too often the lift system was set in one program and not changed, so that when it was time to leave the building the passengers had to fight elevators that had been left on "uppeak" operation.

An option provided by some uppeak traffic control algorithms is the division of the building into two uppeak subzones. This division does not concern landing calls, but only the destination floor of the passengers. Once the uppeak state is selected, some of the lifts are allocated to serve the lower subzone and the remainder to serve the upper subzone. An indicator is illuminated at the main terminal floor, for each lift, stating which floors are served by that lift. Uppeak subzoning is intended to reduce the number of stops per trip made by the lifts, hence increasing the uppeak handling capacity. It works well, if properly tuned, but it is very sensitive to the subzone partition. If the uppeak traffic flow is subject to large fluctuations in the destination of passengers, it may be that one subgroup of lifts is saturated and the other is under used. A later chapter examines this and other techniques to improve traffic handling during uppeak.

During uppeak any passenger wishing to travel up from floors other than the main terminal floor should have little difficulty, as the lifts are frequently stopping at the floors, whilst travelling upwards, to discharge passengers. However, passengers wishing to travel down may find a restricted service or no service at all during the 10−15 minutes of heavy uppeak demand. In the case of no service, an astute passenger can be serviced by registering an up call and entering the first upwards travelling lift stopping at their floor. With restricted service, down landing calls are served by a single lift at fixed time intervals (30 s to 3 minutes) or, alternatively, when they reach the highest priority level.

Although it has been said all control algorithms behave in a similar way for uppeak traffic, ie: lifts fill at the main terminal and distribute to the upper floors, the above are some important control features, which aid efficient handling.

10.5.2 Down Peak Service

During down peak, lifts working on the collective principle, or on a scheduled basis present a problem: the lifts travel too high up the building, tend to fill up to capacity at the upper floors and provide a poor service to the lower floors. Therefore, a major feature of a down peak traffic control algorithm should be the provision of a proper cycling of lifts in order to ensure an even service for all the floors. However, group control systems frequently include a means to detect down peak traffic situations, employing similar methods as those used for uppeak detection and considering heavy loaded lift arrivals at the main terminal floor. Whilst the down peak condition applies, these systems will restrict the service provided to any up traffic and cancel the allocation of lifts to the main terminal, whilst the traffic condition lasts.

Fixed sectoring lift allocation systems are particularly adaptable to this traffic pattern, as the lower sectors are sure to be allocated a lift and do not need to rely on the spare capacity of lifts descending from the floors above. Thus, such systems do not need to provide a separate control algorithm for down peak. Priority timed systems, on detection of a fully loaded down travelling lift, might reduce all the up sector priorities for a preset time, during which preferential service is given to down calls.

Dynamic sectoring schemes, where the sectors are defined according to the status, position and direction of travel of each lift, do not cope very well with down peak traffic. Most lifts, after becoming available at the main terminal and finding no up landing calls registered, are assigned to the top landing calls. Thus, such systems suffer from the same problems as directional collective systems, namely very high average reversal floors and poor service to the lower floors. A separate down peak control algorithm is therefore necessary.

One such system detects heavy down traffic terminating at the main terminal floor. The traffic algorithm then groups the down landing calls into dynamic sectors and assigns lifts to serve call groups in the sectors in a "round robin" fashion. In order to provide a proper cycling of lifts and a systematic and equalised service for down traffic, each time a lift stops in response to a down landing call, the floor is blocked to ensure that each floor is served only once during one service cycle.

10.5.3 Load Bypass

When a lift fills up to capacity it should not stop in response to further landing calls, as such stops would be useless and particularly annoying to the passengers already in the lift. A device is usually available to provide such a feature. However, landing calls should not be cancelled.

It is important that the car load detection is set correctly. As has already been explained, the larger lifts cannot accommodate the rated capacity as indicated in the standards. Consider Example 7.7 again. This refers to a 2500 kg lift. If the load detection were to be set to 60% this would equate to 1500 kg. If an average person weighs 75 kg this is 20 persons and this is just a little larger than the Design Capacity of 19.2 persons. Therefore the lift will frequently stop unnecessarily. The real 60% value would be equivalent to 60% of 24 persons (the Actual Capacity), ie: 1080 kg.

10.5.4 Heavy Demand Floors

Heavy floor demands can occur, for example, at the closing of a meeting or lecture. It is plainly justifiable to bring extra lifts to the floor to deal with such peaks of demand.

In order to cater for local heavy demands, the number of landing calls in each sector can be evaluated by the traffic analyser and compared with the average number of landing calls per sector. A particular sector exceeding the average value by more than a predefined quantity can be set up as a heavy traffic sector. The measure of heavy traffic can be individually adjusted for each sector, allowing preferential service for some selected sectors. A heavy traffic sector is considered to be vacant by the control system, even if one or more lifts are already serving the sector. Extra lifts are brought to this sector, bypassing the landing calls at other sectors.

Another method is to detect at individual floors or specially designated floors that a fully loaded lift has left that floor and a new landing call has been registered within 2 s or so, for the same direction of travel. The traffic controller can then send free lifts to the heavy demand floor.

10.5.5 Lobby and Preferential Floor Service

The lobby or main terminal floor in a building is normally of great importance, owing to the steady flow of incoming passengers. Preferential service is usually provided for these passengers by parking a lift at the main terminal prior to any other sector. Lifts are usually parked with their doors closed at the main terminal floor and can be assigned as the "Next" lift to leave. The lobby floor preferential service implies that a slightly poorer service is provided to the remaining floors in the building. The feature is highly undesirable under certain traffic conditions, such as down peak. A good down peak control algorithm must therefore override such a feature. Another arrangement is that when a lift has parked at the main terminal for a period of time, which can vary from 5 seconds to 3 minutes, it becomes available to satisfy demands in other parts of the building.

Various techniques are used to bring a lift to the main terminal whenever it is vacant. The parking algorithm can preferentially park an idle lift at the main terminal rather than at any other vacant sector. Also, a dummy call can be inserted to force a lift to the main terminal. In a priority timed system, it is the assignment algorithm that allocates an idle lift to the main terminal floor, although it allows the lift to pick up high priority calls in its way down to the main terminal.

A feature related to car preference is the "director" service, which gives special service to the floors, where senior executives or directors are installed. The lift system may be made to recognise landing calls at such floors and to deal with them with higher priority, or key operated switches may be available at these preferred landings which cause a lift to travel direct to the executive floor bypassing all other landing calls, or a lift may be completely segregated out of the bank of lifts for director's service. It is obvious that this sort of preferential treatment can seriously affect the efficiency of the service as a whole, and it should be avoided whenever possible.

10.5.6 Parking Policy

Under light to medium traffic conditions, a lift frequently has no calls to answer. The lift is then free for allocation, and if no further demand exists it might be parked at its current position, or a convenient floor, or sector in the building zone. The parking procedure is mainly intended to distribute the lifts evenly around the building, "in anticipation of traffic demand". A proper parking policy is essential for good lift system performance.

The parking sectors can be the same as the demand sectors in fixed sectoring controllers. In the DS algorithm, the free lift control algorithm provides a parking policy by dividing the building into fixed parking sectors to redistribute any free lifts over the whole building zone.

The parking of a lift at the main terminal floor has preference over parking the lift in other sectors, thus providing preferential service to the main terminal floor. If the lift stationed at the main terminal departs in response to a car or landing call, the nearest parked lift could be immediately despatched to fill its place.

10.5.7 Basement Service

Basement service usually requires special considerations. Even a small basement demand under uppeak and down peak conditions can severely degrade the lift system performance. For this reason, most group control systems will seriously restrict basement service. These floors frequently contain services that are of secondary importance during peak periods, and a restricted service may be sufficient. For example, basement floors may be grouped into one sector, which is given very low priority. Some systems will accept a car call, but not a basement landing call; other systems will allow one single designated lift to reach basement level. Some systems may simply provide no basement service at all whilst a heavy traffic demand prevails. In some cases it may prove preferable for a building designer to provide alternative vertical transportation systems to cope with basement service, thus allowing the more expensive above ground system to serve the remainder of the building.

Section 8.3.3 deals with basement service in more detail.

10.5.8 Car Preference

When a lift is taken out of normal passenger control to be exclusively operated from the inside of the lift, it is said to be in car preference, also referred to as independent service, emergency service or hospital service. The transfer is made by a key operated switch in the lift, which causes the doors to remain open until a car call is registered for floor destination. All landing calls are bypassed and car position indicators on the landings for the lift are not illuminated. The removal of the key, when the special operation is complete, returns the lift to normal control.

Car preference may be useful to give a special personal service, or for an attendant to have complete control on the lift, whenever it is required. A typical example is in hospital buildings, where lifts for carrying beds and stretchers require a car preference switch.

10.5.9 Automatic Shut Down

On older systems, where motor generator sets are employed, the traffic control system might shut down some lifts when the traffic demand is very light, in order to save wear, tear and power. Today, this feature is not required on modern drives. However, a feature to cause a different lift to answer calls in a sequence during light traffic periods is desirable to avoid one lift servicing all calls.

10.5.10 Other Features

A number of other facilities available in group supervisory control systems include the provision of indicators of lift arrivals at landings and a number of switches such as maintenance switches to switch off a lift while maintenance work is in progress and fire switches, whenever required by the fire authority, to enable firefighters to take over the complete control of one or more lifts in a group. Another useful feature is the provision of anti-nuisance devices to ensure that a lift does not answer car calls, if it is empty. This avoids unnecessary car trips and stops due to a practical joker, who register car calls sometimes pressing or touching all the car buttons and then leaving the lift.

10.6 SUMMARY

Lift control systems presenting most of the foregoing features are classically implemented with relay logic or solid state fixed logic. A digital computer is used today to control a group of lifts, but still utilise some of the principles and features discussed in this chapter. The facts and problems of conventional and digital computer implementation of lift group traffic control systems are considered in the next chapter.

CHAPTER ELEVEN

Computer Traffic Control

11.1 DIGITAL COMPUTER CONTROL

The concept of centralised supervisory control systems for buildings, known as: Building Automation Systems (BAS), Building Energy Management Systems (BEMS), etc. using digital computers[1] is already well established. It might include lift group supervisory information as part of the comprehensive information system for a whole building, which might also include other facilities such as employee identification, security control, fire control, environmental control, water treatment and data logging. However, it is not sensible to include the task of lift traffic control in any centralised building control system. Thus, a lift should have all aspects of its control managed independently of other building systems.

The opportunity exists with a computer to program complex tasks to assist the call allocation process, which are impossible to achieve with fixed program systems. This chapter first reviews conventional lift traffic control systems and then discusses the general principles of computer based systems. To avoid the chapter being overburdened with detail, three computer control systems are provided in Control Case Studies CS15, CS16 and CS17 to illustrate many of the principles discussed.

11.2 AN ANALYSIS OF CONVENTIONAL CONTROL SYSTEMS

11.2.1 Four Primary Tasks and Five Rules

There are two levels of traffic control for lift systems. The lower level simply commands individual lift cars to move up or down, to stop or start and to open and close the doors. Closs (1970) showed that once a single car is given a set of car and landing calls to answer, the best procedure is to answer them in a simplex collective manner. This generally results in calls being answered in a different order to their registration.

The higher level of traffic control has the function of coordinating the activity of a group of lift cars, by means of a set of logical rules (the lift control algorithm) defined by the lift designer. The four primary tasks of a group traffic control system in serving both car and landing calls should be:

(a) to provide even service to every floor in a building
(b) to minimise the time spent by passengers waiting for service at a landing
(c) to minimise the time taken by passengers to move from one floor to another
(d) to serve as many passengers as possible in a given time.

There must be conflicts between some of these tasks. For example: an even service may result in some passengers waiting longer.

---

[1] Throughout this chapter the word "computer" will mean digital computer.

A lift system has to follow a set of rules, five in number, which the passengers accept and understand. Two rules (Rules 1 and 2) have been presented in Chapter 9 and three other rules (Rules 3, 4 and 5) can now be introduced:

Rule 1 Car calls *always* take precedence over landing calls.

Rule 2 A lift must *not reverse* its direction of travel with passengers in the car.

Rule 3 A lift *must stop* at a passenger destination floor (it must not pass it).

Rule 4 Passengers wishing to travel in one direction *must not to enter* a lift committed to travel in the opposite direction.

Rule 5 A lift must *not stop* at a floor where no passengers wish to enter or leave the car.

Rule 4 could be violated by absent-minded passengers and Rule 5 could be violated by the search procedure for the optimum path stopping a car and then reversing direction, hence also violating Rule 2.

Closs also showed that the performance of a lift system can be optimised by using a computer. The optimisation was performed by minimising total car journey time (a minimal cost algorithm), considering all possible paths for each car given the arrival and destination floors of each passenger. However, not all of the possible paths are acceptable in a practical lift system, if the five rules are to be obeyed.

11.2.2 The Nature of Passenger Demand

Problems are encountered in attempting to achieve the four primary tasks owing to the random nature of the time of call registration and the landing at which passengers arrive and request service and their ultimate destinations. Extensive observations of actual buildings tend to dispute the traffic pattern depicted in Figure 4.2. Beebe (1980) performed a study of logged data from two commercial buildings and observed the following:

(a) Both uppeak and down peak exist, but last longer than indicated by Figure 4.2 and at lower intensity levels.
(b) During uppeak a considerable volume of down traffic exists.
(c) Similarly, during a down peak a significant level of up traffic exists.
(d) Algorithm switching caused by an incorrect detection of a specific traffic condition can seriously degrade performance.

Unlike traffic analysis, which is often based on pure traffic patterns of up traffic only (pure uppeak), down traffic only (pure down peak) and random traffic with no dominant pattern (pure interfloor), traffic controllers have to deal with the real world of mixed traffic patterns. It is probably best to consider all traffic to be an interfloor traffic pattern with various significant overlaying patterns. If this premise is adopted then it should be possible to deal with the observations (a), (b) and (c). So an efficient traffic controller should deal well with the pure interfloor traffic pattern, but respond to any dominant trends. The latter point (d) is well known, where time clocks have caused the uppeak traffic algorithm to be operating during (say) balanced interfloor traffic.

The main conclusion to be drawn is that the traffic control algorithm must be able to follow changes in passenger demand at all times.

11.2.3 Characteristics of Classical Traffic Control Systems

Chapter 10 discussed four classical control algorithms, which have been embodied into proprietary controllers in various ways. They were: nearest car (NC); fixed sectoring, bidirectional sectors (FSO); fixed sectoring, priority timed unidirectional sectors (FS4) and dynamic sectoring (DS). Although the NC algorithm does possess a limited computational ability (Hirbod, 1975), it is intended for low rise, duplex/triplex installations and it will not be considered further. The other three traffic control systems exhibit a number of common characteristics.

The systems are "multi-algorithmic" with different algorithms or sets of rules for the different traffic patterns. The default operating mode is the interfloor algorithm. Uppeak requires a different algorithm in all cases. Down peak is served by the interfloor algorithm in the FSO and FS4 systems with the minor change of cancelling allocation of cars to the main terminal.

Where a different algorithm is used to handle uppeak or down peak traffic, the changeover is usually abrupt. This can cause disruption of the current traffic, although the severity of the effect depends on the intensity of the current traffic. When the detection is incorrect, the effect is most severe. It would appear that much could be gained by the use of a single control algorithm (Smith and Peters, 2002), which can adapt to different traffic conditions and provide a "bumpless" transfer.

During uppeak the actions of the traffic control algorithms are identical. All a control system can do is to bring cars, which have completed serving their car calls, back to the main terminal immediately. The preferential service given to the main terminal has the benefit of handling minor uppeaks that occur from time to time. Efficient uppeak detection is important as a late detection produces a poor uppeak service.

All systems should allow partly loaded cars to collect landing calls in their path, until they are at least 80% loaded. This is achieved on a directional collective principle. This characteristic is of considerable assistance in conserving handling capacity.

The allocation procedures employed by FSO and FS4 control algorithms allocate free or uncommitted cars to demand sectors. Once the car answers a landing call in its allocated demand sector, it continues to answer all other landing and subsequent car calls in simplex mode and the group controller loses control of the individual cars. Thus committed cars work virtually independent of one another, defeating the objectives of group control.

The FSO and DS systems have no concept of time and do not time landing calls. Thus, they cannot give urgent attention to long wait calls. Only the FS4 system has a concept of time.

11.2.4 Improving a Classical Traffic Control Algorithm

The concept of using existing traffic control algorithms designed to be implemented in relay or solid state logic is a quick way to obtain a computerised traffic controller. However, is this the correct approach, when the computer has such greater computational power available ? Lim (1983) decided to test this thesis and selected the DS algorithm as the best "all round" classical control system. He rationalised his choice by the observations: that static sectoring algorithms generally allocate a car to a sector without checking if there is a better placed car; they allow a loaded car to serve the sector ahead of it without checking if another car already has this duty; and that the DS algorithms focus attention on individual floors.

Having selected the best conventional algorithm Lim decided to introduce modifications one at a time, and compare the performance to the unmodified system, described in Section 9.4. Four poorly performing characteristics were selected, these were: the optimum counter capacity, free car allocation algorithm, waiting time counter and car journey time. Modifications were then added to the original algorithm and the results observed by simulation techniques, see Chapter 16.

Only slight improvements were possible (Barney and Dos Santos, 1985). It would appear that the DS algorithm[1] is an attractive design for a fixed logic controller, and the improvements attempted by Lim were not very effective.

11.2.5 Summary

The discussion so far allows a number of comments to be made:

(a) Passenger arrivals at floors and passenger destination floors are random.
(b) During any traffic period there is never a total trend towards the classical patterns.
(c) It is important that any control algorithm follows the various changes in demand.
(d) A truly optimal solution is not possible.
(e) The modification of any existing control algorithm is not likely to be profitable.

Thus, it has been seen that conventional algorithms possess limitations and their adoption to computer use restricts the flexibility possible with a computer. Quite rightly this suggests that an entirely new approach is necessary if the computational, logic, programming and data storage abilities of a computer are to be exploited in full. The design of such an algorithm must meet the four primary tasks given in Section 11.2.1 and also obey the five rules.

11.2.6 A New Approach to Computer Lift Traffic Control

The control of lift systems is not an isolated area for the application of digital computers to control problems. Computer control has been applied widely, since the late 1950s, in the process and power generation industries. The application of computers to lift traffic group control was an important step in lift technology. For a long time, relay logic had been used to control lift systems. Digital electronic techniques were initially tried in the 1960s by the lift manufacturers, unsuccessfully, as at that time only discrete component circuit implementation was available, with its inherent low reliability, complexity and noise problems.

The lift industry moved successfully in the 1970s from relay logic controllers to those employing digital computers. Digital computers use integrated circuits and provide very high reliability and allow considerable versatility in the type of controller that can be implemented. The facilities and features contained in a traffic control algorithm implemented in a computer are only limited by the imagination and ingenuity of the algorithm designer, the speed of operation of the computer and by financial considerations. Their maintenance does require a higher degree of knowledge than the normal lift maintenance technician possesses, in order to diagnose and repair faulty components. This difficulty can be overcome by the use of software diagnostic routines leading to on-site board exchanges with further repairs carried out off-site.

---

[1] The proprietary implementation of this algorithm was said to have been designed by the use of simulation.

The first approach by the lift industry to the computer control of lift systems was the software implementation of the existing conventional control algorithms. This has been shown not to be the best approach, as the features presented by conventional algorithms are naturally limited by their fixed logic implementation.

Another approach was to consider a newly devised control system. By 1973 a lift manufacturer had developed and implemented a computer based control system, where the landing calls were assigned to lift cars according to the time each car was estimated to take to answer the call. A number of other features were also included, such as the definition of priority levels, the consideration of high activity floors and a priority service for long waiting calls. See Control Case Study CS15.

Such early attempts at lift programming suffered from the major weakness that lift manufacturers were unable to evaluate fully the practicability and performance of new control algorithms without implementing these algorithms and installing them on the lift system. By then, it was too late to make extensive modifications. The availability of simulation programs solved this problem.

11.3 FEATURES TO BE INCLUDED IN A COMPUTER BASED TRAFFIC CONTROL SYSTEM

This section indicates some of the desirable features which a computer based control system should offer. These features are often impossible to achieve with fixed logic systems.

11.3.1 Computing Car Journey Times

The cornerstone of most computer based traffic controllers is the requirement to estimate the journey time for each lift to reach each landing call as part of the call allocation process. Control Case Studies CS15 and CS16 illustrate this feature.

Definition 11.1: Car journey time[1] is the time it takes a lift to travel to a landing call.

The car journey time is estimated according to certain arithmetic and logical rules. The procedure to evaluate car journey times for a normal landing call considers the car calls, which are already registered and must be honoured, and the landing calls to which the car is currently assigned. These journey times may be calculated on the basis of a direct trip to a landing, or an indirect trip where the car stops at intermediate landings on its journey, in order to service car and/or landing calls. The number of such calls between the lift and the calling floor is important. A car journey once assigned to a call then becomes the estimated system response time for that call.

A car journey time consists of several components:

(1) Interfloor flight time (including acceleration, deceleration, levelling and travel at rated speed).
(2) Door operating times (opening and closing).
(3) Car call and landing call passenger transfer times.
(4) Any door dwell time periods likely to exceed the passenger transfer times.

---

[1] This is *car* journey time, not *passenger* journey time.

For component (1) when estimating the car journey time for moving cars, a correction must be made in respect of the acceleration time.

Each time a car stops it is not known how many passengers will enter or leave the car. This has the effect of making the passenger transfer time component (3) difficult to evaluate. If a lift is picking up landing calls at floors in addition to, or in instead of at, the main terminal, it implies it is during an off peak traffic situation. For reasonable levels of interfloor demand, it will be shown in Chapter 14 (Figure 14.4) that the number of car call stops to landing call stops is about 1.2. Put another way, at 20% of the landing stops, more than one passenger enters the lift. Little error would therefore occur in the estimation procedure if one passenger transfer time is assumed for each car or landing stop.

The destination floors of waiting passengers are not known until they enter the car. This leads to the possibility of car calls being registered between that landing call and another landing currently allocated. The discussion regarding passenger transfer times lead to the conclusion that there could be one car stop for each landing call served — but where ? Clearly the call must be between the landing call floor being considered and the last floor of the building in the direction of service. Under balanced interfloor traffic the average floor number could be half way between the landing floor and the last floor of the building. It would therefore be reasonable to make the assumption of positioning a *fictitious car call* every time a landing call is assigned to a lift, half way along its possible journey.

It is inevitable that a high degree of uncertainty must exist in the positioning of a fictitious car call, however, the effect is likely to be small. When traffic is light there is more chance of a larger error, but it can be tolerated owing to the low demand. Whenever the traffic is heavy a car may have a large number of allocations making an incorrect deduction less significant. In an uppeak traffic situation the feature can be turned off as it is of no benefit. It is in down peak where the error would be largest as the destination floor is not half way down the building, but at the main terminal and the feature should be turned off.

11.3.2 Unbalanced Interfloor Traffic

The random arrival of passengers requesting service gives rise to the possibility of the traffic arrival profile at all floors becoming momentarily unbalanced, although the average traffic could still be considered as balanced. This situation can arise when consistently more passengers arrive at a particular floor than any other floor. It is desirable to still maintain an even service distribution in order to meet the requirements of the ideal control system. The handling of an unbalanced situation requires two actions: detection and correction.

Detection requires the estimation of the number of passengers entering the car at a particular floor. This can be achieved by load weighing. Assume car loads can be weighed to an accuracy of 10% (ie: 10 values over the load range). Then the number of passengers boarding at a particular floor can be detected as the extra load over the original load in the car on arrival at the floor. However, when a car stops at a floor with a landing call as the result of a car call, passengers will leave the car. As has been shown previously this is likely to be one person, hence a realistic estimate of extra load can be made. (An exception is when a car stops at its last car call floor, when all passengers are assumed to exit.) It is now possible to estimate the average number of passengers expressed as a percentage of a fully loaded car entering any car when a car stops for a landing call. This information can be used to correct the condition.

11.3.3 Reduction in the Number of Stops

During balanced interfloor traffic, a car serving a landing call will, in general, make at least one further stop (car call) before becoming free. The number of stops made by cars per unit time indicates the level of loading and any reductions in the number of stops will clearly conserve system capacity. There are two ways in which the total number of stops might be reduced:

(a) By serving passengers wishing to travel to the same floor in the same car.
(b) By serving a landing call where a car call already exists.

The first method is only possible using destination registration stations (see Section 11.4.5). The second method is possible by inserting a special procedure into the basic allocation system. The procedure is activated whenever a landing call being considered coincides with a known car call being served by a particular car. If the expected value of landing call waiting time is not too much longer than the original allocation then the landing call should be allocated to the particular car even if any other allocations are smaller. This has the effect of conserving lift system capacity to serve other demands better.

11.4 ADVANCED CONTROL TECHNIQUES

11.4.1 Fundamental Limitations

It is important to point out that although computer based traffic control systems can allocate lifts more efficiently than the classical traffic control systems, there is a limit to what can be done. The main limit is the finite handling capacity resource of the underlying equipment to handle the traffic demands. This is more dependent on good equipment, which is properly set up, than advanced control systems. Once the major inefficiencies have been removed such as: single button calling; stopping full cars; faulty detection of car loads; inefficient door operations; etc., then it is only possible to "trade" one parameter against another. This means that one passenger's shorter waiting time is another passenger's increased waiting time. The effect on the second passenger could be so small that its deterioration is unnoticed, but the effect on the first passenger could be significant.

This section looks at some advanced control techniques that can be applied individually or in concert.

11.4.2 Minimal Cost Functions

Calls are often allocated to a suitable car using the concept of minimum cost, ie: a cost function.[1] This concept operates by performing a trial allocation to all available cars and allocating the call to the car presenting the lowest cost. There are criteria for selecting a suitable cost function. These can, for example, be based on either Quantity of Service, or Quality of Service, or both. In general terms, the Quantity of Service is a measure of

---

[1] "Cost function" is optimal control theory terminology and its inverse, the "performance index", is sometimes quoted.

the lift capacity consumed to serve a specific set of calls, indicated by the total of the journey times of all the cars. This could be minimised by keeping passengers waiting in a lobby until there were enough passengers to make a trip worthwhile. Airlines apply this principle. The Quality of Service is indicated by the average value of either the passenger waiting time or the passenger journey time (waiting time plus in-car travel time).

The minimisation of waiting time implies putting passengers into the first lift that arrives. This would result in no change from the usual procedure.

The minimisation of the total car travel time implies using the smallest system capacity, which is equivalent to using the smallest possible number of cars. The result of this policy would be very large passenger waiting times, a result which would not be acceptable. This criterion alone is thus not suitable as a cost function.

The minimisation of average passenger journey and waiting times are more acceptable objectives. Both times are interrelated and the minimisation of one might be achieved at the expense of the other. An accurate calculation of passenger journey time can only be achieved if passenger destinations are known at landing call registration time. As conventional two button signalling systems are being considered in this discussion, only passenger waiting time can be minimised. In practice, of course, passenger waiting time cannot be measured. The period of duration for a particular landing call only represents the time the first passenger at that landing has to wait. All subsequent passengers benefit from the first registration and actually wait for less time.

Thus the control algorithm can only (practically) reduce the cost of the system response time to service a landing call. In Chapter 14 the passenger average waiting time (AWT) is shown to be very nearly equal to the lift average system response time ($ASRT$) for balanced interfloor traffic. So for the balanced interfloor traffic condition (the main one) this cost function is very suitable. This cost function is less relevant for unbalanced traffic conditions. For example, during uppeak the $ASRT$ is related to the interval (INT) by values of car load. For down peak traffic the AWT is more linearly related to demand (Figure 13.3). Special measures must therefore be taken to deal with unbalanced traffic conditions.

There are other cost functions that can be employed, for example to minimise energy usage (Peters and Mehta, 1998). The hall call allocation strategy has been shown (So and Suen, 2002a) to use more energy than its simple predecessor. Peters and Mehta show that a parking policy uses almost twice as much energy over one that does not have a policy. As indicated in Section 11.4.1, it is only possible to "trade" one parameter against another.

11.4.3 Stochastic Traffic Control Systems

Control Case Study CS16 is a stochastic[1] based traffic control system, named CGC. It was developed by Lim in 1983 and published (Barney and Dos Santos) in 1985. Observations of classically controlled lift systems have indicated that the response times to answer landing calls follow a curved shape similar to the Exponential Distribution curve of Figure 11.1(a). This distribution curve has a large number of calls answered in zero time or during the first time band. However, there is a long tail to the distribution with some calls waiting very long periods of time.

---

[1] Usually the term "statistically based", meaning "facts systematically collected", is used. In this case it seems more appropriate to use the term "stochastic based", meaning "aim at a mark, guess" !

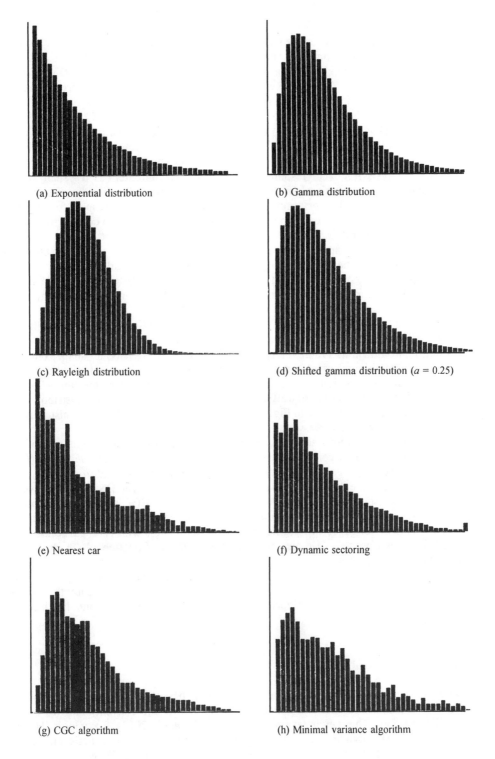

(a) Exponential distribution

(b) Gamma distribution

(c) Rayleigh distribution

(d) Shifted gamma distribution ($a = 0.25$)

(e) Nearest car

(f) Dynamic sectoring

(g) CGC algorithm

(h) Minimal variance algorithm

Figure 11.1 Statistical distributions (after Halpern, 1995)
The x-axis is units of average system response time. The y-axis is frequency.

The underlying premise of the CGC algorithm design was to bring the tail closer to the average and to sacrifice the "instant" collection of some calls by moving the exponential away from the origin to a Gaussian shape similar to the Rayleigh Distribution curve of Figure 11.1(c). The effect would be to give a more even and more consistent service to passengers, by trading the instant response calls for the long tail calls.

What Lim did has subsequently been analysed by Halpern (1992, 1993, 1995). Halpern in his 1992 paper entitled "Variance analysis" introduced a method of analysis based on a Utilisation Factor and a Variance Performance Factor. He showed that a classical traffic control system behaved as a Poisson process. In the 1993 paper ("Variance analysis of hall call response times") he used a computer based system, where he had control of the traffic algorithm. By altering the cost function he was able to improve, and worsen, the performance of the traffic control with respect to response times. He was able to trade off response time against variance, ie: the consistency of the responses. This control he termed Minimum Variance.

In Halpern's 1995 paper ("Statistical analysis ... of call response times") he analyses two classical systems: nearest car (NC) (termed "closest call" by Halpern), and dynamic sectoring (DS), and compares them to three computer based systems: computer group control (CGC) in normal and down peak modes and the Minimum Variance Algorithm (MVA).

It was indicated above that the response time distribution for a classical traffic control system was likely to be exponential, as illustrated by Figure 11.1(a). The ideal distribution would be as Figure 11.1(c), which inclines to a Gaussian shape, but also shows the inevitable tail exhibited by real world systems. A better distribution shape could be a Gamma distribution, as shown in Figure 11.1(b). Here the tail is longer. However, if any computer based traffic control system were to produce a distribution of response times similar in shape to Figure 11.1(d), which is a shifted Gamma distribution, this would be very acceptable.

The other distributions shown in Figure 11.1 relate to Halpern's analysis of the different computer traffic control systems: (e) the nearest car, (f) dynamic sectoring (DS), (g) stochastic (CGC) algorithm and (h) the minimal variance algorithm (MVA). The nearest car algorithm exhibits the classical exponential shape. The DS algorithm, which uses fixed logic, shows a trend towards the ideal. Both the CGC and MVA algorithms more nearly meet the ideal response distribution. Halpern concludes that the computer based systems no longer follow a Poisson process, associated with classical control, but a shifted Gamma process. He also confirms the premise of Section 11.4.1 of a finite (handling capacity) resource.

Thus the intuitive approach by Lim to the CGC algorithm has subsequently been proved mathematically. Control Case Study CS16 gives details and indicates the success of this approach. The CGC algorithm has been implemented by one lift installer (Godwin, 1986).

11.4.4 Other Control Techniques

There are a number of advanced techniques, which can be classified as Artificial Intelligence (AI) techniques and which can be applied to the conventional two button signalling system. These include: expert systems (Qun *et al.*, 2001); fuzzy logic (Ho and Robertson, 1994); dynamic programming (Chan and So, 1996); genetic algorithms (Siikonen *et al.*, 2001; Miravete, 1999); knowledge based systems (Prowse *et al.*, 1992); neural networks (Barney and Imrak, 2001) and optimal control (Closs, 1970). The references are representative of many published papers on this subject and each gives a background to the technique. All these techniques depend on large numbers of events in order to operate satisfactorily. To put these techniques into perspective, consider two buildings, where the uppeak provision is satisfactory.

Building 1 has 8 lifts, serving 16 floors with a total population of 1600 persons. The lift system can handle 15% uppeak arrivals, ie: 240 persons in 5-minutes. This is a significant building, probably prestigious, with a large scale of activity. Statistically, an arrival rate of 0.8 persons per second at one floor (the main terminal) is sufficient justification to claim that the calculations of H and S, in the determination of the handling capacity and interval, are reasonable.

Now consider that 25% of the building population, ie: 400 persons, uses the lifts in a balanced interfloor fashion over a period of one hour. This is an arrival rate of 25 persons, per hour, per floor, or an arrival rate of 0.007 persons per second. This is less than one hundredth of the uppeak arrival rate at the main terminal floor. This level of arrival activity is not sufficient to allow any statistical consideration. Even if it is assumed that 50% of the population use the lifts every hour, ie: an arrival rate of 0.014 persons per floor per hour, it is still not statistically significant.

Now consider Building 2, which has 3 lifts, serving 8 floors with a total population of 400 persons. The lift system can handle 15% uppeak arrivals, ie: 60 persons in 5-minutes. This is a small office building, at the low end of the market place, with an adequate handling capacity. The arrival rate of 0.2 persons per second at the main terminal floor is still sufficient to claim justification for the calculations of H and S.

If 25% of the building population, ie: 100 persons, uses the lifts per hour in a balanced interfloor fashion. This is an arrival rate of 12.5 persons per hour per floor or an arrival rate of 0.003 persons per second. This is just over one hundredth of the uppeak arrival rate at the main terminal floor. Once again this level of arrivals is not sufficient to allow any statistical consideration. At 50% of the population using the lifts every hour, ie: an arrival rate of 0.007 persons per floor per hour, again, not statistically significant.

The above discussion indicates that a statistical approach to a balanced interfloor traffic pattern is not sustainable: simply there are not enough passenger events. The two examples, from each end of the building quality spectrum indicate that if the uppeak sizing is satisfactory then the balanced interfloor activity should provide a similar satisfaction. As the balanced interfloor activity moves into some dominant traffic situation, ie: more traffic at some floors, then the statistical justification improves. Also the demand on the traffic controller lessens as it is more efficient to serve fewer floors with more intending passengers than most floors with one passenger per stop.

Some of the techniques use data gathered over time to predict traffic demand. This would be satisfactory if the demands imposed on a lift system could be relied on to occur reliably each day at the same time. As the number of events is so low, statistically, these techniques are unlikely to improve traffic performance.

11.4.5 New Signalling Methods

The computational potential of the digital computer could be further exploited in order to approach optimal computer control. To achieve this, the supervisory system needs to be presented with additional information related to the passenger destinations. Consider four cases, which are illustrated in Figure 11.2.

Case 1. A car travelling towards a new landing call with a car call for that landing and one car call after that landing, which the new passenger wishes to travel to.

Case 2. A car travelling towards a new landing call with a car call for that landing and one car call after that landing, which the new passenger does not wish to travel to.

Case 3. A car travelling towards a new landing call with a car call after that landing, which the new passenger wishes to travel to.

Case 4. A car travelling towards a new landing call with a car call after that landing, which the new passenger does not wish to travel to.

Case 1 is most efficient as no extra stops are introduced, but the controller does not know the new passenger destination until boarding. Case 2 is efficient at picking up the landing call, but a new car call is introduced, which the controller does not know about until boarding. Case 3 has similar efficiency to Case 2, as a new landing stop is introduced, but the destination is coincident to the existing car call. The controller did not know this until boarding. Case 4 is the least efficient as the controller introduces a new landing stop to pick up the landing call only to discover on boarding that a new car stop will also be required.

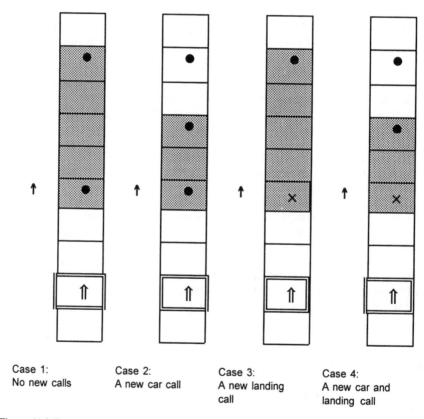

Case 1:   Case 2:   Case 3:   Case 4:
No new calls   A new car call   A new landing   A new car and
                                              call   landing call

Figure 11.2 Four cases when answering a landing call three floors away from an up moving car
Shaded areas indicate original route allocated to the lift.

In all these cases the destination of the passenger making the landing call is unknown, only the direction. A clever traffic supervisor can help with Cases 1 and 2, ie: by using a car already stopping at the landing. Cases 3 and 4 always consume system capacity.

The usual two up/down pushbuttons only give direction information to the traffic controller not ultimate destinations. Could the controller have some intelligence ? Barney *et al.* (2001) suggests that a neural network could help. Neural networks are good at analysing a lot of data over a long period. If a log is kept of the destinations that arise from landing stops, then a weighted table can be produced, which the traffic controller could use to determine which is the best car to send to a landing call. To illustrate this consider Table 11.1.

Table 11.1 Down landing call at Floor 5

0	1	2	3	4	5	6	7	8	9
0.51	0.13	0.02	0.23	0.11		n/a	n/a	n/a	n/a

This shows the likely destinations for a down call at Floor 5 in a 10 floor building. It shows that there is a 51% chance of the call being for the main terminal, a 23% chance of the call being for Floor 3 and so on. Probably a fuzzy logic algorithm would be required to produce the weighted table to deal with fuzzy passenger behaviour such as: not registering a car call as they enter, not entering a car call at all, entering an incorrect car call, getting out at the wrong floor, etc.

11.4.6 Hall Call Allocation

It would be much more useful if the traffic controller knew the intended destination of each landing call. This information could be obtained by replacing the conventional up/down buttons by a panel of passenger destination buttons at each landing, but this needs different fixtures on the landings and a different attitude by passengers. However, most people today are familiar with the digital keypad on telephones, bank teller machines, calculators, etc., so their use should not present a problem. As for the passenger attitude, it has been pointed out that the call registration procedure is not necessarily more difficult to understand than the operation of the usual up/down buttons (Christensen, 1988), and it is probably less of a step in this technological age than when the attendant was made redundant by automatic control and passengers had to operate the lifts by themselves !

The idea of call buttons on the landing was first proposed by Leo Port in 1961 (1961, 1968), but he only had relay logic in which to implement it and could not provide dynamic allocation, only fixed allocation. Installed in two buildings in Australia it functioned in one for some 20 years or more. A dynamic (ACA) system was first described by Closs in 1970 and implemented by a major lift company in 1990 (Schroeder, 1990c), when computer technology had caught up with the ideas. Now installed in many buildings, it has gained acceptance across the world as efficient. Other manufacturers have now applied the technique (Hikita *et al.*, 2001; Smith and Peters, 2002).

Hall call allocation gives the opportunity to track every passenger through from registration to destination. This has great advantages during uppeak as passengers can be grouped to common destinations, as there are large numbers of them. The individual waiting time may increase, the travel time may decrease, but there would be an overall reduction in journey time. During down peak there is no advantage as the destination floor is known. During reasonable levels of balanced interfloor traffic there is little advantage as most landing calls and car calls are not co-incident (see Figure 14.4). However, during an uppeak with some down travelling traffic, or a down peak with some up travelling traffic, there are possible benefits. This leads to a conclusion that an optimum cost (money) system would have a full call registration station at the lobby and other principal floors and two button stations at all other floors. The control algorithm can go into "simple" mode, when dealing with the two button stations by knowing the direction and guessing the destination.

Control Case Study CS17 illustrates in detail the principles underlying this technique.

11.4.7 Conclusions on Advanced Control Techniques

Many of the advanced control techniques employ complex mathematics and involved programming, which makes the practical implementation of the traffic controllers difficult. Also the proper understanding and correct adjustment on site by installation and service persons is doubtful and there is also an increased risk of system unreliability. Powell (2001) states "... the added complexity involved in creating these (neural) networks and putting them into production could not be justified on the (slightly) expected gains in dispatching performance ... over less complicated techniques".

The use of any of the techniques during a dominant traffic flow, such as uppeak or down peak, is unlikely to improve traffic handling over a minimum cost algorithm. The provision of additional destination information, as with call allocation, is unnecessary during light traffic conditions, ie: balanced interfloor, and becomes most effective for heavy traffic situations, particularly uppeak. Then passengers for common destinations can be assembled to travel together. The technique improves the handling capacity for uppeak, but does not assist down peak or interfloor traffic handling (Barney 2000a, 2000b), see Section 12.8.4.

Once a computer is employed to implement the control strategy, the final algorithm is limited only by the imagination and ability (see Section 11.2.6) of the program designer. For example, the search for a "bumpless" transfer of control strategy (see Section 11.2.3) can be dealt with by having one algorithm able to adapt the changing traffic conditions. Also the Hall Call Allocation (see Section 11.4.6) algorithm becomes the Adaptive Call Allocation by detecting when to switch from a waiting time to a journey time cost function. The stochastic algorithm CGC (see Section 11.4.3) could easily be married to the Hall Call Allocation to restrict the allocation of landing calls to those that have waiting for a threshold period of time (see Control Case Study CS16). Learning algorithms can be added to "predict" outcomes and learn to improve the calculation processes such the estimated time to reach a landing call.

All these techniques allow the use of the underlying resource (handling capacity) more effectively for the benefit of all passengers. An added advantage is the systems become more consistent in their response to passenger demands.

11.5 THE ROLE OF THE COMPUTER

To complete this chapter, a review of the computer requirements for traffic controllers is appropriate.

11.5.1 Computer Hardware Requirements

Traffic control is concerned with car movements made in response to passenger demands. To efficiently allocate landing calls to the lift cars, the computer needs to be informed of the position and status of each car. For example, it must know if the doors are open or closed, the car is moving, stopped or out of service, etc. This, together with the landing and car call information, means that the computer must collect a large amount of data. The quantity and variety[1] of the required data required to control a six car group serving 24 floors amounts to some 500 digital input/output signal lines (Barney and Dos Santos, 1977).

To communicate real world events to a computer requires the use of two computer science techniques: interrupt or data scanning. The former applies an electrical signal into the computer hardware. This links with the operating system software causing a cessation of the computer's current activity, in order to deal with the real world event. In the latter technique the computer operating system regularly interrogates (scans) a number of data lines to see if an external event has occurred. Both techniques have their advantages and disadvantages. As the data activity in a lift traffic control system is low and not urgent, ie: very little happens in one tenth of a second, the scanning system is preferred.

The large number of input/output lines makes it necessary to use methods to reduce the number of signal lines necessary to transmit these data bits to and from their points of action. This is best achieved by the use of computer networking techniques.

The choice of computer to control a group of lifts should not present any major difficulties in their selection. Reliability, robustness and cost are the principal factors to take into account. However, the computer to be used should possess a number of important characteristics, namely:

(a) a comprehensive instruction set and good data manipulation facilities to enable efficient programming
(b) a processing speed sufficient to allow the input data to be scanned $5-10$ times per second and all control actions to be carried out
(c) a good input-output and interrupt structure to accommodate the connections to the interface
(d) an expansible configuration to allow additional hardware to be added.

This last characteristic would be necessary should additional algorithms or tasks be included in the system, for instance memory may be added to allow data logging.

11.5.2 Computer Software Structure

The general software structure necessary for the computer control of lifts is indicated in Figure 11.3. The System Supervisor may incorporate, or is a modification of, the computer operating system and will handle all standard peripheral devices. In addition, by accessing the data base, the System Supervisor can determine which control algorithm to select and will initiate each data scan. At every scan, the input signal lines provide information about landing and car calls, car status, etc., and output lines return control commands to the lift system, via the hardware interface. Special software routines must be provided to select the appropriate input/output channels and transfer information between the hardware interface and the data base. An optional traffic monitoring and analysis sub-program can be included to provide run logs of system behaviour and to record incidents on the system for maintenance and other purposes. The structure should be modular, in order to facilitate the alteration or replacement of any modules, as required.

---

[1] Car status (position, direction, load, moving/stationary, door status, car calls), landing calls, etc.

The majority of the software provided to program the structure in Figure 11.3 is standard to most computer control schemes. The design of the standard input-output (I-O) drivers, interface I-O drivers, data base, traffic monitoring and system supervisor sub-modules can be found in many computer and process control texts.

For the lift application area, specific software must be written to provide the traffic control algorithms. A choice of programming language has to be made here. The first traffic control algorithms were programmed in high level languages, such as BASIC or FORTRAN. Nowadays the object orientated languages, such as C++, are the likely choice. It may be necessary, however, to sometimes make use of machine or assembly language code, where speed is required.

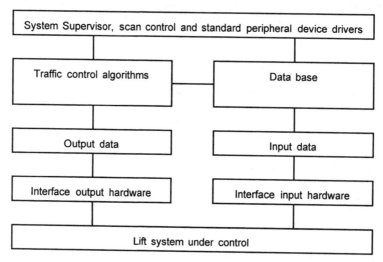

Figure 11.3 Software structure for lift traffic control
Note: signal flow is anticlockwise.

An important advantage of computer lift control is the ease of data collection and logging. The computer can be made to store information on the lift system behaviour and performance, and to print out this information periodically. The performance data can even be used by the supervisory control system to adapt some control algorithm parameters to the traffic requirements.

Nowadays it is possible for any individual car controller to act as a group controller. This is achieved by setting up a network of all individual car controllers. Each receives all the same information (landing calls, car calls, car status, etc.). The possibilities then exist for each individual to "bid" for a new landing call and the best placed car to "win". An additional benefit is that if an individual car goes out of service, or is taken into maintenance or independent service, thus removing it from the group, all the other cars continue to provide a (gracefully) degraded service.

CONTROL CASE STUDIES

Contents

Estimated Time of Arrival (ETA) Traffic Control System

CS15.1 GENERAL PHILOSOPHY

An Estimated Time of Arrival (ETA) digital computer based traffic control system allocates lifts to landing calls, based upon computed car journey times, ie: how long it will take a lift to arrive (Definition 11.1). Early systems of this type, developed in the 1970s, substituted relay or solid state fixed logic by a truly programmable computer. This technique was an obvious one to use once programming facilities were available. The ETA technique remains the underlying basis of many computer based systems on the market. The general philosophy below is that of an actual implementation and gives the opportunity to indicate the various advantages and disadvantages of such a technique. Proprietary implementations will differ and often offer additional features.

CS15.2 OPERATIONAL DESCRIPTION

A primary characteristic of the ETA system is the fast, continuous, collection of lift data and the use of the computational abilities of the computer to process this data, in order to make the control decisions. To do this the system scans the registered car and landing calls, and the position, direction of travel and status of each lift. The scan rate does not need to be faster than every tenth of a second, as little happens on a lift in this time. An estimation of the number of passengers in each lift, based on weighing devices, should be obtained. The time elapsed for every landing call since call registration should be recorded and all the landing call allocations re-evaluated periodically. The frequency of updating can be every second, as this allows sufficient time for re-allocations, except if a lift is already in its slow down sequence.

The procedure for allocation can be as follows. Newly registered landing calls are allocated to the lifts committed to move towards the call in the same direction as the call and also for any uncommitted lifts. Any other lift is regarded as being "in service" and is not considered for allocation. To decide the passenger transfer times, a fixed time of (say) 3 s for each stop due to a landing call is assumed. For car calls, the estimated number of passengers in the lift and the number of relevant car calls is considered. Twice the larger of these two numbers would be assumed as the transfer time in seconds. The landing call is assigned to the lift presenting the shorter car journey time, calculated according to a procedure similar to that described in Section 11.3.1. No account is taken of possible new car calls between the current position of the lift and the new landing call.

The storage and processing of performance data is important in order to make decisions. An evaluation (say) every 60 s should be sufficient. Examples of relevant data would include:

(a) the total number of landing calls answered at each floor during the last 60 s
(b) the total time taken to answer these calls at each floor during the last 60 s
(c) the average landing call waiting time at each floor
(d) the maximum landing call waiting time and the floor at which it occurred.

From the data, the overall average waiting time experienced by the landing calls in the last 60 s and an average landing activity factor based on the number of passengers that required service at each floor can be evaluated. Any landing call that has been waiting for service more than (say) three times the current average waiting time could then be treated as a priority call, and a lift is assigned to make a special run to deal with such a call. Also, any floor presenting an activity level at least (say) three times larger than the current average landing activity could be declared as a high activity floor and is given priority for as long as the condition applies. This facility caters for sudden heavy local demands.

It is useful to declare a number of predefined floors as priority floors and give them permanent service priority. Typical examples are the ground floor, restaurant floors and executive floors, where heavier than average traffic is expected or special service is required. These floors could be defined as parking floors if these were required. The number of priority (or parking) floors should not exceed the number of lifts.

Lifts are therefore assigned to landing calls according to a priority structure. The system deals:

(1) with long wait priority calls
(2) with high activity floors
(3) with priority levels
(4) with the remaining normal landing calls.

In all of these cases the car journey times to the landing call are evaluated and the call assigned to the lift providing a shorter journey time.

A lift assigned to service a long wait priority call, a high activity floor, or a priority level is only allowed to stop in response to car calls before reaching the calling floor. Thus, a different procedure must be adopted to evaluate the car journey times, by considering only car calls to determine the stopping floors and the passenger transfer times. It may happen that a lift which is assigned to service such priority calls discharges its last passengers at a floor where some traffic demand exists. In this case, an information signal would be illuminated in the lift to show its special service condition and request passengers not to enter the lift. The landing call is obviously not cancelled.

There are thus three timed levels of data exchange between the real world of the lift and the digital computer. The first is the basic event scanning at some ten times per second. The second is the updating of call to lift allocations at once per second. The third is the determination of priorities once every minute.

CS15.3 PERFORMANCE

Uppeak traffic situations can be detected using the normal heavy demand feature, by detecting when lifts leave the main terminal floor with a specific loading. For example, three times greater than the average floor activity. Whilst the uppeak condition continues, a number of lifts are despatched to the main terminal, the number of lifts being dependent upon the intensity of the peak demand.

The normal operation algorithm can cope with down peak traffic, although the ETA system is liable to provide a poorer service to the lower floors in the building. Down peak is a heavy traffic situation and, as lifts will not be free for redistribution, "leap-frogging" is liable to occur. This is due to the allocation of lifts to landing calls being made individually and independently, rather than to sectors, thus causing the lifts to frequently bypass landing calls and overtake other lifts. The allocation of lifts based on computed car journey times gives no assurance that the lower landing calls will be given the same service as the upper landing calls, once calls are assigned. Although some account is taken of the time landing calls have remained unanswered in the allocation procedure, this is only partially effective.

Although the algorithm has a concept of time, the normal allocation does not consider landing call waiting times; it is only when a landing call has been waiting too long that it is given priority. This is no more than a "forgotten man" (*sic*) feature, where passengers have to suffer a long wait before being considered for attention.

The system should also include a number of extra features, which are usually provided by good supervisory control systems, namely a car preference key operated switch in each car and "anti-nuisance" devices that prevent an empty lift from answering false car calls.

CS15.4 CONCLUSIONS

The ETA control system can be expected to provide a good uppeak performance. By declaring the main terminal floor as a parking and priority floor, cars will be sent down to deal with the incoming traffic.

The system is not, however, particularly suitable for down peak traffic. If a comparison is made with fixed sectoring systems, similar round trip times are to be expected for the two systems, but the ETA system will present a smaller number of stops per trip, thus it is less efficient.

Under light to medium balanced interfloor traffic conditions, when the free lift parking procedure is active, the system behaviour is very similar to a dynamic sectoring system, and good performance is to be expected. For heavier traffic conditions the redistribution of free lifts is less efficient and the system provides no equivalent to the heavy duty call allocation algorithm available in the dynamic sectoring system. Hence, under heavy interfloor demand levels the ETA system can only be said to be reasonably good, presenting a performance lying between the dynamic sectoring and the fixed sectoring system performances.

It should be emphasised that this control system provides two great advantages, which are inherent to computer control. They are the data logging capability and the possibility of changing the control algorithm parameters by simply re-configurating the computer control program. For example, the number and location of priority floors can easily be altered on site and requires no rewiring. A further advantage is that new control programs can be loaded as they become available, obviating the need to take out the old control system as is required with the conventional systems. On request, the computer can calculate, record and print out the lift systems performance data.

Computer Group Control (CGC) Traffic Control System

CS16.1 GENERAL PHILOSOPHY

The Computer Group Control (CGC) Traffic Control System employs a stochastic control algorithm according to the general principles discussed in Section 11.4.3. Its intention is to provide an even service to all floors, where every landing call is given a fair consideration. This means that the landing call that has been waiting the longest should be given the first consideration for service. To achieve this *egalitarianism*, landing calls are considered to form a queue and will generally be served in order of their waiting time. The system described[1] was designed to serve the balanced interfloor traffic condition, with considerations of the other traffic conditions.

CS16.2 OPERATIONAL DESCRIPTION

Landing calls are arranged, using a computer programming technique, into a first-in, first-out (FIFO) queue. Thus the priority of each call is indicated by its position in the queue, the highest priority being at the head end of the queue (Table CS16.1).

Table CS16.1 Landing call queue

Floor	11	10	8	1	15
Direction	UP	DN	UP	UP	DN
Waiting time	25	21	14	9	5
	Head	–	–	–	Tail

A landing call is taken from the head end of the queue and allocated to a suitable lift using the concept of minimum cost. In order to apply the minimisation procedure it is necessary to calculate the car journey times to travel to a landing call. This is achieved by the procedure given in Section 11.3.1. Part of the procedure is to assume a fictitious car call positioned half way along its possible journey of the lift.

The concept of a high threshold time (*HTT*) is introduced. This time is not a constant, but reflects the current demand into the lift system by expressing it in terms of the average system response time *ASRT*.

$$HTT = KTH \times ASRT \tag{CS16.1}$$

where *KTH* is a constant, and *ASRT* is determined over a suitable recent sampling period T_s. Lim (1983) found suitable values of *KTH* to be 3.0 and T_s to be 90 s.

---

[1] A fuller description is given in Barney and Dos Santos (1985).

The conservation of system capacity can be improved by reducing the number of stops a lift makes. The method adopted is generally as described in Section 11.3.3. If the expected value of landing call waiting time is calculated to be less than the high threshold time, then the landing call can be allocated to another lift, even if the other allocation was smaller. As the high threshold time varies with demand, an even service is provided.

From time to time a particular floor may experience increased traffic demands. These can be detected as indicated in Section 11.3.2. Correction of the unbalanced condition is simple to arrange. The heavy duty floor is moved to the head of the landing call queue, where it will receive immediate treatment. For example, suppose the call at Floor 10 shown in Table CS16.1 were to be given heavy duty status, then the first two head entries in the landing call queue would be transposed.

In some buildings it is necessary to declare certain floors as high priority floors, eg: VIP service or Executive Service. This feature can be dealt with (*sic* reluctantly) by setting a lower and a higher time limit on the waiting time at specified landings. A method is to define the lower and higher limits as proportions of the high threshold time, as Table CS16.2. This ensures the other floors still receive a reasonable service

Table CS16.2 High priority floor list based

Floor	10	5
Lower limit	0.5 *HTT*	0.4 *HTT*
Higher limit	0.8 *HTT*	0.9 *HTT*

The computer group control algorithm can be summarised as follows:

(1) A lift is allocated to a landing call according to least cost except:

    (i)   if another landing call has exceeded a specific, but demand varying, threshold value, and is also allocated to the same lift, the call will not be allocated to that lift

    (ii)   if another lift is scheduled to stop for a car call at the landing being considered and the landing call is estimated not to wait longer than a specific, but demand varying, threshold value, it will be allocated to the other lift.

(2) Landing calls are considered for allocation by examination of a landing call queue, landing calls at the head of the queue being allocated first.

(3) The landing call queue is normally arranged on the basis that the longest wait call is at the head of the queue except:

    (i)   a heavy duty floor landing call may be brought to the head of the queue

    (ii)   a priority floor landing call may be brought to the head of the queue.

(4) The landing call queue is arranged so that the various categories are ordered from the head of the queue as follows:

    (i)   Priority call
    (ii)   Heavy duty call
    (iii)   Long waiting call
    (iv)   Normal call.

CS16.3 PERFORMANCE

The CGC algorithm is designed to handle the interfloor traffic condition, as it exists for much of the working day. Uppeak traffic and down peak traffic are considered special cases of interfloor traffic. A comparison to the dynamic sectoring (DS), which has been shown to be the best performing fixed logic algorithm, is shown in Figure CS16.1, which shows that the CGC algorithm answers less calls immediately and the tail is smaller and smoother.

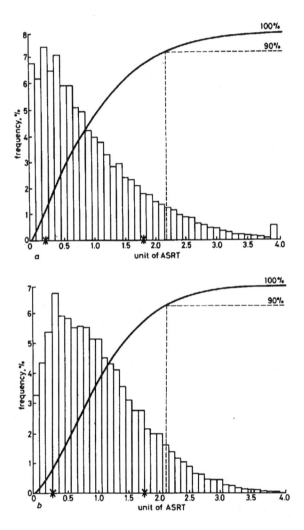

Figure CS16.1 Comparison of interfloor performance DS (upper graph) and CGC (lower graph) algorithms

The CGC algorithm performs in a similar manner to all the other algorithms during uppeak traffic.

The down peak performance of the CGC algorithm is very much poorer than any other algorithm. It would appear that an algorithm designed to perform well under balanced interfloor traffic does not serve down peak adequately and vice versa. It can be concluded that in its attempt to provide an even service to all floors during the down peak period, it does so by sacrificing performance. Optimal performance is thus not achieved. Lim (1983) suggests "that lifts be encouraged to serve adjacent calls". This would require a modified cost function which first minimised the average waiting time and secondly minimised lift journey time between two subsequent landing calls.

CS16.4 CONCLUSION

Lauer (1984) suggests that once lifts respond to demands in a theoretically optimum manner then "further invention or investigation in the area of elevator (lift) supervisor logic would be pointless". As has been shown in the considerations of the design of the CGC algorithm, the consideration of a "bumpless" transfer between different traffic conditions prevents optimality.

CS16.5 EXAMPLE

Consider a 16 floor building served by 4 lifts with lift positions and a landing and car call pattern as given in Figure CS16.2 and a call table as given by Table CS16.1. How will the lifts serve these calls ?

To consider this problem a number of assumptions must be made.

Assume only one passenger with a transfer time of 1.2 s enters or exits at each stop. The single floor flight time is 5.0 s, the door operating time is 3.8 s, and thus the cycle time is 10.0 s. For each floor jumped, add 1.0 s. Delete 1.0 s if a lift is moving.

Assume all calls are being completely re-allocated and calls are taken from the queue shown in Table CS16.1.

Taking each landing call in turn, the service times can be calculated. As an example, take the landing call at the head of the queue, ie: landing call at Floor 11-UP.

Lift 1: $10 + 6 - 1 = 15$ s   (lift moving)

Lift 2: $(10 + 3 - 1) + (10 + 2) + (10 + 8) = 42$ s   (lift moving)

Lift 3: $10 + 3 = 13$ s   (lift stationary)

Lift 4: $10 + 4 = 14$ s   (lift stationary)

Table CS16.3 can now be formed where the bold entries in the double lined cells indicate the allocations made. It is important to note that some later allocations affect earlier ones. For example, the landing call at Floor 15 allocated to Lift 4 will cause the landing call at Floor 10 to wait 25 s.

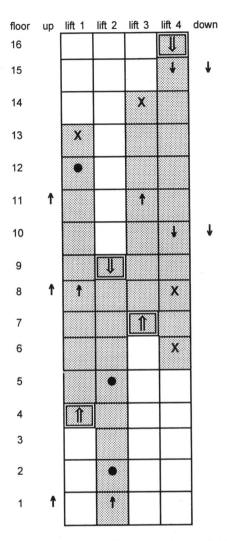

New symbol: **X** fictitious car call. Shaded area shows route taken by lift.

Figure CS16.2 Final disposition of lifts and calls for the example

Table CS16.3 Calculation of landing call times and allocation to lifts

Floor	Wait (s)	Lift 1	Lift 2	Lift 3	Lift 4
11 UP	25	15	42	**13 (1)**	14
10 DN	21	27	41	38	**15 (2)**
8 UP	14	**12 (3)**	39	19	39
1 UP	9	56	**34 (4)**	47	42
15 DN	5	46	66	35	**10 (5)**

Fictitious Car Calls at: (1) Floor 14; (2) Floor 6: (3) Floor 12; (4) Floor 9; (5) Floor 8.

Adaptive Hall Call Allocation (ACA) Traffic Control System

CS17.1 GENERAL PHILOSOPHY

This traffic control system utilises additional information obtained from the passenger as to their destination, not just their direction. A floor calling station must be placed on each floor to allow passengers to select their destination floor.

CS17.2 OPERATIONAL DESCRIPTION

The call allocation algorithm implies a different method of call registration. Instead of the usual two up/down buttons, Figure CS17.1(a), a panel of touch buttons, Figure CS17.1(b), are required at every landing. This can be the familiar telephone key pad. Thus, a passenger arriving, for instance, at Floor 3 and wishing to travel to Floor 11 touches button "1" twice. Within a very short time (<1 s), the passenger would receive an indication on a display beside the call registration station showing which lift has been allocated to their call. No destination buttons are necessary in the car, but an indicator inside the car shows the floors at which the lift is stopping.

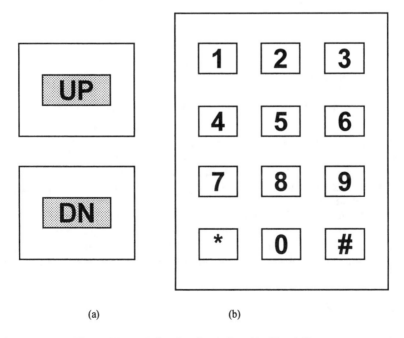

(a)   (b)

Figure CS17.1 Call registration stations (a) old and (b) new

This type of call registration can be adapted to be multi-functional. For example: the * (star) key could be a − (minus) key and be used to enter below main terminal floors. The # (hash) key could be the key to indicate a handicapped person wished to use the lifts and an empty lift could be dispatched. Engineering keys could be programmed, such as pressing the star and hash key together. There are many opportunities.

The computer control algorithm obviously requires that every passenger registers their destination calls. The computer algorithm used to allocate a landing call to a lift is simple to implement. Each time a new call is registered, the computer allocates it in turn to each of the lifts and evaluates the cost of each allocation. The allocation giving the lowest cost is then adopted. Suitable cost functions are passenger average waiting time, passenger average journey time or a combination of both.

CS17.2.1 Waiting Time Cost Function

Consider that a new call is to be allocated to a system of L lifts, each lift (I) with $N(I)$ calls to answer and $WT(I)$ accumulated waiting time for the $N(I)$ calls.

Assume that $NWT(K)$ is the new accumulated waiting time for $N(K) + 1$ calls, when the new call is allocated to lift K.

The average waiting time for the complete set of calls is:

$$AWT = \frac{NWT(K) + \sum_{I=1, I \neq K}^{L} WT(I)}{1 + \sum_{I=1}^{L} N(I)} \qquad \text{(CS17.1)}$$

Which can be written:

$$AWT = \frac{NWT(K) - WT(K)}{1 + \sum_{I=1}^{L} N(I)} + \frac{\sum_{I=1}^{L} WT(I)}{1 + \sum_{I=1}^{L} N(I)} \qquad \text{(CS17.2)}$$

As the two summations in Equation (CS17.2) do not depend on the allocation K, the minimisation of AWT only requires the minimisation of the term $NWT(K) - WT(K)$. This simplifies the evaluation of the cost function, as only this incremental cost is to be evaluated instead of the whole expression for AWT. The quantities $NWT(K)$ and $WT(K)$ are evaluated by simulation.

It should be noted that the incremental cost $NWT(K) - WT(K)$ is made up of several terms. It includes the waiting time for the new call and the increase in the waiting times of calls already allocated to lift K, the extra passenger transfer time resulting from the new call, and any extra stops to pick up and discharge the new passenger.

CS17.2 Journey Time Cost Function

The same concepts apply when average journey time is the cost function. In the equations substitute JT for WT.

CS17.2.3 Average Journey Time with Maximum Journey Time Constraint

A third type of cost function, proposed by Closs (1970), uses average journey time with a maximum waiting time constraint. It operates by costing each allocation against an average journey time cost function, but penalising any solution for which the waiting time of the new call exceeds a predefined value (*MWT*). The algorithm operates as follows:

(1) Evaluate cost of allocation of the new landing call to lift 1:

$$COST(1) = NJT(1) - JT(1) \tag{CS17.3}$$

where $NJT(1)$ and $JT(1)$ have similar meanings to $NWT(1)$ and $WT(1)$, but concern journey times.

(2) Compare the new call waiting time $NCWT(1)$ with the predefined value MWT. If it is smaller than MWT, then $COST(1)$ is not altered, but if it is greater a penalty is added to the cost:

$$COST(1) = COST(1) + \text{penalty} \tag{CS17.4}$$

and the penalty is made up of a fixed value added to a term proportional to the excess of waiting time above MWT. For example:

$$\text{penalty} = 300 + 10 \ (NCWT(1) - MWT) \tag{CS17.5}$$

(3) Repeat the procedure from 1 for all lifts.

The effect of using a penalty is to force the elimination of lifts with a high number of allocations, making it easier to select a more lightly loaded lift. Of course, if all lifts exceed the limit value, a suitable allocation will always be found in terms of minimising the new call waiting time, as the cost (using the same parameters as above) becomes:

$$COST(K) = COST(K) + 300 + 10NCWT(K) - 10MWT$$

$$= COST(K) + 10NCWT(K) + \text{constant} \tag{CS17.6}$$

where $10NCWT(K)$ is the most significant term of those depending on the allocation of K.

A further feature is still necessary in the call allocation control algorithm. After registering the required destination floor and receiving a reply as to which lift will service the landing call, a passenger must walk to the lift. Thus, the allocation procedure must allow sufficient walking time for the passenger to reach the lift from the landing call station when allocating the landing call to a lift.

CS17.2.4 Reduction in Number of Stops

The call allocation algorithm causes calls requesting the same destination floors to be carried by the same lift. This has the effect of reducing the number of car stops. However, in some cases the cost of allocating a new landing call to a lift already stopping at the calling landing or destination floor is marginally greater than the cost to allocate the call to another lift not stopping at either floor. Although the allocation is perfectly proper, it might be better not to allocate the new call to the lift with the lowest cost, as by not doing so capacity is reserved for future calls.

To cater for this idea a penalty $p\%$ is introduced for each extra stop motivated by the new call. To prevent operation of this penalty under low traffic conditions, the penalty is made dependent on the incremental cost of the allocation and is proportional to car load.

$$\text{penalty} = \frac{p}{100} \times \text{incremental cost} \times \frac{\text{load}}{CC} \qquad (CS17.7)$$

where CC is the rated capacity and the load is measured as the average value of the number of passengers inside the lift, or queuing for service. The procedure improves performance for values of p up to 10%. For larger values of p the algorithm is self defeating, as it produces less appropriate allocations.

CS17.2.5 Dynamic Uppeak Subzoning

During an uppeak, the obvious cost function to implement with call allocation is journey time. This is because a waiting time allocation criterion would do no more than allocate every new call to the first available lift at the main terminal which possessed space capacity, in the same way as the collective-distributive algorithm. If journey time is the cost function, calls terminating at the same floor tend to be allocated to the same lift, hence reducing the number of stops per trip and the round trip time. The system handling capacity is increased and the main terminal floor more frequently served. However, a waiting passenger may not board the first available lift, and this may produce increased waiting times. The overall effect is that better journey times are produced, in comparison to conventional algorithms, for the whole range of traffic intensities, but under some circumstances it can result in longer waiting times, mainly for light traffic originating at the main floor.

It will be shown in Chapter 12 that uppeak subzoning is sometimes used by conventional group control systems to improve the uppeak handling capacity. Subzoning is very sensitive to the way in which the zone partition is arranged and should ideally be adjusted for every traffic situation. As in practice a fixed partition is implemented, it cannot respond to the wide fluctuations found in arrival traffic patterns. Knowing the advantages of uppeak subzoning, and the adaptability of a computer implemented algorithm in coping with input traffic variations, a dynamic subzoning concept can be implemented in the ACA system. The building is divided into three subzones, as shown in Figure CS17.2.

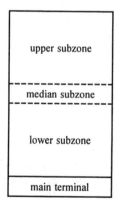

Figure CS17.2 Uppeak subzoning for the ACA traffic control system

The lifts are divided into two subgroups, one for the lower sector and the other for the upper sector. No indication of this partition is given to the passengers. A newly registered landing call is allocated to a lift in the usual way, by evaluating the costs of the allocation of the call to every lift and choosing the allocation giving the lowest cost. However, during the evaluation of the cost, the allocation of a call registered for the lower subzone to a lift allocated to the upper subzone is penalised, and so is the allocation of a call with a destination in the upper subzone to a lift in the subgroup serving the lower subzone. The penalty, which is added to the cost of the allocation, is a function of the load of the two subgroups of lifts, and can be expressed as:

$$\text{penalty} = (1 + \frac{b}{100})k \qquad \text{(CS17.8)}$$

where k is a constant value and b measures the imbalance of lift loads between the upper and lower subgroups as a percentage of the highest subgroup lift load.

The fact that the loads of the two subgroups of lifts are taken into account contributes to equalise these loads. For example, the allocation of a call terminating at a floor in the lower subzone to a lift assigned to the upper subzone can be penalised by a quantity ranging from zero, if all the upper subzone lifts are idle, to $2k$ if the lower subzone lifts are idle.

A call registered to the median subzone can be allocated to either subgroup of lifts, with preference for the subgroup with the smallest load. The allocations to the lifts assigned to the heavier loaded subgroup are penalised by a quantity which equals the absolute value of b multiplied by k.

A correction mechanism allows this technique to deal with extremely unbalanced traffic destinations, as if excessive unbalance between the subgroup loads is detected, the subzone limits are automatically adjusted.

CS17.3 SYSTEM PERFORMANCE

Waiting time has proved to be the most suitable cost function to implement in the ACA control system under various traffic patterns, with the exception of uppeak, for which journey time is used.

The ACA system is able to detect a traffic situation where most traffic originates at the ground floor, independently of the level of traffic intensity, and can automatically switch the cost function from waiting time to journey time. This will provide very good journey time performances, although under light up traffic situations the waiting times may be considered unacceptable, because sometimes a call is not allocated to the first lift to depart. However, it can be considered that waiting times are psychologically more important than journey times, and that under light traffic conditions it is not important to reduce the number of stops, and therefore waiting time is the appropriate cost function. To cater for these considerations, the ACA system can use waiting time as cost function until heavy uppeak traffic is detected, when it switches to journey time.

A lift system under adaptive call allocation control presents increased uppeak handling capacities. As a result, it can cope with uppeak traffic intensities that would saturate a similar lift system under conventional control. Average journey times are improved over the whole range of traffic intensities, and average waiting times are improved for the heavy traffic demands.

The performance of ACA under interfloor traffic is substantially better than for conventional systems.

A similar situation occurs for medium to heavy down peak traffic. Under heavy down peak traffic, where the knowledge of the passenger destination floor is no advantage as most passengers have destinations at the main terminal, the average waiting times presented by the ACA algorithm may exceed those presented by a fixed sectoring algorithm. This reflects the handicap resulting from the necessity of fixing the allocation of a new call at call registration time.

The ACA system presents a powerful capability to adapt to unbalanced and changing traffic circumstances. It also possesses those advantages inherent to computer control, namely the flexibility and ease of algorithm modification without any rewiring being performed.

CHAPTER TWELVE

Uppeak: Dissection and Paradigm

12.1 INTRODUCTION

Uppeak traffic sizing defines the underlying capability of the lift installation. This size should be decided by the requirements of the target building, whether speculative or high prestige. Once the uppeak capability has been decided the quantity and quality of the service in all the other traffic conditions is also decided. This chapter discusses some considerations and techniques that can be applied to the uppeak traffic condition normally encountered during the morning incoming traffic. Some of the sections answer pragmatic questions, but others are simply for curiosity.

12.2 UPPEAK FORMULAE

Two major parameters in the evaluation of the round trip time equation are the probable number of stops (S) and the highest reversal floor (H). Both these parameters are dependent on the number of floors served (N) and the number of passengers in the lift (P). The number of passengers is dependent on their arrival rate and the assumed probability distribution function (pdf) of the passenger arrival process determines how they arrive.

Consider the simplest arrival process, which is the rectangular pdf. This is a pdf where there is a simple linear relationship defining the times of passenger arrivals. The following formulae have been derived and are shown in Table 12.1 for reference.

For equal demand: S (see Equation (5.5), Section 5.3)
H (see Equation (5.12), Section 5.4)

For unequal demand: S (see Equation (7.6), Section 7.3.1)
H (see Equation (7.13), Section 7.3.3)

Table 12.1 Formulae for S and H for rectangular pdf

Demand	Probable number of stops (S)	Highest reversal floor (H)
Equal	$S = N \left[1 - \left(\dfrac{N-1}{N} \right)^P \right]$	$H = N - \displaystyle\sum_{i=1}^{N-1} \left(\dfrac{i}{N} \right)^P$
Unequal	$S = N \left[1 - \dfrac{1}{N}\displaystyle\sum_{i=1}^{N} \left(1 - \dfrac{U_i}{U} \right)^P \right]$	$H = N - \displaystyle\sum_{j=1}^{N-1} \left[\displaystyle\sum_{i=1}^{j} \dfrac{U_i}{U} \right]^P$

These formulae, derived using the rectangular pdf, considered the probability of a particular floor having a specific attraction.

Consider the Poisson pdf. This is a pdf where there is an exponential relationship defining the times of passenger arrivals. The following formulae have been derived and are shown in Table 12.2 for reference.

For equal demand: S (see Equation (5.15), Section 5.6.2)
                              H (see Equation (5.16), Section 5.6.2)

For unequal demand: S (see Equation (7.15), Section 7.3.1)
                              H (see Equation (7.16), Section 7.3.3)

Table 12.2 Formulae for S and H for Poisson pdf

Demand	Probable number of stops (S)	Highest reversal floor (H)
Equal	$S = N\left(1 - e^{-\frac{\lambda INT}{N}}\right)$	$H = N - \sum_{i=1}^{N}\left(e^{-\frac{\lambda INT}{N}}\right)^{i}$
Unequal	$S = N - \sum_{i=1}^{N} e^{-\lambda INT \frac{U_i}{U}}$	$H = N - \sum_{j=1}^{N}\prod_{i=N-j+1}^{N} e^{-\lambda INT \frac{U_i}{U}}$

These formulae, derived using the poisson pdf, considered a time related exponential distribution.

Notice that Equations (5.15) and (5.16) are not presented exactly as shown in Section 5.6.2. The identifying subscript p is omitted (trivial). More importantly the term λT has been replaced by λINT, as the time interval T used by Tregenza (1972) and Alexandris (1976) can be related to the lift system interval INT. This allows another substitution as λINT is P, as defined by Equation (5.17). This brings all the equations into the same form and some further mathematical approximations can be carried out (not shown here) to show their close equivalences.

12.3 LOW CALL EXPRESS FLOOR

When evaluating the round trip time Equation (4.11) various parameters are required such as H, S and P. The parameter H is the high call reversal (HCR) floor, which the lift reaches on an average journey during a round trip. It is important to derive a value for H accurately (Section 5.4) and not assume it to be N for a low to mid rise building zone or $N - 1$ for a tall building zone as some designers do. A pedant will then question:

"Where is the first stopping floor that the lift serves during an uppeak trip ?"

It might be assumed to be the floor next to the main terminal. But it may not be. The reason for the question is that Section 7.4 indicated that a lift may not reach its rated speed after a single floor run, particularly for high speed lifts. Thus an error in the round trip calculation may arise. The question has no great relevance where the served zone is not immediately adjacent to the main terminal, as the time to travel the express zone will be large. But what is the magnitude of the error ?

Kavounas (1992a, 1993a, 1993b) examined this question, terming the first stopping floor the "Lowest Call Express (LCE)" floor. The first paper briefly reported his findings and the latter two give the theory more fully. The problem Kavounas faced is illustrated in Figure 12.1. Here LCE is given by the symbol h. For equal floor populations he first "rephrases" Equation (5.12) as:

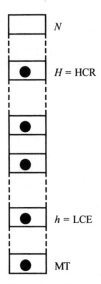

Figure 12.1 Illustration of Lowest Call Express floor

$$H = N - X \qquad\qquad (5.12\text{bis})$$

and then says:

$$h = 1 + X \qquad\qquad (12.1)$$

By following the principles applied in Section 5.4, and after a great deal of mathematics, Kavounas shows that X is:

$$X = \sum_{i=1}^{N-1} \left[\frac{i}{N}\right]^{P} \qquad\qquad (12.2)$$

Kavounas comments that the term "1" represents the lowest possible floor that can be served, ie: the first one above the main terminal, and considers that the term X represents the reciprocal nature of the problem. For unequal populations, Kavounas derives a similar expression:

$$h = 1 + Y \qquad\qquad (12.3)$$

where:

$$Y = \sum_{j=1}^{N-1} \left[\sum_{i=j+1}^{N} \frac{U_i}{U}\right]^{P} \qquad\qquad (12.4)$$

It should be noted that the inner summation range of the equation for Y differs from the corresponding part of Equation (7.13) given below. This is due to the inner summation for H being concerned with the influence of the populations of the lower floors, whereas the inner summation for h is concerned with the influence of the populations of the upper floors. The numerical difference is small.

$$H = N - \sum_{j=1}^{N-1} \left[\sum_{i=1}^{j} \frac{U_i}{U}\right]^{P} \qquad\qquad (7.13)$$

The result indicates that the first stopping floor is as far from the main terminal as the high reversal floor is from the last floor in a building zone. This gives symmetry to the car call pattern, which is of little consequence except for small lifts serving many floors. For example, a 10 person lift serving 22 floors would have a stopping range from Floor 2 to Floor 20 (see Table 5.1).

Kavounas (1993b) concludes:

"Although the LCE consideration always affects some of the components of the *RTT* summation, it is often in ways that neutralise each other. Therefore, the *RTT* may often turn out to be virtually unaffected by the LCE consideration. This is why the LCE principle was originally termed (1992) a third order consideration."

The pedant should now be satisfied.

12.4 THE 80% CAR LOADING FACTOR

Conventional traffic design calculations assume that lifts only load, on average, to 80% of their actual capacity (Definition 7.3). Note that the actual capacity is not the simplistic rated load (kg) divided by 75 (Definition 7.2), which ensures safe operation, but the available space divided by 0.21 m^2 (see Table 7.2). But why has 80% been chosen ?
Some workers (Forwood and Gero, 1971; Tregenza, 1972; Zimmermann, 1973) say it is normal practice based on experience. Strakosch (1967)[1] states that passengers left to themselves "will seldom fill an elevator to more than that (80%) during uppeak" and in later editions (1983, 1998)[1] offers the advice: "Actual loading is less than the weight allowed since people will not crowd that close together". Loading a service facility to 80% is a well known classical statistical technique used in bulk-queue, multi-server facilities, when a good service is desired. It would possible that someone in the lift industry long ago with statistical experience, whose name is lost to posterity, suggested 80% be used. However, can the 80% figure be proved ?
Barney and Dos Santos (1977) set out to investigate this assumption by computer simulation. Figure 12.2 was plotted, as the result of over 400 simulations for groups with from four to six lifts. Here the vertical axis indicates the uppeak performance figure as *AWT/INT* and the horizontal axis shows the corresponding percentage car load. The value for the passenger average waiting time (*AWT*) was obtained from the simulation. The value for the interval (*INT*) and the percentage car load was calculated using the procedure described in Section 6.11. This procedure matches the actual arrival rate to the lift system handling capacity. It thus provides values for the interval and the percentage car load when this match is made. The relationship for the uppeak performance figure *AWT/INT* is chosen in order to normalise the results for all lift systems regardless of their size, floors served, dynamics, etc. Each point plotted (■) represents one simulation.
Some interesting facts emerge at once from Figure 12.2. First, there is a spread of points, due to the randomness of the passenger arrivals and destinations. For example, some simulations approach the 100% performance figure at only a 60% value of car load. Secondly, there is a definite pattern. A weighted average performance figure line can be

---

[1] 1st edition, page 64; 2nd edition, page 74; 3rd edition, page 84.

superimposed on the graph, as shown. It indicates the increase of *AWT/INT* with an increase of car loading. A large number of the simulations present a performance figure smaller than 50% at low car loadings. As the car loads increase, this excellent performance gradually vanishes and at a 60% percentage load the average performance figure is approximately 50%.

For the car loads above 70% there are some runs showing a performance figure in excess of 100%. The number of such runs for percentage loads up to 80% is small and probably reflects the effect of car bunching (see Section 12.5). The average performance is still good in this range. The number of poor runs increases for loads above 80% and for 90% loads the average performance figure exceeds 100%. Above 90% percentage load the situation deteriorates very rapidly and only a few runs give an acceptable performance. Such runs will correspond to the lifts running with a constant headway and large queues building up. Thus, the uppeak lift system performance depends on the car loading and for heavy percentage loads, say above 90%, average waiting times in excess of an interval time must be expected. Figure 12.2 is presented in an idealised form as Figure 6.1, where the spread of values is indicated by dashed lines.

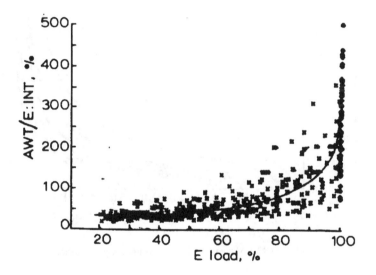

Figure 12.2 Uppeak performance from simulations

Computer simulation is an empirical modelling method: another method of analysis is by mathematical modelling. A notable feature of Figure 12.2 is a "knee" at a system utilisation of 80% (as characterised by percentage car load), above which the performance deteriorates rapidly, leading eventually to intolerable queuing situations. The form of this curve is by no means unique and, as stated above, occurs with bulk-queue, multi-server facilities. This leads to the proposition that a theoretical derivation of Figure 12.2 might be possible.

Alexandris *et al.* (1979) considered a multi-car lift system as a bulk service queuing problem. To perform the mathematical analysis, certain model assumptions were made:

(1) Only the uppeak traffic pattern is to be considered.
(2) Lifts do not leave without passengers.
(3) Lifts return to the main terminal, even when there are no calls.
(4) There is no limit to the length of the queue of passengers waiting for service.

(5) Passenger arrivals obey the Poisson process.
(6) There are no priorities, passengers use whichever lift becomes free.
(7) The queue service discipline is first-in, first-out.
(8) Service is by batches of size no greater than the rated capacity of the lifts.
(9) The service time for each batch is exponentially distributed.

All of these assumptions are reasonable as most proprietary controllers operate on the basis of assumptions (2), (3), (6) and (8), and passengers often behave according to assumptions (5) and (7). Complex mathematical processes (Alexandris, 1977) resulted in the curves shown in Figure 12.3 being produced. This figure shows a facility utilisation factor R (horizontal axis) plotted against mean passenger waiting time (vertical axis). The number of lifts in the groups ranged from one to eight. The mean passenger waiting time was normalised by dividing it by the lift system interval time. The graphs are similar in shape to Figure 12.2.

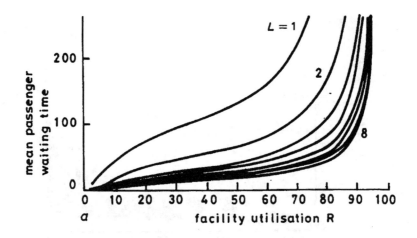

Figure 12.3 Mathematical model of uppeak performance

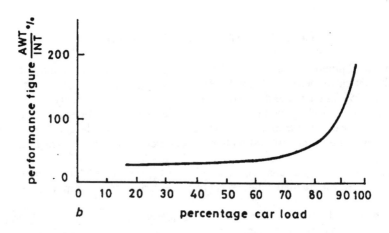

Figure 12.4 Simplification of Figure 12.3

The simulation runs from which Figure 12.2 was constructed assumed groups of from four to six lifts. So, by eliminating those curves for numbers of lifts ranging from 1 to 3 in Figure 12.3 and averaging the rest, Figure 12.4 can be drawn. The shape of this curve is very similar to that of Figure 6.1. Thus, justification for Figure 12.2, which was obtained empirically by simulation, is confirmed by mathematical analysis.

12.5 TRAFFIC ANALYSIS – THE INVERSE S-P METHOD

Often there is a need to know the number of passengers in a lift, for example when deciding whether the sizing is correct, or when an estimate of bunching is needed, or in order to estimate the likely passenger waiting time. Several methods exist to determine the number of passengers in a lift, including: load weighing, photocell signals, sensitive pads, imaging systems and observers. Al-Sharif (1992) suggests another method.

In the design of lift systems the traditional design method is to derive the round trip time from various parameters. One of these parameters is the number of stops (S), which for equal floor populations is given by:

$$S = N \left[1 - \left[\frac{N-1}{N} \right]^P \right]$$

(5.5)

Equation (5.5) can be inverted to produce a formula for P:

$$P = \frac{\ln \left[\frac{N-S}{N} \right]}{\ln \left[\frac{N-1}{N} \right]}$$

(12.5)

Al-Sharif validates the method thoroughly in his paper and shows it has a close correspondence to real systems. Thus, a very simple method of "load weighing" is obtained, which could be easily incorporated into traffic supervisors enabling them to switch algorithms to meet traffic demands. It is interesting to note that Basset Jones (1923) used this formula for another purpose, in order to determine the variance of S from the expected value E(S).

12.6 BUNCHING

Uppeak calculations are based on average values for the various parameters. Deviations from these average values in the actual lift installation will cause deviations from the calculated performance. One phenomenon that prevents optimum performance is "bunching". Ideally, lifts arrive to transport passengers at the main terminal with a separation time equal to an average interval. In practice this does not happen and lifts arrive with an irregular interval. In an extreme case all lifts in a group could arrive and leave simultaneously, like a huge single lift with a capacity equal to the sum of their individual capacities.

Early traffic controllers dispatched lifts from terminal floors with a fixed headway. This was satisfactory provided the lifts returned before the headway time had expired. A disadvantage was that the introduction of a headway larger than the underlying average interval reduced the underlying handling capacity. The technique of using a lobby dwell time during intense uppeak traffic works well, as it ensures lifts leave the main terminal substantially filled to full capacity. There is no degradation in handling capacity, as long as the lobby dwell time is shorter than the time to load a lift. Modern hall call allocation systems, where passengers register their destinations and are then allocated a car, will not, by definition, suffer from bunching.

Al-Sharif (1993) illustrates bunching by considering two lifts with a round trip time of 50 s serving a passenger arrival rate of one person per second. If the lifts arrive uniformly in time, ie: every 25 s, then the average waiting time might, simplistically, be half the interval, ie: 12.5 s and 25 persons will leave in each lift. If the first lift now arrives at 10 s and the second at 50 s, then the first lift transports 10 people ($AWT = 5$ s) and the second transports 40 people ($AWT = 20$ s), assuming it is big enough. The total average waiting time will be the weighted average, ie:

$$AWT = (10/50 \times 5) + (40/50 \times 20) = 17 \text{ s}$$

which is longer than the theoretical 12.5 s.

Suppose now that the lifts do not have infinite capacity, but can only accommodate 25 persons, then 15 persons will be left to wait another 10 s for the next lift. They will be joined by a further 10 persons ($AWT = 5$ s). The weighted average AWT will then be:

$$AWT = (10/60 \times 5) + (25/60 \times 20) + (15/60 \times (20+10)) + (10/60 \times 5) = 17.5 \text{ s}$$

Thus bunching does not matter too much if all waiting passengers can board the lifts to reach their destinations. The average interval may still be as designed but the average passenger waiting time will increase. Provided all waiting passengers are served, the handling capacity of the lift installation is not affected. However, if passengers are left behind then the handling capacity has been compromised.

Schroeder (1990a) considers bunching to be one of the reasons for the rapid increase in passenger average waiting times, when cars load to greater than 50% of their capacity. He suggests estimating bunching by forming the ratio: actual passenger average waiting time divided by the average waiting time that would be obtained if the linear relationship $AWT = INT/2$ applied, ie: $2AWT/INT$. This ratio is unity for low car loadings (<50%) and a large number as car loadings increase to 100%. It would be better to invert the ratio, ie: $0.5INT/AWT$, so that unity represents no bunching and zero represents total bunching. This method requires the acquisition of values for interval and passenger average waiting time.

Figure 6.1 illustrates uppeak performance with the value of INT on the vertical axis being the actual value the lift provides in response to each level of car loading. Table 6.4 tabulates Figure 6.1. Considering all car loadings below 50% to give a value of 0.5, Table 6.4 can be converted to give the inverted Schroeder table, *viz*:

Table 12.3 Bunching factor (after Schroeder, 1990a)

Car loading	50%	60%	70%	80%	85%	90%	95%
AWT/INT (Table 6.4)	–	0.50	0.65	0.85	1.01	1.30	1.65
Bunching ratio	1.0	1.0	0.8	0.6	0.5	0.4	0.3

All figures are rounded

Table 12.3 indicates that, at the usually assumed car loading of 80%, the bunching factor is 0.6. This would appear to be the limit of passenger tolerance under the Schroeder method. Schroeder suggests that, as the number of lifts in a group increases, then the bunching ratio will worsen. However, the irregularities will be less noticeable as generally the interval will be smaller. A single lift can also be considered to have a bunching factor simply manifested as an irregular round trip time.

Al-Sharif (1993) proposed a formula based measure of bunching. If the time difference between lift number i departing and lift number $i + 1$, is defined as $t_{i,i+1}$, then the difference between this time and the ideal time can be taken as a measure of how much bunching exists. The ideal time is the average interval. Thus the time difference is:

$$(t_{i,i+1} - RTT/L) \tag{12.6}$$

This time difference can be either positive or negative, as an early arrival of a lift is as bad as a late arrival. To penalise large deviations from the interval, this time difference formula was squared, a technique which also takes care of the positive and negative values. Using this method, a value of zero represents no bunching and unity total bunching. Because of the reversed valuation, when applying the Schroeder criteria in the method above, a tolerable acceptable value may be 0.4. This method relies on obtaining the times of each lift departure. The full formula is not given here.

Kavounas (1992b) asked if bunching is (a) a self correcting situation, ie: the non-bunched state is one of lesser energy, or (b) a self aggravating situation (like wildfire) which needs early detection and correction, or (c) neither. The foregoing suggests it is (b).

12.7 MORE ON UPPEAK PASSENGER AVERAGE WAITING TIME, TRAVEL TIME AND JOURNEY TIME

12.7.1 Alternative and Alternate Definitions

It is important to check the definitions used, when comparing traffic designs. In Chapter 6 the following definitions were proposed and have been used throughout this book.

Definition 6.6: The passenger average waiting time is the average period of time, in seconds, that an average passenger spends waiting for a lift, measured from the instant that the passenger registers a landing call (or arrives at the landing), until the instant the passenger enters the lift.

Definition 6.8: The passenger average travel time (ATT) is the average period of time, in seconds, which an average passenger takes to travel from the main terminal floor to the requested destination floor measured from the time the passenger enters the lift until alighting at the destination floor.

Definition 6.9: The passenger average journey time is the average period of time, in seconds, measured from the instant an average passenger first registers a landing call (or arrives at the landing), until alighting at the destination floor.

It can be seen that $AJT = AWT + ATT$ precisely. Now consider the following definitions from NEII-2000 (see References), which have been paraphrased a little:

Definition 12.1: The passenger waiting time is the actual time a passenger waits after registering a landing call until the responding elevator door(s) begin to open.

Definition 12.2: The time to destination of a single passenger is measured from landing call registration until the car doors start to open at the destination floor.

Whereas Definitions 6.6, 6.8 and 6.9 are a complete set, Definitions 12.1 and 12.2 are not complete as there is no NEII definition of travel time. The NEII definitions are psychologically sound in the sense that a passenger will consider the waiting time to be over, when the lift doors are opening for boarding (Definition 12.1), however, there is a piece of time missing between the car doors opening and the passenger entering the lift. Definition 12.1 is close to Definition 6.6, except for the one missing time element. The passenger will also consider that the travelling time is over when the lift doors are opening for alighting (Definition 12.2), however, there is a piece of time missing between the car doors opening and the passenger leaving the lift. Definition 12.2 is close to Definition 6.9, except for the one missing time element.

Peters (2001) defines waiting time as Definition 12.1 and transit time as follows.

Definition 12.3: Transit time is the time from the elevator doors beginning to open (at the arrival floor) to the time the doors begin to open again at the passenger's destination (floor).

Definition 12.3 is different to Definition 6.8, as it is measured between door opening events. So the times waiting to enter the lift at the arrival floor and to leave the lift at the destination floor are missing. In practice, these two missing times may well balance out.

The above discussion illustrates the need to agree terminology (Barney *et al.*, 2001; Bates, 1993).

12.7.2 Another Formula for Passenger Average Waiting Time

In Section 6.6.3 it was suggested that passenger average waiting time could be obtained easily from Figure 6.1 and Table 6.4, but no proof was given. Section 12.4 showed the provenance of Figure 6.1 in Figure 12.2. Figure 6.1 can used to complement the conventional design procedures in order to obtain an estimation of average waiting times. The procedure is very simple. For a particular lift configuration and a specified traffic demand, the designer calculates the interval and the percentage load, using the procedure in Section 6.11. From the value of the calculated percentage load on the horizontal axis of Figure 6.1, the value for *AWT/INT* can be read from the vertical axis. As the calculated interval is known, the average waiting time can then be evaluated.

Equation (6.1) represents the *AWT/INT* curve in Figure 6.1 for 50%<*P/RC*<80%.

$$AWT = [0.4 + (1.8P/RC - 0.77)^2]INT \tag{6.1}$$

A more accurate, but more complex curve fit covering a wider range (40%<*P/RC*<95%), was proposed by Kavounas (1993c) as:

$$AWT = 0.2e^X \tag{12.7}$$

$$\text{where } X = 0.235e^{2.27\frac{P}{RC}}$$

12.7.3 More on Passenger Average Travel Time (*ATT*)

Equation (6.4) gives a formula to calculate average travel time (*ATT*):

$$ATT = 0.5Ht_v + 0.5St_s + Pt_p \qquad (6.4)$$

So and Suen (2002b) have suggested a more accurate formula for the travel time. This is shown as Equation (12.8):

$$ATT_s = 0.5 \left[\frac{S + 1}{S} \right] Ht_v + 0.5(S + 1)t_s + Pt_p \qquad (12.8)$$

When *S* is much larger than unity ("1"), the So and Suen equation becomes that of Equation (6.4). The difference between the two formulae is not large.

Example 12.1

Consider the following data:

$N = 16, CC = 16, H = 15.3, S = 9, t_v = 1.0 \text{ s}, t_s = 10.0 \text{ s}, t_p = 1.0 \text{ s}.$

Using Equation (6.4):

$ATT = 65.5 \text{ s}$

Using Equation (12.8):

$ATT = 71.4 \text{ s}$

The So and Suen average travel time values are more pessimistic than Barney's, being some 5 or 6 s longer.

12.8 IMPROVING UPPEAK HANDLING CAPACITY

12.8.1 Rationale

Sometimes the traffic designer specifies too few lifts, or the architect refuses to provide sufficient space for the number of lifts required, or the building population increases and the installed lift system cannot handle the uppeak traffic demand. Several techniques are available to improve the uppeak handling capacity of an installation (Barney, 1992). These techniques, sometimes called "boosters", have to be used carefully as they generally cause other traffic conditions to deteriorate (see Section 12.8.4).

12.8.2 Conventional Uppeak Traffic Control

Under conventional traffic control, once the uppeak traffic condition has been detected (eg: by load weighing, number of car calls registered, etc.) all cars are returned to the main terminal floor after the last passenger has exited at the high call reversal floor (*H*).

Down landing calls are ignored or serviced on an occasional basis. This traffic control system is available from most manufacturers.

Equation (4.11) is used to calculate the round trip time (*RTT*). Using Equation (4.7) the interval (*INT*) can be found and using Equation (4.11) the handling capacity (*HC*) can be found. The passenger service interval (*PSINT*) will be equal to the interval (*INT*) as each passenger enters the first car to arrive.

Definition 12.4: The passenger service interval is the period of time, in seconds, between lift arrivals serving the destination floor of the passenger.

Example 12.2

Consider a 16 floor building served by six lifts with the following data:

$L = 6$, $N = 16$, $P = 10$, $t_v = 1.0$ s, $t_s = 8.0$ s, $t_p = 1.0$ s.

From Table 6.8a: $H = 15.0$, $S = 7.6$, $P = 10$.
Using Equation (4.11) the *RTT*   = 119 s.
Using Equation (4.7) the *INT*   = 20 s.
Using Equation (4.6) the *HC*   = 151 persons/5-minutes.
Using Equation (6.4) the *ATT*   = 48 s.
Using Equation (6.6) the *AJT*   = 58 s.

This example will be used to compare the other four traffic algorithms discussed below. It will be seen that there are more stops compared to the other four traffic algorithms. Also, all the cars must travel to a reversal floor, high in the building.

12.8.3 By Subzoning

In subzoning systems, the building zone is divided into two subzones and the lift group is divided into two subgroups for the duration of the uppeak period. The cars are permanently allocated to a subzone and passengers are directed to the subgroup which serves their floor by illuminated signs. The subzones may not contain equal numbers of floors, nor may equal numbers of lifts serve each subzone. The technique works well with at least six lifts in the group and has been offered by a number of lift manufacturers.

The effect is to reduce the number of floors served and hence the number of probable stops each lift can make. This reduces the value of the middle term in the round trip time equation (4.11) and increases the handling capacity. The disadvantage is that passengers will have to wait longer for a lift serving their subzone. The traffic designer should always attempt to provide similar interval times at the main terminal for each subzone.

The calculation would use Equation (4.11) for the lower subzone. The upper subzone is calculated in the same way except allowance has to be made for the express jump through the lower subzone by using Equation (5.22).

Example 12.3

Consider the building in Example 12.2 with a lower subzone of 9 floors and an upper subzone of 7 floors.

Lower subzone:

From Table 6.8a: $N = 9$, $L = 3$ $H = 8.6$, $S = 6.2$, $P = 10$.
Using Equation (4.11) the RTT = 95 s.
Using Equation (4.7) the INT = 32 s.
Using Equation (4.6) the HC = 95 persons/5-minutes.
Using Equation (6.4) the ATT = 39 s.
Using Equation (6.6) the AJT = 55 s.

Upper subzone:

From Table 6.8a: $N = 7$, $L = 3$, $H = 6.7$, $S = 5.5$, $P = 10$.
Using Equation (5.22) the RTT = 111 s.
Using Equation (4.7) the INT = 37 s.
Using Equation (4.6) the HC = 81 persons/5-minutes.
Using Equation (6.5) the ATT = 49 s.
Using Equation (6.6) the AJT = 67 s.

Table 12.4 Comparison for uppeak subzoning from Example 12.3

Zone	N	L	INT	HC	ATT	AJT
Original building	16	6	20	151	48	63
Lower subzone	9	3	32	95	39	55
Upper subzone	7	3	37	81	49	67

All are figures rounded

Table 12.4 compares the original building (Example 12.2) with the subzoned building (Example 12.3). The table indicates that the handling capacity has been increased from 151 to 176 persons/5-minutes, ie: by 16%. This is the equivalent of one extra lift. The passenger service interval ($PSINT$) will be equal to the interval of the lifts serving the respective subzone. There is an increased passenger waiting time (20 s to 32/37 s) for both subzones as indicated by the increased interval times. But the passengers take less time travelling in the lifts in order to reach their destinations (48 s to 33/37 s). The average journey time for passengers travelling to floors in the lower subzone is smaller and those travelling to floors in the upper subzone will take a little longer time (63 s to 67 s).

12.8.4 By Sectoring

Uppeak subzoning can be extended by dividing the building into more than two subzones or sectors. The number of sectors can be made equal to (or slightly less than) the number of cars. Cars are not permanently assigned to a sector. As cars arrive at the main terminal floor they serve the sectors in a strict "round robin" fashion. Passengers will have to wait longer for service, but the group interval is smaller. Passengers are directed to cars serving their floors by destination signs above the cars and have to be continually scanning the destination panels placed outside each lift entrance until they find their desired destination indicated. Where there are more lifts than sectors, it allows some lifts to be travelling back to the main terminal as the others travel up the building. Each sector generally contains the same number of floors, except the highest may have less floors and the lowest may have more floors. The number of floors in each sector is small, eg: 3/4/5, and consequently the round trip time is reduced and the handling capacity increased. The passenger service interval (*PSINT*) will be equal to the interval of the group multiplied by the number of sectors. One lift manufacturer (Powell, 1992) has proposed this system.

To analyse the system the procedure used for subzoning could be extended and each sector calculated individually. The individual round trip times could then be averaged and an average interval found for all the lifts in the group. This group interval would then need to be adjusted by multiplying it by the number of sectors to obtain the passenger interval. An alternative is to consider a notional sector placed centrally in the served zone and to calculate its round trip time, interval, etc.

Example 12.4

Consider the building in Example 12.2. The building will be divided into 4 sectors of 4 floors each. There will therefore be six lifts "sharing" four sectors. The notional sector for a calculation will be considered to be Floors $7-10$.

The sectored building:

From Equation (5.12): $H = 3.9$, and Equation (5.5): $S = 3.8$, $P = 10$.
Using Equation (5.22) the *RTT* $= 80$ s.
The group interval $= 13$ s.
The passenger service interval will be four times the group interval, ie: 56 s.
Using Equation (4.6) the group *HC* $= 211$ persons/5-minutes.
Using Equation (6.5) the *ATT* $= 37$ s.
Using Equation (6.6) the *AJT* $= 65$ s.

Table 12.5 Comparison for uppeak sectoring from Example 12.4

Zone	N	L	INT	HC	ATT	AJT
Original building	16	4	20	151	48	63
Sectored building	16	4	56[1]	211	37	65

All figures are rounded

The number of stops is reduced as each lift now only serves four floors. Table 12.5 indicates that the handling capacity has been increased by 46% from 151 to 211 persons/5-minutes. This is the equivalent of nearly three extra lifts. The passengers have to wait on average 2.8 times longer, as indicated by the increased interval time (20 s to 56 s) for a lift serving their destination. They, however, spend less time in the lift (48 s to 37 s) and their average journey times are slightly longer (63 s to 65 s).

A sectoring system can be improved by applying dynamic sectoring where the number of floors in a sector is not fixed, but changes according to the number of passengers in a lift (car capacity factor). Thus if lifts travel to a particular sector above a threshold capacity, consideration can be given to changing the sector boundaries. This technique will increase the handling capacity, but could further confuse passengers when they try to locate a lift travelling to their floor. It is also more complex technology.

---

[1] $PSINT = 4 \times 14$ s $= 56$ s.

12.8.5 By Call Allocation: Single Zone

Research in the 1970s (Closs, 1970, 1972) reported by Barney and Dos Santos (1977) and implemented by a lift manufacturer (Schroeder, 1990b, 1990c) provided a different approach. This was a new signalling arrangement whereby passengers registered their destinations before they entered a lift. This system is generally termed Call Allocation, sometimes Hall Call Allocation, Destination Call Allocation, Adaptive Call Allocation, etc.

If a keypad is provided (see Figure CS17.1) at the main terminal floor, passengers can register their destination floor and a more efficient allocation can be made. Cars are not permanently allocated to specific floors. Passengers are notified (on the keypad) which car will take them to their destination, immediately they have registered their call.

The call allocation traffic supervisor allocates passengers to lifts in a way that ensures a smaller number of stops. This causes the round trip time to become shorter, the passenger waiting time to generally become longer and the passenger journey time to generally become shorter.

The basis of the algorithm is to consider not just one lift (conventional procedure) but several lifts at once. The number of lifts considered can vary from two to four. The effect of a larger number is longer waiting times. A compromise is usually made at between two and three lifts. The number of lifts considered is termed "the look ahead factor", k. The conventional calculation of H and S is thus not possible. Fortunately Schroeder (1990d) offers a solution.

The conventional formula for S is given by Equation (5.5) and the modified formula by Schroeder is given by Equation (12.9):

$$S = N \left[1 - \left[\frac{N-1}{N} \right]^P \right] \tag{5.5}$$

$$S = \frac{N}{k} \left[1 - \left[\frac{N-1}{N} \right]^{kP} \right] \tag{12.9}$$

It will be noted that Equation (12.9) implies that each lift will only serve N/k floors, but with a large lift carrying kP passengers. Similarly the conventional formula for H is given by Equation (5.12) and the modified formula by Schroeder is given by Equation (12.10):

$$H = N - \sum_{i=1}^{N-1} \left[\frac{i}{N} \right]^P \tag{5.12}$$

$$H = N - \sum_{i=1}^{N-1} \left[\frac{i}{N} \right]^S \tag{12.10}$$

The value obtained for H in Equation (12.10) is dependent not on the passengers carried (P), as in the conventional case, but on the number of stops (S) made. The passenger service interval (*PSINT*) will be k multiplied by the group interval.

Example 12.5

Consider the building in Example 12.2 with a call allocation traffic controller using a look ahead factor (k) of 2.

Using Equation (12.9) $S = 5.8^1$ and using Equation (12.10) $H = 14.1,^2$ $P = 10$.

Using Equation (4.11) the RTT = 102.6 s.
Using Equation (4.7) the INT = 17 s.
Using Equation (4.6) the HC = 175 persons/5-minutes.
Using Equation (6.4) the ATT = 40 s.
Using Equation (6.6) the AJT = 57 s.

Table 12.6 Comparison for call allocation from Example 12.5

Zone	N	L	INT	HC	ATT	AJT
Original building	16	6	20	151	48	63
Call allocation	16	6	34[3]	175	40	57

All figures are rounded

Table 12.6 indicates that the handling capacity has increased by 16% from 151 to 175 persons/5-minutes, the equivalent to one extra lift. The passengers have to wait 1.7 times longer, as suggested by the increased interval time ($PSINT$), ie: from 20 s to 34 s, for a lift serving their destination. They, however, spend a little less time in the lift (48 s to 40 s) and their journey times are shorter (63 s to 57 s).

12.8.6 By Call Allocation: Subzoning

A further technique with call allocation is to use dynamic subzoning. Here the building is divided into two subzones. The boundary of the subzones can change according to the demand to each of the subzones, determined by the individual car loadings. The intending passengers will be unaware of the changing boundary, as they are always told at call registration which lift they are to travel in. The passenger service interval will be k multiplied by twice the group interval.

The calculation will be the same as that employed for single zone Hall Call Allocation system, but with a reduced number of floors. In the case of the higher subzone, account must be taken of the time to pass through the lower subzone.

Example 12.6

Consider the building in Example 12.2 with a call allocation traffic controller using a look ahead factor (k) of 2 and with a lower subzone of 9 floors and an upper subzone of 7 floors.

---

[1] kP is 20. Use Table 6.8b, look down column for 20 persons and across row 16 and divide value by 2.

[2] Index will be 5.8. Use Table 6.8a, interpolate between columns 5 and 6 to read a value opposite row 16.

[3] $k = 2$, so twice the calculated interval.

Assume three lifts per subzone.

Lower subzone:

Using Equation (12.9) $S = 4.0^1$ and using Equation (12.10) $H = 7.8.^2$
Using Equation (4.11) the RTT $\quad$ = 76 s.
Using Equation (4.7) the INT $\quad$ = 25 s.
Using Equation (4.6) the HC $\quad$ = 119 persons/5-minutes.
Using Equation (6.4) the ATT $\quad$ = 30 s.
Using Equation (6.6) the AJT $\quad$ = 55 s.

Upper subzone:

Using Equation (12.9) S is 3.4^3 and using Equation (12.10) H is $5.9.^4$
Using Equation (5.22) the RTT $\quad$ = 85 s.
Using Equation (4.11) the INT $\quad$ = 28 s.
Using Equation (4.6) the HC $\quad$ = 105 persons/5-minutes.
Using Equation (6.4) the ATT $\quad$ = 36 s.
Using Equation (6.6) the AJT $\quad$ = 64 s.

Table 12.7 indicates that the handling capacity has increased by 50% from 151 to 224 persons/5-minutes, the equivalent to three extra lifts. The passengers have to wait 2.7 times longer, as suggested by the increased interval time (*PSINT*), ie: from 20 s to 50/56 s, for a lift serving their destination. They, however, spend less time in the lifts (48 s to 30/36 s) travelling to their destination floor and their average journey times are shorter or the same (63 s to 55/64 s).

Table 12.7 Comparison for subzoned call allocation system from Example 12.6

Zone	N	L	INT	HC	ATT	AJT
Original building	16	6	20	151	48	63
Lower subzone	9	3	50^5	119	30	55
Upper subzone	7	3	56^5	105	36	64

All figures are rounded

12.8.7 Some Conclusions

The four techniques described above: subzoning, sectoring, call allocation and call allocation plus subzoning, all increase the handling capacity of the underlying lift installation. There is always a penalty to pay in terms of increased waiting times and journey times. The travel times, however, are shorter. Table 12.8 compares the four techniques where, for the subzoned systems, the handling capacities have been added and the intervals, travel times and journey times have been averaged across the subzones.

---

[1] kP is 20. Use Table 6.8b, look down column for 20 persons and across row 9 and divide value by 2.

[2] Index will be 4.0. Use Table 6.8a, extrapolate from column 5 to deduce a value opposite row 9.

[3] kP is 20. Use Table 6.8b, look down column for 20 persons and across row 7 and divide value by 2.

[4] Index will be 3.4. Use Table 6.8a, extrapolate from column 5 to deduce a value opposite row 7.

[5] $k = 2$, so twice calculated interval.

Table 12.8 Comparison of the four "boosters" and the original (underlying) system

System	INT	HC	ATT	AJT
Original building	20	151	48	63
Subzoning	33	176	44	61
Sectoring	52	211	37	65
Basic call allocation	34	175	40	57
Call allocation with subzoning	52	224	33	60

The techniques substantially increase (boost) the handling capacity of the underlying system for the uppeak traffic condition. If a building is underlifted, or the population increases, then the techniques can be used to improve the handling of the morning peak traffic. The passengers will have to endure a longer wait (bad psychologically) in return for not queuing, ie: they board the first lift serving their floor. However, they will spend less time travelling in the lifts (good psychologically) and only take a little more time to reach their final destination.

The subzoning and basic call allocation techniques offer about 16% extra handling capacity with shorter travel times and similar journey times. The waiting times increase however by 65% (interval changes from around 20 s to 33 s). The call allocation technique can hide most of this increased time from the passenger. This is possible as the call registration panel can be some distance from the lift that the passenger will travel in. The passenger will therefore take some time to walk to the lift entrance. This walking time can be used to overlap the actual service time to the passenger. The walking time can be (say) 10 s, but it should not be any longer than the group interval in case an assigned lift is delayed.

The sectoring and call allocation with subzoning offer about 50% more handling capacity, a reduced travel time, but a 25% longer journey time. The waiting time increases considerably by 250% (interval from 20 s to 50 s). The call allocation technique can hide only some of this time from the passenger.

If a lift system has sufficient handling capacity, but the passenger waiting, travel and journey times are unacceptable, then the techniques can be used to reduce these times. For example, if with subzoning the required handling capacity was still to be 151 persons/5-minutes (Table 12.8) then the interval (waiting time), travel time and journey times would fall. A further benefit, in these circumstances, is that the car loads will also be reduced, giving the passengers a more pleasant ride.

Uppeak boosters can improve the overall performance of a system for uppeak, either by increasing the handling capacity or by improving the passenger times. However, such techniques do not improve the performance of the other major traffic conditions. They stay the same. This problem will be discussed further in Chapters 13, 14 and 15.

12.9 DEALING WITH A GROUP OF LIFTS WITH DIFFERENT SPECIFICATIONS

It usual for all the lifts in each group to have the same specification, ie: speeds, capacities, number of floors served, operating times, etc. Occasionally one or more lifts in a group will be different. Examples are:

■ Firefighting lift of smaller and slower specification.
■ Goods/passenger lift of larger and slower specification.
■ Lifts that serve a number of basement floors in addition to the general floors.
■ Lifts that serve a number of penthouse floors in addition to the general floors.

To illustrate the calculation consider two examples.

Example 12.7

In a group of four lifts, one lift is used for goods. The data are:

$N = 12$, $P = 16$, $H = 11.7$, $S = 9$.

Regular lifts: $t_v = 1.0$ s, $t_s = 10.0$ s, $t_p = 1.2$ s.
Goods lifts: $t_v = 2.0$ s, $t_s = 13.0$ s, $t_p = 1.0$ s.

The round trip time, from Equation (4.11), for the regular lifts is 161.8 s.

The round trip time, from Equation (4.11), for the goods lift is 208.8 s.

The average round trip time will be:

$$\frac{3 \times RTT_{reg} + 1 \times RTT_{gds}}{4} = \frac{3 \times 161.8 + 208.8}{4} = 173.6 \text{ s}$$

and the interval will be 43.4 s.

Example 12.8

Peters (1997a) analysed a system of four lifts, two of which served the main terminal and two served one basement (see Section 8.3.3). He found the round trip time of the two lifts not serving the basement was 80 s and the two lifts serving the basement was 90 s. Following the same procedure as for Example 12.7 above, the average round trip time will be:

$$\frac{2 \times RTT_{reg} + 2 \times RTT_{bas}}{4} = \frac{2 \times 80 + 2 \times 90}{4} = 85 \text{ s}$$

The interval at the main terminal will be 85/4, ie: 21 s. However, the interval at the basement will be 90/2, ie: 45 s.

12.10 A FINAL IMPORTANT WORD

Uppeak design depends on how many passengers (P) a lift can transport when it is serving the uppeak traffic condition (see Sections 4.5.2, 5.2, 5.13, 6.6.3 and 7.8). The procedure is the same regardless of whether the design is for a new lift or an existing lift. The design procedure is to match the number of arriving passengers with the handling capacity of the installed lift system. This is usually carried out using an iterative process, such as that described in Section 6.11. Having found the number of passengers (P) to be carried, the car size must be selected. The first consideration is to assume a random arrival process. To cater for this, the value obtained for P should be taken as 80% of the car size. Now the designer must decide the best size of car which will accommodate the passengers from the ranges offered by the lift manufacturers. These will generally be to the ISO 4190 (1999) standardised sizes, although some manufacturers may offer other useful sizes such as 900 kg and 1800 kg rated loads.

To select the correct size of lift, should the designer take criterion (a) or (b) to decide on the value for the number of passengers P ?

(a) the Rated Car Capacity (CC), ie: rated load divided by 75 kg
(b) the Actual Capacity (AC), ie: the platform area divided by 0.21 m² (see Table 7.2).

The decision can depend on where in the world the lift is to be installed. Originally the assumed passenger weight was 150 pounds, ie: 68 kg, in the UK and the USA with the passenger standing on 2.0 square feet (0.19 m²). The assumed average weight of a passenger in Europe now is 75 kg. This may be suitable in parts of Europe and North America. However, in Scandinavia and Russia it could be 80 kg and probably is only 65 kg around parts of the Asia-Pacific rim. The area occupied by a passenger will also change from 0.21 m².

Barney (2000a) provides a table giving suggested occupancy levels for several assumed average passenger weights, as an adaption of Table 7.3 by adding nominal rated load. This is shown as Table 12.9.

Table 12.9 Occupancy levels according to assumed weight and area of occupancy

Available car area (m²)	Nominal rated load (kg)	Desirable number of passengers according to weight/area occupied			
		65 kg/0.18 m²	68 kg/0.19 m²	75 kg/0.21 m²	80 kg/0.22 m²
1.30	450	7	7	6	6
1.66	630	9	9	8	7
2.00	800	11	11	10	9
2.40	1000	13	13	11	11
2.90	1275	16	15	14	13
3.56	1600	20	19	17	16
4.20	2000	23	22	20	19
5.00	2500	27	26	24	23

All passenger figures are rounded

The comfort considerations (Day, 2001a, 2001b), which includes cultural differences, must also be considered carefully. This is very subjective and guidance is not given here.

The selection of the right car size or the value for P is very important. When the selection is made, the reasons for it should be clearly documented.

CHAPTER THIRTEEN

Down Peak: Dissertation and Hypothesis

13.1 EARLY WORK

13.1.1 Background

At the end of a working day the occupants of a building wish to leave as quickly as possible. Most, if not all, of the lift traffic during this outgoing or down peak traffic period terminates at the main terminal and the demand for lift service is usually very high. Surprisingly very little attention is devoted to down peak traffic, and the evaluation of a lift system for this traffic pattern is far from common practice. The conventional lift traffic designer is happy to conclude that if a system is properly designed to cope with the uppeak traffic then it will be able to move the down peak traffic satisfactorily and this is generally so. The hypothesis is that a lift system will be designed to serve the requirements of the uppeak traffic according to the type of building, whether it is high prestige or low end speculative. The assumption then is that if the lift system provides the necessary Quantity and Quality of Service required in uppeak, then a matching Quantity and Quality of Service will be obtained during down peak. This conclusion will be shown to be a criterion based on Quantity of Service rather than on Quality of Service.

The question is "Can down peak be analysed in a similar way to uppeak ?". And "Can a concept of a round trip time be used ?". The answer is yes to both questions as can be seen by examining Figure 4.8, which shows a spatial plot of car movements during down peak traffic. There is a similarity to the uppeak traffic patterns in two respects. The first is that all trips pass through the main terminal floor. The second is the regular staircase stopping pattern. This pattern has a reverse slope to the uppeak, but with less stops. It suggests that an analysis similar to the uppeak analysis is possible.

13.1.2 Basic Down Peak Analysis

Two authors give some indication of the down peak traffic intensities (quantity criterion) which need to be served. Strakosch in 1967[1] established a requirement that a lift system should be able to evacuate the population of a building in $15-30$ minutes, which he later refined in 1983[2] to $25-40$ minutes as an update to modern practice. He says that for office buildings, the 5-minute down peak of traffic "may exceed any other traffic peak by $40-50\%$". Zimmermann (1973) says that, from his experience, the evening "crush" lasts $7-9$ minutes in most buildings, moving $40-50\%$ of the population. He also suggests a peak demand of 25% of the population for 5 minutes.

Consider the down peak traffic condition. Lift cars discharge passengers at the main terminal floor, travel back up the building to the floors above, fill up at a number of stops and express back down to the main terminal. This is almost the reverse of the

---

[1] Page 102. [2] Page 112.

uppeak traffic pattern, except ... !   Where there is a suitable traffic control system ensuring a reasonably equal service to all floors, the average reversal (from up direction to down direction) occurs at a lower floor than the uppeak high call reversal floor. See Figure 4.8 for an illustration of this effect. So (H_D) is smaller. The number of stops is observed to be smaller than during uppeak, as cars fill at three, four or five floors. This is owing to the intensity of the passenger demand at "going home time", even in flexitime workplaces, which fill the lifts at a smaller number of floors. Hence a much lower number of stops (S_D) than during uppeak. There is also a tendency for the cars to fill nearer to the maximum available car capacity (AC, Table 7.2), whenever there is still space available, the "no touch" syndrome being abandoned at "going home time" ! This could lead to more efficient loading and unloading of passengers and shorter passenger transfer times (t_p). So there will be higher car loads, which are more quickly transferred. Thus the round trip time will be smaller than for other traffic patterns and, consequently, the down peak handling capacity will be inherently greater than for uppeak, but by how much ?

If the expected number of stops and the expected highest reversal floor were known, and if cars are assumed to fill to maximum capacity, then the down peak round trip time could be calculated using a modified version of Equation (4.11), *viz*:

$$RTT_D = 2H_D t_v + (S_D+1)t_s + 2ACt_p \tag{13.1}$$

Both Strakosch and Zimmermann describe empirical rules for down peak round trip time evaluations. Strakosch (1967) presents a method where he considers that the lift cars nearly always reach the top floor, as "a conservative measure", and the number of stops may be determined from the knowledge of the population of each floor and the nature of that population, for example: whether the cars could fill at a single floor. If the exact nature of the occupancy is not known, then he estimates that the down peak expected number of stops per trip (S_D) is approximately 75% of the probable number of stops in the uppeak situation (S).

Zimmermann (1973) uses 75% of the number of floors (N) serviced above the main floor as the highest reversal floor (H_D) and he assumes that the expected number of stops is equal to the ratio of the number of floors and the number of cars (L). Table 13.1 summarises these ideas of down peak evaluation.

Table 13.1 Evaluation of down peak parameters

Parameter	Strakosch	Zimmermann
Expected number of stops S_D	$S_D = 0.75S$	$S_D = N/L$
Expected highest reversal floor H_D	$H_D = N$	$H_D = 0.75N$

These ideas are applied in Example 13.1.

13.1.3 Example 13.1

Consider a system of four, 16 person lifts serving 16 floors above the main terminal floor with lift dynamics of $t_v = 1.6$ s, $t_s = 8.4$ s, ($T = 10$ s), $t_p = 1.0$ s.

Using the ideas of Table 13.1 and Equation (13.1), it is possible to evaluate the down peak interval of cars at the main terminal floor, the down peak handling capacity and the down peak to uppeak handling capacity ratio. Table 13.2 summarises the results.

Table 13.2 Illustrative example

Parameter	Uppeak	Strakosch	Zimmermann
P	12.8	16	16
H	15.3	16	12
S	9.0	6.75	4.0
INT	39.6	37.1	28.1
HC	97	129	171
DNPHC/UPPHC ratio	n/a	1.3	1.8

All figures are rounded

Table 13.2 indicates a wide disparity between the two authors.

13.1.4 Effects of the Traffic Controller

There is another important aspect of lift behaviour during a down peak traffic situation that must be considered, this is the effect of the traffic control system. If the high call reversal technique is applied, then a lift reaching the top of the building may fill to capacity in a small number of stops, owing to the heavy down peak traffic demands. It will then travel to the main terminal floor, bypassing a number of landing calls at the lowest floors. Thus if each car, after discharging the passengers at the main terminal, is allowed to answer landing calls registered from the top of the building, it is possible that the lowest floors will never obtain service. To avoid this possibility, the supervisory control algorithm applying to the down peak traffic pattern must allocate the cars in some "round robin" fashion to guarantee a balanced service to all floors in the building. This is best achieved by dividing the building into sectors (Definition 9.4). All modern group supervisory control systems provide such a facility.

13.1.5 Quality of Service

The evaluations of Table 13.2 only consider the Quantity of Service. Bearing in mind the effect of the traffic control system cycling the cars to groups of floors, it is now possible to consider the Quality of Service. Strakosch (1967)[1] states that:

"service should be available at every floor at intervals no longer than 60 s".

This statement emphasises a very common confusion between the interval of cars at the main terminal floor and the service period at a particular floor. Indeed, if the lift cars fill up at a few floors, they cannot call at all floors during each trip, and the average time interval between successive services at a particular floor (*FINT*) is longer. One way to estimate this floor interval is:

$$FINT = N/S_D \times DNPINT \qquad (13.2)$$

Considering the lift system given in Table 13.2, the frequency of service at a particular floor according to the Strakosch's postulation is:

$$FINT = 16/6.75 \times 37.1 = 88 \text{ s}$$

---

[1] Page 103.

and from Zimmermann's postulation gives:

$FINT = 16/4 \times 28.1 = 112$ s

Another way to calculate the floor interval is to divide the building into the same number of sectors as there are lifts. This gives:

$FINT = L \times DNPINT$ \hfill (13.3)

Strakosch's values become:

$FINT = 4 \times 37.1 = 148$ s

and Zimmermann's values become:

$FINT = 4 \times 28.1 = 112$ s

All the floor service intervals above break Strakosch's rule of a service time of less than 60 s.

An interesting conclusion can be drawn from Table 13.2 and the results for *FINT*. The values obtained from the two methods and the two authors are so different that they offer no degree of design confidence. Thus, what values should be used for the various parameters ?

13.2 FINDING VALUES FOR S_p, *DNPINT* AND *DNPAWT* DURING DOWN PEAK

Barney and Dos Santos (1977) simulated over 2000 lift installations covering a wide range of variables and different traffic control strategies. They defined a number of parameters.

13.2.1 Down Peak Demand

Definition 13.1: Down peak demand (α) is expressed as a percentage of the number of potential passengers (λ) arriving at a lift system and requiring service during a 5-minute peak period, with respect to a reference value, *UPPHC*, which is the uppeak handling capacity of the same lift configuration for 80% uppeak car loadings.

Definition 13.1 is given as Equation (13.4):

$$\alpha = \frac{\lambda}{UPPHC} \times 100\% \hfill (13.4)$$

For a particular lift configuration, which is properly designed for uppeak (from a conventional point of view), the uppeak rate of passenger arrivals matches the uppeak handling capacity *UPPHC*. A reasonable range of values for the down peak demand placed on the lift system in terms of passenger arrival rates could be from a value just below *UPPHC*, say 80% of *UPPHC*, through 100%, where uppeak and down peak handling capacities are equal, to a value well above *UPPHC*. If the information on peak demand values during down peak provided by some authors (eg: Strakosch, 1967; Zimmermann, 1973) is used, then it is necessary to consider demand levels exceeding *UPPHC* by 40−50%. This parameter, which acts as an independent variable, should therefore cover a range of values from 80% to over 150% of *UPPHC*.

13.2.2 Down Peak Number of Stops

Definition 13.2: Down peak percentage number of stops is the average number of stops per trip (*DNPSTPS*) during a down peak traffic situation, measured as a percentage of a reference value (*UPPSTPS*), which is the uppeak expected number of stops per trip for 80% uppeak car loadings.

Figure 13.1[1] shows the ratio of number of down peak stops to uppeak stops against passenger demand. The average curve (solid line) shows a ratio of 40% at a low demands and, at the more likely demand level of 150%, a ratio of 50%. Thus the number of down peak stops is about half the uppeak stops. This is lower than Strakosch and higher than Zimmermann indicated. A compromise !

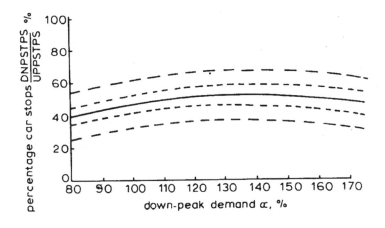

Figure 13.1 Down peak stops compared to uppeak stops

13.2.3 Down Peak Interval

Definition 13.3: Down peak percentage interval (*DNPINT*) is the interval of cars at the main terminal during a down peak traffic situation, measured as a percentage of a reference value, (*UPPINT*), which is the uppeak interval of cars for 80% uppeak car loadings.

Figure 13.2 shows the ratio of down peak interval to uppeak interval against passenger demand. The average curve (solid line) shows a value of about 66% at a demand of 150%. Thus the down peak interval is about two thirds of the uppeak interval. The conclusion is that the down peak handling capacity will exceed the uppeak handling capacity, in a similar ratio, ie: by approximately 60%.

---

[1] Figures 13.1, 13.2 and 13.3 are a summary of the detailed results shown in Barney and Dos Santos (1977). The curves show the overall probable limit of values (long dashed line) and the possible range of controller influence (short dashed line).

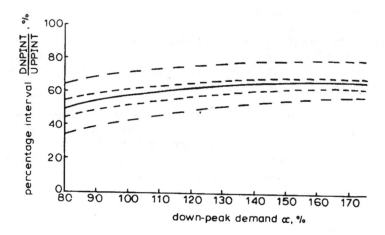

Figure 13.2 Down peak interval compared to uppeak interval

13.2.4 Down Peak Performance

Definition 13.4: Down peak performance figure is the measure of the quality of service provided by a lift system during a down peak traffic situation, expressed as a percentage of average down peak passenger waiting time (*AWT*) and of the reference value *UPPINT*.

Figure 13.3 shows the performance figure of average down peak passenger waiting time (*AWT*) normalised by dividing it by the uppeak interval (*UPPINT*).

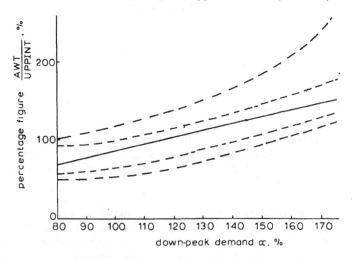

Figure 13.3 Down peak performance compared to uppeak performance

A straight line approximation can be drawn on Figure 13.3, and results in the equation:

$$DNPAWT = 0.85\alpha UPPINT \qquad (13.5)$$

but from Equation (6.1):

$UPPAWT = 0.85UPPINT$

and so:

$DNPAWT = \alpha UPPAWT$ (13.6)

If a down peak handling capacity of 150% of uppeak is assumed, then:

$DNPAWT = 1.5 \times UPPAWT$ (13.7)

Thus, although the down peak handling capacity is improved, for example in this case, by 1.5 times, the passenger average waiting time deteriorates by the same value, ie: by becoming 1.5 times longer.

In terms of the uppeak interval and using Equation (13.5):

$DNPAWT = 1.5 \times 0.85 \times UPPINT = 1.275 \times UPPINT$

This means the typical down peak average passenger waiting time, in this case, is 27.5% longer than the calculated uppeak interval.

Example 13.2

Suppose a lift system has an uppeak handling capacity of 100 persons/5-minutes at an uppeak interval of 30 s. What will the average passenger waiting time be for a down peak arrival rate of 140 persons/5-minutes ?

$\alpha = 140/100 = 1.4$

Using Equation (13.5):

$DNPAWT = 1.4 \times 0.85 \times 30 = 35.7$ s

13.3 REVISITING EQUATION (13.1)

13.3.1 Average Number of Stops (S_D)

What is the value for the down peak number of stops S_D ?

Schroeder (1984), after making some assumptions, formed a mathematical expression for the optimum number of stops for a range of lifts and rated capacities. He concluded "the optimum number of stops is rather low", but did not quantify "low". The value for S_D, which has been found by simulation in Section 13.2.1 as 0.5S, concurs with this statement.

13.3.2 Average Highest Reversal Floor (H_D)

What is the value for the down peak highest reversal floor H_D ?

Schroeder (1984) suggests H_D might be $0.5N + 0.5S_D$. However, the highest reversal floor will depend largely on the traffic controller. If it operates in such a way as to cycle cars around the building into "sectors", in order to ensure an even service, then H_D will be somewhere near to the central floor of the served zone. It could $0.5N$. However, as the traffic controller will require the lifts to reverse direction at the highest call in each sector, H_D may well be higher, say $0.5N + 0.5S_m$, where S_m is the number of floors in a service sector. If a lift has not filled to capacity in its assigned sector, a good traffic controller will permit the lift to stop for calls registered in the next sector encountered. This will be particularly true where sectors contain a small number of floors. Thus there will be a tendency for the highest floors in a sector to have already been serviced, except, of course, for the highest sector. This may well cause the value for H_D to be lower. Perhaps a best estimate for H_D is $0.5N$.

13.3.3 The Average Number of Passengers

What is the value for the down peak number of passengers P_D ?

If lifts tend to fill nearer to their actual capacity during down peak, then the transfers will be more efficient. Although the value of P will increase, the value for t_p will decrease. Thus the value for $P_D t_p$ could remain constant at Pt_p, ie: the same as during uppeak.

13.3.4 The Down Peak Round Trip Equation

Equation (13.1) can now be written as:

$$RTT_D = 2 \times 0.5Nt_v + (0.5S+1)t_s + 2Pt_p \qquad (13.8)$$

$$RTT_D = Nt_v + (0.5S+1)t_s + 2Pt_p \qquad (13.9)$$

The down peak interval (*DNPINT*) can be obtained by dividing RTT_D by the number of lifts. The down peak handling capacity (*DNPHC*) will then be:

$$DNPHC = 300P/DNPINT \qquad (13.10)$$

Reconsider Example 13.1. Using Equation (13.9) an extra column can now be added to Table 13.2 to give Table 13.3.

Table 13.3 Illustrative example – Table 13.2 extended

Parameter	Uppeak	Strakosch	Zimmermann	Barney
P	12.8	16	16	12.8
H	15.3	16	12	8
S	9.0	6.75	4.0	4.5
INT	39.6	37.1	28.1	24.3
HC	97	129	171	157.7
DNPHC/UPPHC ratio	n/a	1.33	1.76	1.63

All figures are rounded

Examination of Table 13.3 shows that the interval is 24.3 s, the handling capacity is 157.7 persons/5-minutes and the DNP/UPP ratio is 1.63. This is approximately half way between Strakosch (pessimistic) and Zimmermann (optimistic), and indicates the down peak handling capacity is some 63% larger than during uppeak.

Is the improvement in down peak handling capacity consistent for a different number of floors and rated car capacities? Applying Equation (13.8) to the "four corners of the lift world and its centre of gravity" (see Section 5.12.3) results in Table 13.4.

Table 13.4 Comparison of down peak to uppeak handling capacity

Number of floors	Rated car capacity	*DNPHC/UPPHC* ratio
10	10	1.60
10	24	1.54
16	16	1.62
20	10	1.62
20	24	1.61

Table 13.3 confirms the hypothesis that an installation during down peak traffic inherently has at least 60% more capacity then in uppeak. As a rule of thumb, traffic designers can assume the ratio *DNPHC/UPPHC* is 1.6.

13.3.5 Quality of Service

How will the Quality of Service change? What will the floor interval be and what will the passenger average waiting time be?

Again considering the installation in Example 13.1, assume that with the four lifts that there are four sectors of four floors. If a sector is served once every four (main terminal) intervals, then the floor interval is:

$$L \times DNPINT = 4 \times 24.3 = 97.2 \text{ s}$$

Using Equation (13.5), the down peak passenger average waiting time will be:

$$0.85\alpha UPPINT = 0.85 \times 1.63 \times 39.6 = 54.9 \text{ s}$$

This is much better than the floor interval would suggest, because it is an average value, whereas the floor interval is a maximum value. Also, because the traffic controller can be programmed to pick up landings calls ahead, whilst the lift still has available capacity, some high floors in a sector will have already been served before a lift is assigned to the sector. If Strakosch's rule of service within 60 s is applied to passenger waiting time and not floor interval, his criterion is met.

To achieve this Quality of Service requires the careful setting up of the load weighing system and the number of floors in a sector. If the load weighing is set at say 60% of Actual Car Capacity (see Table 7.2), then a lift with just over 60% loading will not stop for further landing calls ahead. As it is considered that passengers will squash closer together when going home, the load weighing could be set to about 75% during down peak. This may mean some lifts may become loaded above the conventional 80% value on occasions.

It is important that the number of floors in each sector is large enough for cars to fill in a sector. It is also important to ensure that the floor interval is not too long. The design range for floors per sector is probably from three to five floors.

Consider Example 13.1 again. The down peak handling capacity is:

158 persons/5-minutes

This is an arrival rate for all floors of:

158/300 = 0.53 persons/second

At a single floor this is an arrival rate of:

0.53/16 = 0.033 persons/second

If the floor interval is 97.2 s, then, as each lift arrives, the number of passengers waiting, on average, will be:

0.033 × 97.2 = 3.2 passengers

To fill the car this will require only four stops. For this system dividing the building into four sectors of four floors is ideal. If there were more cars available, six say, and four sectors were retained, then the floor interval would improve by 33% to 64.8 s and the passenger average waiting time would similarly improve.

13.4 CALCULATION PROGRAM

Equation (13.9) can be easily programmed. Table 13.5 shows the calculation for the re-evaluation of Example 13.1, as discussed in Section 13.3.4 and reported in Table 13.3.

Table 13.5 Re-evaluation of Example 13.1

```
**********************************************************************************
Lift traffic calculations          by Dr Gina Barney          Down Peak Estimate
**********************************************************************************
Design name: Down peak                     Run No.: 1              Date: 2002
................................................................................
{N}umber of floors          = 16      Number of {P}assengers = 12.8
Number of {L}ifts           = 4       Building {POP}ulation   = 647
Number of {SEC}tors         = 4
Interfloor {D}istance (m)   = 4       {E}xpress {J}ump (m)    = 0
{P}ass. {T}ransfer time (s) = 1       Speed (m/s) {v}         = 2.5
{AC}eleration (m/s2)    = 1       {J}erk (m/s3) = 1.5  {S}tart {D}elay (s) = 0.4
Door times (s): {C}lose = 3       {O}pen        = 2    {AD}vance           = 0.12
................................................................................
Flight time (s) = 4.72             Performance time(s) = 10.00
................................................................................
Average high reversal floor =  8.0     Average number of stops =  4.5
RTT =    97.4  Interval =  24.3  Handling capacity = 157.7  %POP =   24.4
Floor interval =  97.4                 DNP/UPP ratio = 1.63
**********************************************************************************
```

13.5 ESTIMATING THE DOWN PEAK HANDLING CAPACITY AND AVERAGE PASSENGER WAITING TIME

The traffic design procedure should always be to satisfy the uppeak requirements. Equation (13.9) should enable a lift designer to estimate the down peak round trip time (and hence the handling capacity) in relation to an uppeak design. For example, using a *DNPHC/UPPHC* ratio of 1.6, a 12% system may have a 5-minute down peak handling capacity of 19.2% and a 17% system may have a down peak handling capacity of 27.2%.

Over a 10-minute outgoing traffic period this represents from 38% to 54% of the zone population leaving the building. These figures relate well to the experience expressed by Strakosch and Zimmermann given in Section 13.1.2 of this chapter.

Using Equation (13.5) the designer can estimate the down peak average passenger waiting time.

Example 13.3

Consider Example 13.1, which has an uppeak handling capacity of 15%, ie: 97 persons/5-minutes and an uppeak interval of 40 s. What is the down peak passenger average waiting time, the down peak handling capacity in persons/5-minutes and the percentage of the building population that can leave in 10 minutes ?

Assuming the cars are loaded to 80% capacity, the uppeak passenger average waiting time (see Table 6.4) is:

$UPPAWT = 0.85 \times 40 = 34$ s

From Table 13.5 the *DNPHC/UPPHC* ratio is 1.63, and using Equation (13.5) the down peak average passenger time is:

$DNPAWT = 1.63 \times 34 = 55$ s

The uppeak handling capacity is 97 persons/5-minutes.

The down peak handling capacity is $97 \times 1.63 = 158$ persons/5-minutes.

In 10 minutes 316 persons can leave, which is 49% of the population.

13.6 A WARNING

The methods used to improve or boost the uppeak handling capacity do not improve down peak handling capacity. This is still dependent on the underlying installation performance.

To understand this dependence, consider Example 13.1 again. The underlying handling capacity is 97 persons/5-minutes, the original building population is 647 persons and the percentage population handled is 15%. As Example 13.3 shows, in down peak this becomes 24.5%. Now suppose that the population increases to 1000 persons. The original system now has an uppeak percentage population of 9.7% (97/1000). An uppeak booster is applied, which brings the uppeak handling capacity back to 15%. However, the down peak handling capacity will still only be 1.63 times the underlying uppeak handling capacity, ie: 15.8%. Under down peak conditions only 31.6% of the increased building population will be able to leave over 10 minutes.

Thus the uppeak handling capacity can be improved, but the down peak handling capacity remains unchanged. If a system were to be designed with uppeak boosters, to meet the uppeak design criteria, the down peak criteria will not be met. Siikonen (2000) suggests if uppeak boosters are employed, that the uppeak criterion for handling capacity should be increased by 20−30%. This only goes part of the way. In the example above the uppeak handling capacity criterion would need increase by 50%.

13.7 CONCLUSIONS

A detailed examination of the results of a large number of simulations allowed two objectives of the down peak analysis to be fulfilled, namely: the evaluation of the Quantity of Service provided by a lift configuration under down peak traffic conditions, and an estimation of the corresponding Quality of Service. The examination, like that for the uppeak examination, has been for a pure down peak traffic demand with no demand elsewhere.

The Quantity of Service is not dependent on the control policy. The average interval of cars at the main terminal is slightly below two thirds of *UPPINT*. Thus the down peak handling capacity exceeds the uppeak handling capacity by over 60%. If it could be assumed that the lift cars can be loaded to 100% of Rated Car Capacity during down peak, then a down peak handling capacity as high as 200% of *UPPHC* could be obtained.

However, the interval of cars at the main terminal does not mean that a particular floor is served at the same frequency, since each car only serves a few floors on each trip. The simulation analysis shows that the down peak lift performance is heavily dependent on the control policy in this case. The conclusion is that, for most lift systems, in order to make full use of the down peak handling capacity, it is necessary to allow for higher average passenger waiting times than those encountered during uppeak traffic. These waiting times are generally longer by the same ratio as the handling capacity is improved. That is if the down peak handling capacity is 1.6 times larger than the uppeak handling capacity, then the passenger average waiting time during down peak will also be 1.6 times longer than those endured during uppeak.

It may happen that the designer does not know the lift supervisory control system that will be installed and how the floors are served. Consequently, there is a difficulty in deciding what the performance might be. It is interesting to note that the best down peak algorithm is similar to the uppeak sectoring "booster" algorithm (Section 12.8.4). In down peak the landing calls are collected in a sector and the passengers are taken to the main terminal, ie: the opposite action to that of the uppeak operation.

It should also be noted that the figures presented are average figures and that particular configurations will deviate from the average, as illustrated in Figures 13.1, 13.2 and 13.3. Also, the situations considered are for equal floor demands. Thus, care must be taken in the interpretation of any evaluations where this is not so. Simulations will be necessary in complex situations.

Down peak traffic handling is always dependent on the underlying uppeak handling capability.

CHAPTER FOURTEEN

Interfloor Traffic: Debate and Pragmatism

14.1 INTRODUCTION

Once a building is occupied, its population will move around the building about their normal business randomly (in the statistical sense). A small percentage of occupants will leave the building and be replaced by new arrivals. A typical time to observe such an interfloor traffic pattern is during the mid-morning or the mid-afternoon periods. The traffic demand is frequently balanced and there is no dominant pattern of arrival at any floor or any dominant destination floor. Passengers using the lift system at this time eventually return to their original departure floor, hence the use of the term balanced interfloor traffic. This is illustrated in Figure 4.9, which shows a typical spatial plot of lift movements around a building during an interfloor traffic period. The plot can be viewed upside down, or back to front, and still displays a random pattern ! At other times there may be some unbalanced interfloor movements for example to/from refreshment floors, hotel lobbies, etc. These traffic flows are not as dominant as those found during pure uppeak or down peak traffic, and not as random as those found during balanced interfloor traffic. A more significant traffic pattern occurs at mid day and this will be considered in Chapter 15.

The analysis of lift systems under interfloor traffic conditions is rare and it is very seldom mentioned by lift writers. It is however the traffic pattern that exists for most of the working day. A few authors have developed methods to evaluate the round trip time for interfloor traffic. Strakosch (1967),[1] based on lift engineering experience, considered interfloor traffic and the expected number of stops by intuitive means and provided a step by step method to calculate a round trip time. Tregenza (1972) presents a formula for the evaluation of the round trip times and deduces relationships for the expectancies of the high reversal floor and the expected number of stops per trip. The cars are assumed to circulate uniformly around the building, but always returning to the main terminal floor.

The mathematical evaluation of interfloor traffic is even rarer than an intuitive analysis. Alexandris *et al.* (1979) developed formulae for the most general traffic condition of balanced interfloor traffic using statistical analysis. He considered a Poisson probability distribution function and produced equations for both H and S. They are very complex and are not reproduced here. He proved their validity by reducing them to the classical formulae for H and S for uppeak traffic. Barney and Dos Santos (1985) report these results. Peters (1997a) provides other formulae of equal complexity as his General Analysis formulae. These formulae are summarised in CIBSE Guide D: 2000. Both the Alexandris generalised formulae and the Peters General Analysis formulae only apply when there is a dominant traffic flow, ie: when lifts pass through the main terminal every trip.

In practice, during balanced interfloor traffic, lifts do not make a round trip as defined for both the uppeak and down peak traffic patterns. Examining Figure 4.9, where there are three cars, shows 251 reversals of direction of which only 87 occurred at the

---

[1] Page 87.

main terminal. Each round trip contains two reversals in the circuit. This would account for 174 of the reversals and therefore indicates nearly one third of all trips do not include the main terminal. Thus calculations of round trip times using values of H and S derived from the Alexandris/Peters formulae will provide results which have no meaning and this approach is not recommended.

The methods of evaluation proposed by Strakosch and Tregenza, and the mathematical formulae proposed by Alexandris and Peters, could all be suitable for the now discarded scheduling systems, where cars are despatched at regular time intervals from the terminal floors, but they are not applicable to modern non-scheduling (on demand) lift systems. Modern systems work on an allocation basis, and the cars are frequently taken out of their normal cycles to make an express run to particular floors or sectors, either as a parking strategy or to provide extra service to heavy traffic demand sectors.

14.2 INTERFLOOR DEMAND

A further characteristic of balanced interfloor traffic is that there are larger demands made on the lift system in terms of the number of stops made by each lift. Whereas during uppeak and down peak traffic there are a number of passengers boarding or alighting together, during balanced interfloor traffic there is a stronger likelihood that each individual passenger requires one stop to be picked up and one stop to be set down. A good indicator of overall activity during interfloor traffic may well be the number of stops a lift is making.

With the exception of Tregenza's work, no mathematical consideration is made on the intensity of traffic flow. It is simply predicted that passenger average waiting times will deteriorate with an increase in the passenger demand. What is demand ?

Strakosch defines three categories of demand: light, medium and heavy. Light traffic is where the number of passengers requesting service is no more than two to three times the number of lifts in service. Medium traffic is defined when lifts fill to less than 50% of their capacity and heavy demand is when lifts fill to over 50% of their capacity. Heavy demand is unlikely as cars will not fill to 50% of their capacity during balanced interfloor activity. Lift car loadings at this level would generally indicate a dominant flow of some sort. Bedford (1966), analysing a fixed sectoring proprietary supervisory control system similar to that described in Section 10.4.3, considers a system busy, if the number of stops made per car per minute is about 2.25.

14.3 QUALITY OF SERVICE: PERFORMANCE FIGURE

In the balanced interfloor traffic situation, the handling capacity of a lift system would be a less important parameter than for the uppeak and down peak traffic conditions, even if it had a meaning. This is because the number of passengers being transported is generally smaller. It is most unlikely that all the occupants of a building would require to use a lift during each hour of the working day. However, the lower demand is offset by a greater diversity in destination requirements. Thus, Quality of Service is more significant under interfloor traffic conditions, as the lift system is expected to provide an excellent service in order to save occupants' time.

The Quality of Service may be measured in terms of passenger average waiting times or in terms of passenger average journey times. From an employer's point of view, journey times are more important, as they measure the total time spent by an employee obtaining lift service. From the user's point of view, waiting times are psychologically more important. It is considered that, once a passenger enters a lift car, they will not mind if the car takes a certain time to reach their destination floor, as long as that time is not too long. However, passengers will become impatient if they have to wait too long for service. This is particularly so at the main terminal floor, which may receive special treatment during interfloor traffic by, for example, parking a free lift there whenever possible.

14.4 INTERFLOOR PASSENGER DEMAND

The lift system performance during interfloor traffic is certainly dependent on the rate of passenger arrivals. A traffic flow can be measured by the number of passengers requiring service per unit time and would represent different demand levels for different lift configurations. It would, however, lack normalisation. Thus, in order to provide normalisation and also to relate balanced traffic performance with the uppeak design, it is sensible to present the balanced interfloor passenger demand as a percentage of the uppeak handling capacity (*UPPHC*). In Chapter 13 down peak demand was related to *UPPHC* and this enabled the down peak handling capacity to be directly compared to the underlying uppeak handling capacity. Following the same principle, a definition of passenger demand could be:

Definition 14.1: Balanced interfloor 5-minute demand (β) is expressed as the number of passengers requiring service (λ) from the lift system during a 5-minute period, expressed as a percentage of a reference value, *UPPHC*, which is the uppeak handling capacity of the lift configuration for 80% car loadings.

$$\beta = \frac{\lambda}{UPPHC} \times 100\% \qquad\qquad (14.1)$$

Thus, each lift configuration should be considered under a variety of passenger arrival rates, from very low traffic intensities to demand levels that saturate the lift system. No precise indication could be found in the lift literature of when a demand level was considered as a light or a heavy demand load, except those by Strakosch indicated earlier in this chapter. Thus a wide range of demand levels is used in the analysis, varying from 10% to 100%.

Again, as for the down peak study, the passenger average waiting time will be presented as a percentage of *UPPINT*, in order to define normalised performance figures, which makes them less dependent on particular lift configurations, and also allows this traffic condition to be related to the uppeak design. A definition can now be provided for balanced interfloor performance.

Definition 14.2: The balanced interfloor performance figure is the measure of the quality of service provided by a lift system under balanced interfloor traffic conditions, expressed as a percentage of the passenger average waiting time to a reference value, *UPPINT*, which is the uppeak interval of cars for 80% uppeak car loadings.

As with the analysis of uppeak traffic (Section 6.6.3 and Section 12.4) and down peak traffic (Section 13.2), simulation techniques are best employed to study interfloor traffic. Barney and Dos Santos (1977) together with Lim (1983) conducted over 2000 simulations covering a wide range of variables and different traffic strategies. The balanced interfloor traffic performance figures plotted in Figure 14.1 show performance deteriorating with demand. The performance is also dependent on the supervisory control policy over the whole range of demands. The performance figures for the different control algorithms at low demands are very similar, except for the Adaptive Call Allocation (ACA) algorithm (Control Case Study CS18), which will be discussed later.

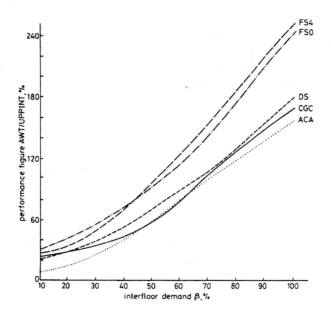

Figure 14.1 Interfloor performance graphs for several traffic control algorithms
Legend for traffic algorithms:
FS4 Fixed sectoring priority timed
FSO Fixed sectoring common sectors
DS Dynamic sectoring
CGC Computer group control
ACA Adaptive call allocation

At the 10% very low demand level, they present average waiting times around 30% of *UPPINT*. With an increase in demand, the performance degrades at an increasing rate and two groups of traffic control algorithms appear on Figure 14.1. The fixed sectoring algorithms FS4 (Section 9.4.2) and FSO (Section 9.4.3) perform badly compared to the dynamic sectoring (Section 9.4.4) and the computer based algorithms (Section 10.4.3 and Control Case Study CS18). The similarity of performances for the FS4 and FSO algorithms shows that the allocation of cars on a priority timed basis presents no advantage over a policy of redistribution of free cars around the building, complemented by a sector heavy load detection mechanism as used in the DS dynamic sectoring algorithm. This observation confirms the manufacturer's statement that the dynamic sectoring algorithm was designed specifically for the interfloor traffic condition.

The ACA algorithm, as depicted in Figure 14.1, appears to provide a superb performance at low traffic levels. This is due to a slightly different method of representation. As passengers into an ACA system register calls at a landing destination station and then walk to an appropriate car, the ACA system only starts to register passenger waiting time after some 5 s of walking time has elapsed. Hence the ACA curve is a little overstated. A re-calculation would show, that up to mid range values of β (40% $\leq \beta \leq$ 60%), the ACA system and the CGC system give virtually identical results after which the CGC algorithm is worse than the ACA algorithm.

Figure 14.1 clearly shows that fixed sectoring systems are less effective for balanced interfloor traffic. At very high demands all the lift systems saturate and at 100% demand the passenger average waiting time is 1.4 to 2.5 times the uppeak interval (*UPPINT*), dependent on the traffic algorithm used.

Figure 14.1 can be simplified to Figure 14.2 by drawing a waited average line through all the different control algorithms. The long dashed lines indicate the overall probable limit of possibilities and the short dashed lines represent the possible limit caused by the influence of the control algorithms.

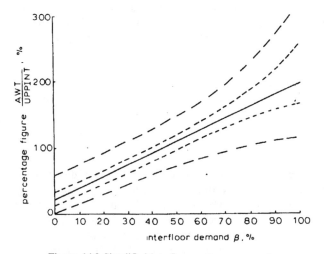

Figure 14.2 Simplified interfloor performance graph

An equation can be derived for the straight line in Figure 14.2 to allow an estimation of interfloor performance to be obtained, as:

$$IFAWT = UPPINT(0.22 + 1.78\beta/100) \tag{14.2}$$

Example 14.1

Consider an interfloor demand of 30% made on a lift system with an uppeak interval of 34 s. What is the likely passenger average waiting time ?

$$IFAWT = 34[0.22 + 1.78 \times 30/100] = 25.6 \text{ s}$$

14.5 ACTUAL PASSENGER DEMAND

The demand levels used to derive Figure 14.1 range from 10% to 100% as defined by Definition 14.2 and Equation (14.1). How do these demand levels relate to actual demand ?

In uppeak and down peak traffic situations there are short periods of five to ten minute peaks of demand for the lift system to handle. Lift system sizing is determined by the 5-minute handling capacity during uppeak, which if correctly chosen, generally means that the lift system also meets the longer periods of down peak demand. During balanced interfloor traffic the passenger demand is continuous over longer periods of time, ie: in excess of one hour. The demand levels are thus lower, but the activity of the lifts is not necessarily smaller as the number of stops is higher. There is a need to relate the demand indicated in Equation (14.1) with a longer term passenger activity.

Assume that each morning period is four hours long between occupant arrival (uppeak) and the mid day break, and that the afternoon period is four hours long between the mid day break and the departure time (down peak). Then in the morning and afternoon periods there will be about a three hour period when the building occupants are working. The occupants will be at their desks or in meetings and will be much less active. It might be reasonable to assume that each occupant uses the lift system once every morning period and once in every afternoon period. The demand on the lift system would thus be about one third of the building population using the lift system every hour. This demand parameter will be defined.

Definition 14.3: Balanced hourly interfloor demand (γ) is the number of passengers requiring service from the lift system during a one hour period, expressed as a percentage (x) of the building population (U).

$$\gamma = \frac{xU}{100} \tag{14.3}$$

Relating this demand to the uppeak demand. If 30% of the building population uses the lifts during one hour of interfloor activity this is equivalent of only 2.5% demand over 5 minutes. Similarly if 36% of the population uses the lifts during one hour of interfloor activity this is equivalent to 3.0% demand over 5 minutes. These interfloor demand levels are considerably smaller than the uppeak values, which are typically from 12% to 17%, ie: interfloor demand is about one fifth of the uppeak demand.

It is convenient to relate Definitions 14.1 and 14.3 by considering the following:

From Definition 14.3 the number of passengers handled in 5 minutes is:

$$\gamma = \frac{xU}{100 \times 12} \tag{14.4}$$

Substitute into Equation (14.1):

$$\beta = \frac{xU}{100 \times 12} \times \frac{100}{UPPHC} = \frac{xU}{12UPPHC} \tag{14.5}$$

but:

$$\%POP = \frac{UPPHC}{U} \times 100 \tag{14.6}$$

Rearranging Equation (14.6) and substituting in Equation (14.5):

$$\beta = \frac{xU}{12} \times \frac{100}{U\%POP} = \frac{x100}{12 \times \%POP} \tag{14.7}$$

To illustrate the relationship, suppose a lift system has an uppeak $\%POP$ of 12% and a 36% interfloor one hour demand, then the value for β will be:

$$\beta = \frac{36 \times 100}{12 \times 12} = 25\%$$

Similarly a lift system with an uppeak %*POP* of 15% and a 30% interfloor one hour demand will have an interfloor demand (β) of:

$$\beta = \frac{30 \times 100}{12 \times 15} = 17\%$$

Thus the likely range of usage of, say, one third of the building occupants using a lift every hour is represented by the low end of interfloor demand β at less than (say) 25%. At this range of usage the performance figures shown in Figures 14.1 and 14.2 are smaller and the passenger average waiting times are unlikely to exceed 50% of the calculated uppeak interval.

Example 14.2

For a lift system designed to handle a 15% uppeak traffic demand (%*POP*) with an uppeak interval (*UPPINT*) of 34 s, determine the likely balanced interfloor performance when 50% of the building population use the lifts each hour. Compare the fixed sectoring and computer based systems, stating approximate performance times.

Using Equation (14.7):

$$\beta = \frac{50 \times 100}{12 \times 15} = 28\%$$

Using Figure 14.1 for fixed sectoring systems:

performance figure = 50%

AWT = 0.5 × 34 = 17 s

Using Figure 14.1 for computer based systems:

performance figure = 30%

AWT = 0.3 × 34 = 10 s

14.6 NUMBER OF STOPS

Part of the philosophy of interfloor traffic patterns was that lifts made more stops, whilst handling fewer passengers. Thus, it may be of interest to record the number of stops as evaluated from the simulation runs. Figure 14.3 shows the number of stops made for an increasing interfloor demand. The number of stops made by each car shown in the figure increases with demand, until a saturation level is reached. The saturation value, of the order of 4.5 stops/car/minute, is not very dependent on the lift configuration or the control algorithm. This is a predictable result, because a saturated system causes the cars to stop at almost every floor and the number of stops per minute is consequently limited by the floor cycle time.

The number of stops can be estimated for the lift systems simulated using Equation (14.8).

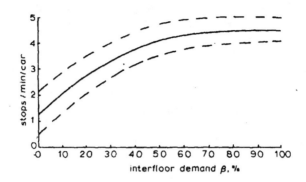

Figure 14.3 Number of stops during interfloor traffic

Limit to number of stops car/minute:

$$\frac{60}{T + (P_l \times t_l) + (P_u \times t_u)} \tag{14.8}$$

where:

    T is the single floor performance time

    P_l and P_u are the number of passengers entering and leaving the car at each floor

    t_l and t_u are the single passenger loading and unloading times.

This formula assumes that the passenger transfer times exceed the door dwell times. To evaluate the formula, a number of assumptions can be made. The single floor performance time can be taken as 10.0 s and the individual passenger transfer times taken as 1.0 s. At saturation assume two passengers leave and two passengers enter at each stop. This determines the denominator terms as 14.0 s. Then the number of stops is 4.3. A similar result to that obtained by simulation.

Siikonen has indicated that the limiting number of stops may be even lower, as any systems that employ long door dwell times, without passenger detectors, increase the cycle time as these times are longer than the passengers require to board or alight. She puts the limiting value at 3.3 stops/car/minute.

It is interesting to note that Bedford (1966) said a lift system was busy with 2.25 stops per car per minute. This suggests that a busy system is below that of a saturated system.

Lim (1983), as part of his work to compare the dynamic sectoring (DS) and computer group control (CGC) algorithms, plotted the different types of car stops against interfloor demand. This is shown as Figure 14.4.

Figure 14.4 shows that at low levels of demand there are very few common landing and car call stops (CLSTP). Even at high demands, common landing and car call stops only amount to about 20% of the total car stops (CSTP). This observation is confirmed by examining the curve showing the ratio of car call stops to landing call stops (CC/LCSTP). This is almost constant at a figure of 1.2. This supports the earlier suggestion that each prospective passenger causes a car stop to be picked up and introduces a new call to be set down. There are very few algorithmically induced movements.

Taking the Bedford "busy system" of 2.25 stops per car (CSTP) to Figure 14.4 shows an interfloor demand range of 17% to 25%, which corresponds to the 30% to 36% values of passenger activity mentioned in Section 14.5.

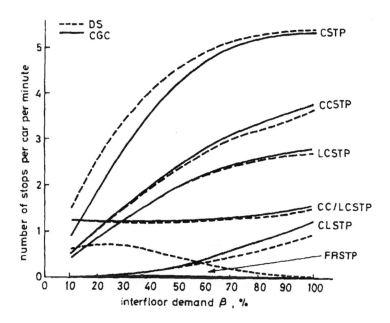

Figure 14.4 Comparison of lift car stopping activity
Legend:
CSTP = total car stops
CCSTP = car call stops
LCSTP = landing call stops
CC/LCSTP = ratio car calls to landing calls
CLSTP = common landing and car call stops
FRSTP = algorithmically induced stops
DS = dynamic sectoring
CGC = computer group control

14.7 RELATING PERFORMANCE

During uppeak and down peak traffic the usually applied performance measure is passenger average waiting time, except this parameter cannot be measured. So the obvious and measurable parameter is the interval. During interfloor traffic the concept of the round trip and an interval is not viable. So what can give an indication of performance ?

Figure 14.2 gives a theoretical relationship between the passenger average waiting time and the underlying uppeak interval. Lim (1983) carried out a number of investigations to find a better performance criteria for interfloor traffic. This was to use ninety percentiles (90%) values of passenger waiting time. This idea fits well with the one hour view that must be taken of interfloor traffic. This moves the debate away from averages, which only satisfy half the passengers, towards a 90% value, where only 10% of the passengers are disadvantaged.

Figure 14.5 gives a comparison between average waiting time and a ninety percentile waiting time for the range of interfloor demand. If these two parameters are expressed as a ratio, the graph of Figure 14.6 is obtained. This graph shows an almost constant ratio of 2.2 for interfloor demands over 30%. However, this ratio is not useful as passenger average waiting time cannot be easily measured.

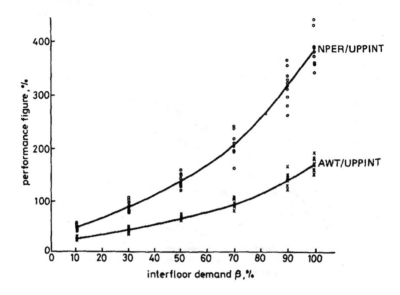

Figure 14.5 Comparison between passenger average waiting time and the ninety percentile

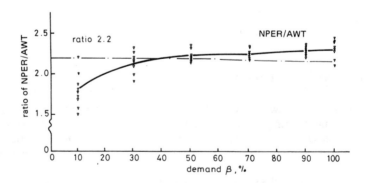

Figure 14.6 Comparison between passenger average waiting time and the ninety percentile

Because Lim was simulating the lift systems he was able to plot, in Figure 14.7, the ratio of the ninety percentile to the average system response time (see Definition 6.7), which can be measured.

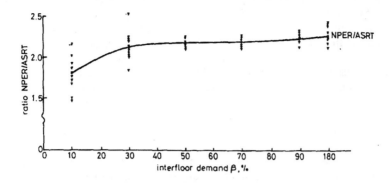

Figure 14.7 Ratio of the ninety percentile to the average system response time

Figure 14.7 shows a ratio of approximately 2.1 between the ninety percentile and the average system response time for interfloor demands greater than 30%. It can therefore be concluded that 90% of passengers wait no longer than twice the average system response time. It should be noted that at the more likely demand levels of less than 30%, the ratio becomes smaller to approximately 1.8.

From Figure 14.6, $NPER = 2.2AWT$

From Figure 14.7, $NPER = 2.1ASRT$

So $AWT = 0.96 \times ASRT$

Thus AWT and $ASRT$ are almost equal.

14.8 CONCLUSIONS ON BALANCED INTERFLOOR TRAFFIC

A large number of simulation runs for a variety of lift configurations, demand levels, number of floors and supervisory control systems have allowed the assessment of the Quality of Service provided by a lift system under balanced interfloor traffic conditions.

The lift performance degrades severely with demand and is dependent on the type of supervisory control system implemented. A dynamic sectoring system is more efficient than the fixed sectoring systems, but the adaptive call allocation computer control algorithm presents the best performance.

A "busy" lift system can be defined as one where cars make 2.25 stops/minute and handle about one third of the building population in one hour. Interfloor demands higher than 50% tend to saturate the lift system, causing the cars to stop at most floors thus degrading the quality of service. At the saturation level of demand, lifts stop over twice as often as during a busy period.

The system response time provides a measure of Quality of Service as it is related to ninety percentile of all passengers not waiting longer than twice the average system response time.

The concept of a round trip time has no meaning during balanced interfloor traffic. Once interfloor traffic becomes unbalanced then it may transform into a dominant traffic pattern such as uppeak or down peak, which are analytical. The traffic pattern could mutate into an unbalanced traffic pattern such as that experienced at lunch time. This will be considered in the next chapter. The underlying handling capacity of the lift system is never utilised during balanced interfloor traffic. Siikonen (2002) suggests this underlying handling capacity might be 40% larger than during uppeak. If all this handling capacity were to be used, each person in the building would be using the lift twice every hour and waiting times would be very large. Because only some 25−30% of the underlying capacity is used, the passenger waiting times are generally excellent.

14.9 A WARNING

The information presented in this chapter enables a lift designer to estimate the balanced interfloor lift performance in relation to an uppeak design. However, it should be noted that the figures given here refer to a balanced demand per floor. If a building is unequally populated, it is known from practical simulations carried out that waiting times can become extended.

Most methods employed to improve, or boost, the uppeak handling capacity do not improve balanced interfloor performance as the capability of the underlying lift installation remains the same. The effect on the performance during interfloor traffic of a "boosted" system will be less obvious, as the system utilisation is much lower. The only traffic control technique that could be used to boost interfloor performance would be a call allocation system. This system could serve the calls more efficiently as both the calling floor and destination floors are known the controller. But as the passenger demand is much lower than the system capability, it is unlikely that the expense is justified over a two button system.

CHAPTER FIFTEEN

Review of Traffic Patterns (including Mid Day Traffic)

15.1 PREAMBLE

In the previous chapters, uppeak, down peak and interfloor traffic patterns have been analysed and discussed. These were all considered as pure patterns, ie: with NO mixed mode traffic. The mid day traffic pattern is defined in Section 4.4.3 and other traffic patterns were touched upon in Section 4.4.5. This former pattern can be analysed mathematically. Other patterns, especially where there are mixed modes of demand, are best examined using simulation techniques (see Chapter 16).

15.2 MID DAY OR LUNCH TIME TRAFFIC

15.2.1 Background to the Traffic Condition

The mid day or lunch time traffic is today regarded as a most severe test of the capability of a lift system to handle the passenger demands. It is a traffic condition that makes use of all of a lift systems capability. This period of time can last over two hours depending on the business arrangements in a building. During this period passenger demand builds up as passengers depart from their "home" floors and then return some time later. A simple case would be a building that does not contain refreshment or other facilities and the traffic generated would include:

(a) passengers leaving their home floor to travel to the main terminal
(b) passengers returning from the main terminal to their home floor
(c) passengers moving between floors (interfloor traffic).

In a more complex case, where there are refreshment or other attractions, these may be placed on a "facilities" floor. If the facilities floor were the same floor as the main terminal its effect would be similar to the simple case described above. More likely the facilities floor will not be the main terminal and might simply be a restaurant. It may, however, include other facilities such a travel agent, banking facilities, shops, leisure centre, etc. Also it may not be situated in the lower part of the building and could be the highest floor. The facilities floor can therefore be very attractive to the building occupants. Besides traffic types (a), (b) and (c) listed above, other traffic generated would include:

(d) passengers leaving their home floor to travel to the facilities floor
(e) passengers leaving the facilities floor to travel to their home floor
(f) passengers leaving the facilities floor to travel to the main terminal
(g) passengers leaving the main terminal to travel to the facilities floor.

The character of the traffic changes throughout the mid day period. Initially, in the first 30 minute period, persons begin to leave their home floors, ie: traffic types (a) and

(b) in the simple case, and Traffic types (a), (b) and (d) in the more complex case. These persons eventually travel away from the main terminal floor creating traffic type (c) in the simple case and traffic types (e), (f) and (g) where there is a facilities floor. But at the time that the first period persons begin to travel, new persons will start their mid day break. This counter traffic flow can create a very high demand on the lifts. The lower demand at the beginning of the mid day break thus becomes a complex high demand in the middle period of the mid day break, which reduces to a lower demand similar to the first period at the end of the mid day period. The most severe demand of the whole mid day period is the central portion and the lifts can load to their full capacity.

How is this demand to be modelled and what should its value be ? In uppeak traffic there is a benchmark of a pure up demand, in down peak there is a benchmark of a pure down demand and in interfloor traffic there is a benchmark of a completely random demand. Can a benchmark be suggested ?

Peters and Sung (2000) suggest that if the uppeak handling capacity were to be (say) 15% then there could be a 5% uppeak traffic, a 5% down peak traffic and a 5% interfloor traffic. This suggests the modal split of traffic would be three ways and would be a value related to the uppeak handling capability of the lift system.

Siikonen (1998), based on some traffic measurements across the world, quantifies the values at 40% incoming, 40% outgoing and 20% interfloor traffic. She does not specify the demand level, but indicates that whatever it is, the proportions are as stated above.

These two authors use different terminology: Peters uses uppeak/down peak traffic and Siikonen uses incoming/outgoing traffic. Both these terminologies suggest the simple case rather than the complex case. It might be better to term the traffic as up traffic and down traffic, as this will deal with both cases. The Siikonen scenario would seem more likely for the middle part of the mid day period and that the traffic would divided as 40% up traffic, 40% down traffic and 20% interfloor traffic. A further step can be taken by assuming that the interfloor traffic is divided equally in the up and down directions. The best benchmark could therefore be 50% up traffic and 50% down traffic.

15.2.2 Calculation of Mid Day Traffic

Can the mid day traffic pattern be calculated ? To establish a formula for the round trip time during the mid day period a number of assumptions need to be made.

(a) There is a restaurant floor and main terminal floor near together, ie: in the lower part of the building.

(b) Lifts leave the lower part of the building with a car occupancy of 80% of the rated capacity (Table 7.2), ie: with P passengers.

(c) Lifts stop S times on the way up and reverse at the highest reversal floor H.

(d) Lifts leave the highest reversal floor and stop S times on the way down and arrive at the lower part of the building, loaded to 80% of the rated capacity, ie: with P passengers.

(e) The lifts exhibit a round trip, ie: they call at the main terminal every trip.

Initially the main terminal will be an extra stop, but as the traffic pattern establishes itself the lifts will make $2S$ stops every round trip. The time consumed for stopping is therefore $2ST$ (where T is the performance time).

The number of passengers that board and exit on both the way up and the way down the building is P passengers. The time consumed for passenger transfers is therefore $4Pt_p$.

The lift will pass through $(H - S)$ floors both on the way up and the way down the building (see Equation (4.10) and Discussion 4.5.3). The time consumed to travel past these floors is therefore $2(H - S)t_v$.

$$RTT_M = 2(H-S)t_v + 2ST + 4Pt_p$$

$$RTT_M = 2Ht_v + 2St_s + 4Pt_p \tag{15.1}$$

where from Equation (4.14): $T - t_v = t_s$

The handling capacity can be determined by considering the number of passengers transported over the whole round trip. This is $2P$ passengers, giving the formula:

$$MIDHC = 300 \times 2 \times L \times P/RTT \tag{15.2}$$

No account has been taken of interfloor traffic. It might be reasonable to assume that under this heavy demand situation any interfloor traffic boards and alights at the stops serving the main traffic flow and that the interfloor traffic does not introduce any extra stops.

It is also assumed that the traffic controller can deal with a restaurant floor adjacent to and below the main terminal floor, or conversely the restaurant floor adjacent to and above the main terminal floor, when lifts might fill to capacity at the first floor visited and therefore be unable to accommodate any further passengers at the next adjacent floor. The controller would need to detect skipped floors and send lifts to service them. The traffic controller should also employ load weighing to avoid stopping when there is no capacity available for new passengers to board.

Where the restaurant floor is at the top of the building it would be assumed that the cars fill to 80% capacity prior to departure. Similarly cars would be assumed to fill to 80% capacity prior to departure from the main terminal floor.

Table 15.1 shows the results of applying Equation (15.1) to the "four corners of the lift world and its centre of gravity" (see Section 5.12.3).

Table 15.1 Comparison of mid day to uppeak handling capacity

Number of floors	Rated Car Capacity	MID/UPP ratio
10	10	1.34
10	24	1.23
16	16	1.24
20	10	1.29
20	24	1.17

The table shows a range of values during the mid day period. The average handling capacity is some 1.3 times the underlying uppeak handling capacity. These calculated figures do not take into account any incidental interfloor traffic handled. Siikonen (2002) has shown by the simulation of a small number of systems that the mid day to uppeak ratio is 1.2-1.4. Using the same technique, she also states that she considers the down peak to uppeak ratio is 1.6-1.8, compared to the value of 1.6 derived using Equation (13.9). The work reported in Chapter 13 relies on the extensive simulation of over 2000 systems and should encourage a designer to use the lower value of 1.6. The differences between the Siikonen values and those reported here could be accounted for by the different calculation and simulation techniques, the number of systems considered and the traffic control systems employed. However, the values are very similar.

15.2.3 Mid Day Demand

Peters and Sung (2000) related their modal split of traffic to the underlying uppeak handling capacity. Table 15.1 also relates demand to uppeak capacity. But what is the demand? Can this demand be modelled and simulated? Again no benchmark of likely demand exists. Factors to be considered:

(a) What is the length of the mid day period?

(b) What is the length of each person's mid day period?

(c) What percentage of floor occupants use the lifts during the mid day period?

(d) How long do they stay away from their home floor?

(e) In a restaurant how long do they stay (residence time)?

(f) What is the departure/arrival profile of passengers?

(g) What is the modal split between a main terminal and a facilities floor (if any)?

These are broad assumptions and will have a number of other factors that affect them. For example how many covers are there in the restaurant, how efficient is the service in the restaurant, are there snack/self service kitchen facilities on, or close to, each floor, etc.? The following are practical suggestions based on experience:

(a) 2 hours

(b) 30 minutes

(c) One third per hour; two thirds per mid day period

(d) 30 minutes

(e) 30 minutes

(f) In a 30 minute period there are six, 5 minute periods. Assume the departure profile is similar to down peak and that all the demand is over 10 minutes, ie: two 5 minute periods. Also assume that the return profile occurs at the start of the next 30 minute period and also last 10 minutes. Note the co-incident demand.

(g) Two thirds to the facilities floor, one third to the main terminal.

Assumption (f) relates well to the evening down peak passenger behaviour. Then building occupants wish to leave the building as quickly as possible. The heavy demand lasts for 10 minutes and uses all the underlying capacity of the lift system. The underlying capacity of the lift system was shown in Chapter 13 to be 1.6 times the underlying uppeak handling capacity. Building occupants are in a similar situation during the mid day period as they are in their own time and want refreshment or to leave the building and therefore it is reasonable to assume that they make similar demands on the lift system.

The scenario indicated above may be more severe, if the mid day period is shorter, or if the numbers of persons using the lifts is larger. The demand may, however, be smaller, if the demands are more random and do not occur simultaneously as indicated in the scenario.

It is possible to relate assumption (f) to a demand level. Each floor has a 100% population. There are four, 30-minute mid day periods, thus 25% of the population is available to use the lifts every 30-minute period. Only two thirds of them use the lifts (16.7%) and then the demand is spread over two, 5 minute periods. The demand in each of the 5 minute periods is thus 8.3%. This is a demand well below any ever considered for an uppeak design. There is a difference, however, as when the down traffic demand coincides with the up traffic demand then the lifts must provide twice this requirement, ie: 16.7%. Using the mid day/uppeak ratio of 1.3, the underlying uppeak handling capacity must be at least 13%.

15.2.4 Example 15.1

Consider a building with 11 floors above the main terminal. Floor 1 is a restaurant floor. The single floor transit time is 2.0 s, the performance time is 10 s and the passenger transfer time is 1.0 s. There is a population of 180 persons on Floors 2−11. For a 14% uppeak arrival rate (252 persons/5-minute), what is a suitable system ?

With 8 lifts of a rated capacity of 1800 kg the design capacity is 18.6 persons (Table 7.2) and with an occupancy of 14.9 persons this provides the 14% uppeak arrival demand. A calculation using the LTD program gives an interval of 17.5 s. Other parameters from the calculation are: $H = 9.7$, $S = 8.6$, $t_v = 1.6$ s, $T = 10.2$ s, $t_p = 1.0$ s.

What is the mid day traffic handling capacity ?

Using Equation (15.1):

$$RTT = 2 \times 9.7 \times 1.6 + 2 \times 8.6 \times 8.6 + 4 \times 14.9 \times 1$$

$$= 238.6 \text{ s}$$

The mid day interval *MDINT* is:

$$MDINT = 29.8 \text{ s}$$

Using Equation (15.2):

$$HC = 300 \times 2 \times 8 \times 14.9/238.6$$

$$= 300 \text{ persons/5-minutes}$$

The uppeak handling capacity was 252 persons/5-minutes, giving a mid day to uppeak ratio of 1.19.

What would be the worst case scenario ? Following the suggestions in Section 15.2.3.

Population of each floor	= 180 persons
Proportion who use lifts in mid day period	= 120 persons
Floor population who use lifts in one hour	= 60 persons
Floor population who travel to the main terminal in one hour	= 20 persons
Floor population who travel to the facilities floor in one hour	= 40 persons
The departure profile from each floor over two hours:	= 120 persons
The departure profile from the main terminal over two hours:	= 400 persons
The departure profile from the facilities floor over two hours:	= 800 persons

The departure pattern is shown diagrammatically in Table 15.2.

Table 15.2 Representation of passenger departures from each floor during a mid day period of 2H:10m

Fl	First mid day hour												Second mid day hour												Third ...	
	1	2	3	4	5	6	7	8	9	10	11	12	13	14	15	16	17	18	19	20	21	22	23	24	25	26
10	15	15	0	0	0	0	15	15	0	0	0	0	15	15	0	0	0	0	15	15	0	0	0	0	15	15
9	15	15	0	0	0	0	15	15	0	0	0	0	15	15	0	0	0	0	15	15	0	0	0	0	15	15
8	15	15	0	0	0	0	15	15	0	0	0	0	15	15	0	0	0	0	15	15	0	0	0	0	15	15
7	15	15	0	0	0	0	15	15	0	0	0	0	15	15	0	0	0	0	15	15	0	0	0	0	15	15
6	15	15	0	0	0	0	15	15	0	0	0	0	15	15	0	0	0	0	15	15	0	0	0	0	15	15
5	15	15	0	0	0	0	15	15	0	0	0	0	15	15	0	0	0	0	15	15	0	0	0	0	15	15
4	15	15	0	0	0	0	15	15	0	0	0	0	15	15	0	0	0	0	15	15	0	0	0	0	15	15
3	15	15	0	0	0	0	15	15	0	0	0	0	15	15	0	0	0	0	15	15	0	0	0	0	15	15
2	15	15	0	0	0	0	15	15	0	0	0	0	15	15	0	0	0	0	15	15	0	0	0	0	15	15
Rest	0	0	0	0	0	0	100	100	0	0	0	0	100	100	0	0	0	0	100	100	0	0	0	0	100	100
MT	0	0	0	0	0	0	50	50	0	0	0	0	50	50	0	0	0	0	50	50	0	0	0	0	50	50

Table 15.2 shows the maximum demand occurs in Periods 7−8, 13−14, 19−20 and 25−26. In each of these 5 minute periods the total up plus down demand is 300 persons. As the lifts can handle 300 persons/5-minutes, this means the mid day period can be satisfied in terms of handling capacity.

15.2.5 Observations on the Mid Day Period

The lift system may be able to handle the demand, but what will the passenger waiting times be ? The uppeak interval for the underlying lift system in Example 15.1 was 17.5 s and the mid day interval was 29.8 s. The lifts were assumed to load to the same occupancy level in both cases. This means that the passenger average waiting time would be longer during the mid day period by the ratio of the interval values. The ratio in this case is 1.7, ie: almost twice as long. The likely formula for the mid day passenger average waiting time would therefore be:

$$MIDAWT = 0.85MIDINT \tag{15.3}$$

In practice, the mid day demand may not be as large as discussed above. Two other scenarios are shown in Simulation Case Study CS21. Unlike Example 15.1, these use a lower demand and a skewed overlap. This might be a more likely scenario.

Whatever the assumptions that are made, the procedure above gives a possibility of a calculation analysis. A simulation, however, allows a more realistic judgement to be made.

15.3 REVIEW

The primary traffic condition is pure uppeak. This traffic condition is usually the condition to be satisfied in the traffic design. It is analytic and formula can be derived. The pure down peak and mixed mode, mid day traffic patterns are usually satisfied by the sizing carried out for the uppeak traffic pattern. Both of these patterns are analytic if assumptions are made. The only traffic pattern that defies a mathematical examination is the interfloor traffic pattern. Fortunately a lift system subjected to this pattern is never called upon to deliver its full capability. Interfloor demands are modest compared to the other three traffic patterns. Table 15.3 indicates the formulae derived.

Table 15.3 Review of traffic patterns

Traffic pattern	Round trip time equation	Handling capacity equation	Passenger average waiting time equation
Up-peak	$RTT = 2Ht_v + (S+1)t_s + 2Pt_p$ Equation (4.11) - Page 99	$UPPHC = 300P/UPPINT$ Equation (4.8) - Page 97	$AWT = [0.4 + (1.8P/RC - 0.77)^2]INT$ Equation (6.1) - Page 134
Down peak	$RTT_D = Nt_v + (0.5S+1) + 2Pt_p$ Equation (13.9) - Page 330	$DNPHC = 300P/DNPINT$ Equation (13.10) - Page 330	$DNPAWT = 0.85\alpha UPPINT$ Equation (13.5) - Page 328
Inter-floor	No equation available	No equation available	$IFAWT = UPPINT (0.22 + 1.78\beta/100)$ Equation (14.2) - Page 339
Mid day	$RTT_M = 2Ht_v + 2St_s + 4Pt_p$ Equation (15.1) - Page 349	$MIDHC = 300 \times 2P/MIDINT$ Equation (15.2) - Page 349	$MIDAWT = 0.85MDINT$ Equation (15.3) - Page 353

All the analysis presented in this book is mainly based on the classical traffic pattern, depicted in Figure 4.2. It has been used to describe how people use lifts in a building and to size lift systems successfully for over half a century. But it is largely a figment of the imagination, as it probably has never existed. Siikonen (1998, 2002), Peters and Sung (2000), Powell (2002) and many others all report considerable differences to Figure 4.2. Just because Figure 4.2 has been "discredited" it does not mean it should be abandoned as a valuable tool. As a "benchmark" it is generally accepted world wide. Countless buildings have been designed to its "illusion" and the designs work ! Why does it work ? The answer is that the uppeak design provides an underlying capacity, which sets the performance of the three other major traffic conditions of down peak, interfloor and mid day traffic.

The previous chapters have looked at the ratio of inherent handling capacity for these three traffic patterns compared to uppeak handling capacity. With uppeak considered to be unity, the ratios are:

uppeak   1.0
down peak   1.6
mid day   1.3.

These ratios assume that all the handling capacity can be utilised.

The interfloor handling capacity is never utilised, as the demand is about one fifth of the uppeak demand. In fact, lift system performance under interfloor traffic demand has an inherent capability as high as 1.4 times the underlying uppeak capability (Siikonen, 2002):

interfloor   1.4.

All the formulae given in Table 15.3 relate to pure traffic patterns. These are unlikely to be seen in practice. However, the tradition of applying the classical approach has generally proved satisfactory. Chapter 16 deals with the more complex traffic patterns by using the simulation technique. This technique allows a full understanding of a lift system's capability.

CHAPTER SIXTEEN

Simulation and Computer Aided Design

16.1 SIMULATION AND CAD DEFINED

Chapters 4−8 have used calculation methods to design and evaluate the performance of lift installations. These calculation methods are based on mathematical models of the lift system. Other methods that have been discussed in Chapters 12−15 are based on the computer modelling of the lift systems to provide empirical formula. This computer modelling was carried out using a simulation program operating in batch mode.[1] All calculation methods rely on simplifications and assumptions in order to make the calculations possible. Such simplifications and assumptions are discussed in Chapter 7. Other simplifications need to be made to deal with special situations, see Chapter 8.

Where the traffic design requirements are more complicated, computer simulations may need to be carried out. Simulation allows all traffic conditions to be examined more exactly. They allow a check to be made of the underlying basic traffic design and a comparison to be made with the more realistic representation obtained by simulation. The realism of the simulation depends on the richness of the model used. The simulated results may be quite different from those obtained by calculation. This is because all calculations are based on statistics and generally deliver an average answer to a design, not a specific one.

The computer simulation of engineering processes is particularly appropriate where the study of the actual process is difficult or dangerous, too costly, would take too much time, or would be inconvenient. Existing lift systems fall into this category. In the case of a new lift installation it does not even exist.

Definition 16.1: Simulation is the development and use of models to aid in the evaluation of ideas and the study of dynamic systems or situations.

Digital computers are most suitable for the simulation of discrete systems (but not continuous systems[2]) that can be described by sets of logical equations. A lift system is a discrete system:

- Each individual passenger arrival is a discrete event.
- Each individual passenger departure is a discrete event.
- Each lift is a discrete unit.
- Each floor is a discrete entity.
- Each car movement is a discrete occurrence.
- Each door operation is a discrete occurrence.
- etc., etc.

---

[1] Batch mode is where a large number of simulations are carried out, without human interaction, against an input file of data, writing the data to output files for later analysis.

[2] Continuous systems are usually analogue systems and are described by differential equations.

Digital computer simulation programs can be either event based, ie: the model is updated every time something happens, or time based, ie: the model is updated at regular intervals. A lift system is sparse in the number of events that occur (see Section 11.4.4), compared to some engineering systems. Most events also do not require immediate action, and some events initiate identical actions, making it efficient to service them at the same time. It is therefore sensible to select the time based method, with an update interval chosen to service all events in a reasonable time.

Engineering design involves the appreciation of shape, form and relative values, thus the graphical presentation of data allows the designer to appreciate a design quickly. The designer should submit the input data, receive the results back quickly, appreciate them and resubmit the design, if required. This is computer aided design (CAD) and the process is illustrated in Figure 16.1.

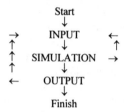

Figure 16.1 Phases of a computer aided design

Definition 16.2: Computer aided design is where a designer interacts directly with a complex computer process and in so doing closes the design loop.

The necessity for the computer aided design of lifts was foreseen by Jackson (1970) who wrote:

"a real need . . . is a computer program to simulate the likely performance of proposed lift systems . . . Different numbers, speeds and groups of lifts should be considered, as well as different control systems . . . the results would show designers the performance of several proposals . . . [and allow] . . . rational decisions".

The CAD of lift traffic systems would follow the pattern shown in Figure 16.2.

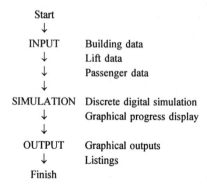

Figure 16.2 Phases in the CAD lift traffic systems

16.2 ABOUT PC-LSD AND ELEVATE

Most lift companies and many consultants have calculation programs and some have simulation programs. The information regarding their internal structure is not generally available. This chapter looks at the traffic design of lift systems using simulation within a CAD process, illustrating the discussion by examples from two programs available in the public domain: PC-LSD and ELEVATE.

PC-LSD (Personal Computer - Lift System Design) was developed by Dos Santos (1972) at the University of Manchester Institute of Science and Technology. Programmed in FORTRAN IV for a main frame (now called a server) time sharing computer (Digital Equipment PDP10), it was later transferred to a DEC minicomputer (PDP11) and in 1985 to the first IBM compatible personal computers, running under the DOS operating system. After dos Santos, the design suite was further enhanced by others (Mousellati, 1974; Hirbod, 1975; Swindells, 1975; Lim, 1983). All development, except bug fixes, ceased in 1985.

The programming facilities available in 1972 were such that PC-LSD is modular in design and suffers from a very poor user interface for the input of data. It can, however, simulate a wide range of traffic algorithms, which include: two simple collective algorithms (COL, MCO), one duplex/triplex algorithm (THV), three fixed sectoring algorithms (FSW, FS4, FSO), two dynamic sectoring algorithms (DS, SCH) and four computer control algorithms (CGC, ACA, CCU, MAS). PC-LSD is rich in graphical displays of passenger waiting and journey times, percentile plots, call response graphs, car spatial movements, etc. These graphical outputs can be printed using screen capture programs. Full numerical print outs are available.

PC-LSD was supplied to over 20 enterprises in the UK, Europe, North America and Asia, including lift companies, standards bodies, consultants and educational establishments. Difficult to use it never became an industry standard, but is still available for the enthusiast.[1]

ELEVATE is much younger and has the look and feel of most modern computer applications with a simple interface for the input of data and easy navigation around the suite. It does not have many in-built traffic control algorithms. It does, however, provide a facility for users to incorporate their own algorithms. ELEVATE's standard output is limited to some simple percentile graphs and numeric print outs. However, as large amounts of data are collected which can be output to a spreadsheet application program, users can design their own graphs. The simulation phase is illustrated by a display of lift movements, as the simulation proceeds, and is much appreciated by architects. The ELEVATE suite also includes calculation programs which range from the simple methods, such those based on Equation (4.11), as detailed in Appendix 2, through those based on the generalised analysis procedure (Peters, 1990) to full simulation.

Some 500 copies of ELEVATE[2] are in operation world wide and it is under continuous development and many of its shortcomings are being addressed. It is likely to become the industry standard for the 21st Century.

In general, PC-LSD standardises the format and style of the data entry and presentation for the traffic designer whereas, ELEVATE does not presume to do this and allows the traffic designer freedom to produce outputs of their choosing.

In the discussion that follows, features of PC-LSD are identified as {P} and ELEVATE as {E}.

---

[1] www.liftconsulting.org

[2] www.peters-research.com

16.3 UNDERLYING STRUCTURE OF A DIGITAL COMPUTER SIMULATION PROGRAM FOR LIFT TRAFFIC DESIGN

16.3.1 Time Slices

Section 16.1 explained that it is sensible to select the time based method of digital simulation. Thus an update interval must be chosen to service all events in a reasonable time. The choice of the update interval, or time slice, is important if the accuracy of the numerical values obtained is to be maintained.

One method {P} is to have a relationship with the speed of a lift. The simulation program would automatically select the update interval. If (say) the update interval was one tenth of the single floor transit time then most events would be accurately represented. For a lift with a high rated speed (say 5 m/s) travelling a short distance (say 3.0 m) the update interval would be small (0.06 s). Of course a slow speed lift (say 0.25 m/s) with a long travel distance between floors (say 10 m) would have a long update interval (4.0 s) and this could mean a loss of accuracy. This loss of accuracy can be overcome by keeping all times as an integer number of update intervals plus a remainder.

Another method {E} is to select a fixed update period (say 0.1 s) and to keep real values of all numbers. The maximum error with this method would then be no more than 0.1 s.

16.3.2 Display Update

The display presented during the simulation phase could be updated at every time slice. This is unnecessary as the simulation will be running many times faster than real time (<60 times). A suitable display update could be every ten update intervals {P}. If the user does not want this detail then the display update could be made a variable {E} and altered to suit the user's requirements.

16.3.3 Random Number Generation: Arrivals and Destinations

The arrivals and destinations must be selected. A common approach is to use a random number generator, which on a digital computer will not be produced from "white noise", but from a pseudo-random binary sequence (PRBS) subroutine. PRBS subroutines need a starting point known as a "seed". Sometimes this seed is fixed {P} or can be different for each design run {E}.

Some programs {E} use a PRBS number generator to randomly introduce passengers into the simulation and to select their destination. Other programs {P} use a PRBS number generator for destinations, but a Poisson process for arrivals.

16.3.4 Number of Simulations

A digital computer always gives the same answer in response to the same input. If computer time is at a premium (as it was in 1970) then one simulation per design would be carried out {P}. This has been shown to be accurate (Dos Santos, 1972). However if a much more authoritative result is desired and computer resources are available (as they are in the 21st Century), then a number of simulations can be performed {E} and an average obtained.

16.3.5 Simulation Period

Lift systems have to serve four main traffic conditions: uppeak, down peak, interfloor and lunch time. These are best studied as distinct time bands {P} of one hour of activity. If different time bands for study were desired {E} then the start and finish time of each period would need to be defined. It has been established (Section 4.5.2) that 5 minutes is a realistic minimum time period in which to analyse data. Longer periods can be considered for the other traffic conditions. It therefore follows that in any simulation period that data are accumulated over 5 minute periods and analysed over one or more 5 minute periods. These 5 minute periods can be called data collection periods.

16.4 SIMULATION AND DESIGN OF LIFT SYSTEMS

16.4.1 Input Phase

Figure 16.2 indicated there were three sets of input data required to describe a simulation: the building; lift; and passenger data sets. These data sets are described in Section 5.1. Before data entry starts there should also be an identification header for the work.

One approach to data entry is an interactive dialogue of question and answer {P}. Here all the data for a design are entered in a logical order and all input data must be entered before the designer is allowed to proceed to the simulation. Any errors can be corrected or data changed by entry to associated sequences of data. Figure 16.3 shows an extract from such an approach.

```
Design Name:    Test
Run Number:     23
Date:           21st Century

Which module to run next (INP/SIM/OUT) ? INP

What next (type ? for help) ? T

Speed of lifts (m/s) = 2.5

Flight time (acc = 1.0 m/s2):

    1       2       3       4       5
    4.6     6.0     7.3     8.5     9.6

OK/acc/enter ? OK

What next (type ? for help) ? E

Which module to run next (INP/SIM/OUT) ? SIM

Display arrivals at floor = 10
```

Underlines indicate data entered.

Figure 16.3 Data entry by question and answer (PC-LSD)

Nowadays computer users are more used to a screen entry approach {E}. Here each screen permits a set of associated data to be entered. Checks need to be made that all valid data are entered before a simulation is performed. Figure 16.4 shows a screen {E} using this approach.

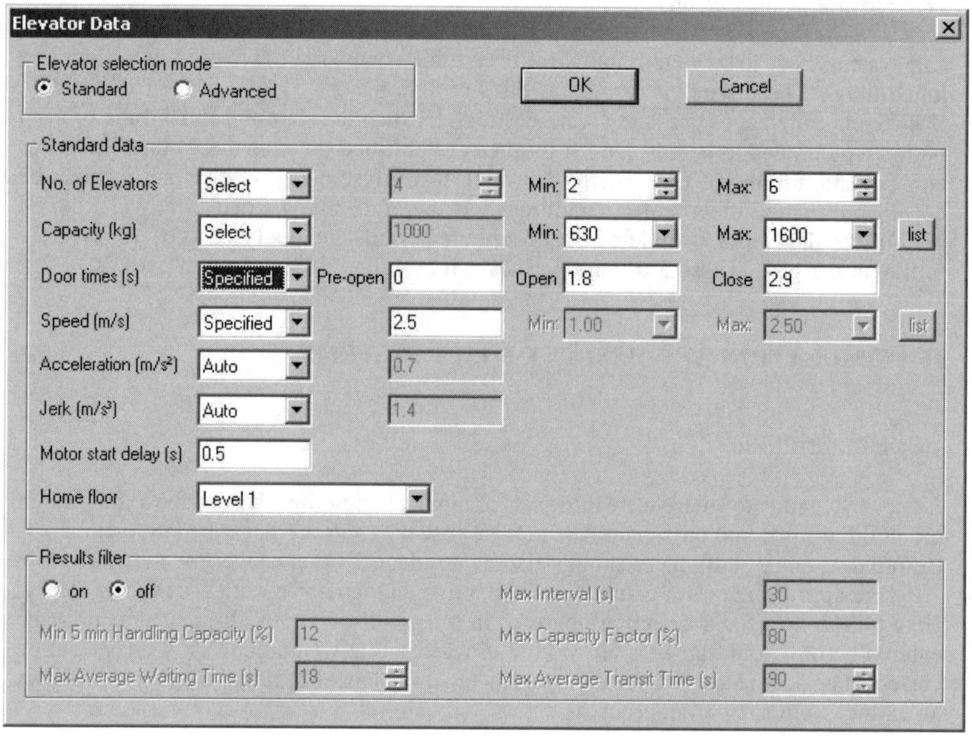

Figure 16.4 Screen based data entry (ELEVATE)

16.4.1.1 Building data set

The building data set includes:

- Number of floors
- Interfloor distances
- Express jump.

The simplest case is where a lift system serves a main terminal and a number of floors above. In this case the easiest description is to provide the total travel distance {P}. Even where the interfloor distance varies, for example to accommodate the different heights found at the main terminal floor, service floors, etc. the effect of these difference is not usually significant, as Section 7.4.3 indicates. To be able to enter {E} the interfloor distance of each floor will give a more accurate result.

Where there is service to an upper zone then the first floor served (the Express Zone Terminal) must be entered, as the travel time through it can be large. Some programs {P} are arranged for this to be defined at the start of data entry.

The identification of each floor for the purposes of analysis can be simply to give them numbers {P}, or to give them any identification including names {E}. In the former case a translation table may need to be provided. The main terminal {P} or home floor {E} should be defined.

16.4.1.2 Lift data set

The lift data set comprises:

- Number
- Capacity
- Speed
- Acceleration and jerk
- Single and Multi-floor flight times
- Door opening time
- Door closing time
- Advance door opening time
- Start delay
- Floors served
- Traffic control algorithm.

The whole purpose of a traffic design is to select the number and capacity of the lift installation and its associated parameters. It is possible to allow the simulation program {E} to auto-select the parameters, or they can be defined at the start of the study {P & E}. An advantage with auto-selection is that a tyro designer, who by definition is unskilled, can be "trained" to use appropriate values. A disadvantage is that the tyro may accept an inappropriate selection. Regardless how the parameters are selected, whether by the designer's conscious decision or by auto-selection they should be checked carefully.

The number of lifts may be known because the architect or owner has defined them or because it is an existing building. Some simulation programs {E} will try several combinations of number and capacity of lifts whilst searching for a solution.

The capacity of a lift is always debatable (see Section 7.9). Some programs {P} require a maximum number of persons permitted in the lift to be specified. Other programs {E} allow the rated load (kg) to be specified and allow the designer to specify the average weight of the passengers. This then defines the maximum number of passengers considered in the design.

Speed can be selected from Table 5.3 {P}, or auto-selected {E}, according to the travel distance. The average interfloor distance should be provided and if the acceleration (possible) and the jerk (unlikely) are known, then single and multiple floor flight times can be calculated using the analysis of Appendix 1. This calculation can be embedded into the program {P & E} or the designer could enter specific floor data {P}. This latter data may have been obtained by measurement (see Appendix 3).

Door opening, closing and advance opening (or pre-opening) times can be selected from Table 5.6 {P}, or auto-selected {P & E}. Again these times might be those obtained from measurements.

What is important is the performance time (T). This includes all door times, the flight time and other times, such as start delays. Generally the designer is wise to adjust the component parameters, in order to achieve the target performance time whether as a design parameter or from a site measurement.

Some installations restrict service to certain floors or do not physically serve certain floors, ie: basements. It should be possible to indicate this in the design data {E}. Some installations have lifts of different capacities and speeds included in the group. Simulation of such circumstances is advantageous {E}.

The traffic control algorithm is the most important feature of a simulation program. Several generic types should be available including: collective {E & P}, timed collective {P}, nearest car {P}, fixed sectoring {P}, dynamic sectoring {P}, estimated time of arrival (ETA) {E & P}, stochastic {P} and hall call allocation {E & P}.

Each algorithm will have parameters that require selection. These include sector boundaries, load detection levels, landing call timers, priority floors, parking floors, etc. All these parameters must be adjustable {P}.

16.4.1.3 Passenger data set

The passenger data set includes:

- Arrivals
- Destinations
- Transfer times
- Disruption factors
- Stair usage factor.

The passenger arrivals can be specified by an envelope defining the rate of arrivals in calls per 5 minutes. A simple, easy to define envelope is the gaussian shape shown in Figure 16.5(a). This requires three parameters to define the shape: the peak value, the base value and the time period for which the arrival rate exceeds the base rate, as Equation (16.1).

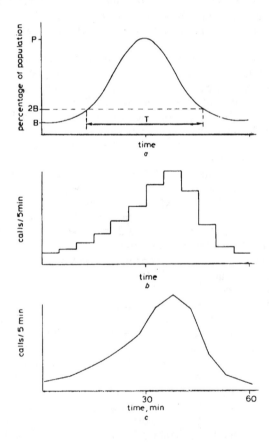

Figure 16.5 Arrival profiles (a) standard (b) step (c) sloped

The peak rate is defined for a 5-minute period and the base rate is defined for a 30 minute period. The shape can be used to represent uppeak and down peak, by adjusting the thinness or fatness of the peak. It can also be used to simulate an interfloor arrival process, which starts and ends at low values and builds up to a larger value at the middle of the simulation period, eg: a lunch time period.

A more practical envelope is a stepped shape, as shown in Figure 16.5(b). The steps could have been obtained from observations on site, for example. Input of this data would involve entering 12 values. Entering 12 values for each floor in the simulation would be tedious and error prone. Facilities to assist include a common profile {P} technique applied to all floors, or the use of copy, cut, paste and fill techniques {E} familiar to spreadsheet users.

The envelope of Figure 16.5(c) is the envelope of Figure 16.5(b) but smoothed by sloping the lines connecting different 5 minute periods.

It is useful at the end of the entry of passenger arrivals to be informed {P} of the total population arriving in one hour. This can be used to relate to actual daily arrivals compared to total building population (see Section 6.5).

The definition of the destination floors for the arriving passengers is important. In uppeak all floors might be assumed equal {P}, which simplifies data entry. In down peak the main terminal floor will have the greatest attraction. In all other traffic conditions the attraction of one floor to another will vary from 5 minute period to 5 minute period.

One method to enter data {P} is to assume all floors are equal and worth 100. To make a floor more attractive 100 is added, making 200, and to make a floor attract no traffic 100 is subtracted, making zero. Table CS18.1 illustrates an uppeak traffic situation where Floors 21–30 have equal attraction and Floor 7 (the facilities floor) has 50% attraction. Table CS20.1 shows an interfloor traffic pattern where Floors 21-30 have an equal attraction and Floors 0 (main terminal) and 7 have twice the attraction. The advantage of simplicity is sometimes lost by the need for a complex set of off line arithmetic to work out the attraction of each destination.

Another method {E} is to be able to define for each simulation period the percentage attraction of each floor relative to the other (Peters, 2000). All attractions in any simulation period must add up to 100%.

Level 14	↑ 0%	↑ 0%	↑ 0%		↑ 17%
Level 13	↑ 17%	↑ 17%	↑ 50%		↑ 5%
Level 12	↑ 0%	↑ 0%	↑ 0%		↑ 5%
Level 11	↑ 17%	↑ 17%	↑ 0%		↑ 3%
Level 10	↑ 0%	↑ 0%	↑ 0%		↑ 8%
Level 9	↑ 17%	↑ 17%	↑ 0%		↑ 7%
Level 8	↑ 0%	↑ 0%	↑ 0%		< 233
Level 7	↑ 17%	↑ 17%	↑ 0%		↓ 5%
Level 6	↑ 0%	↑ 0%	↑ 0%		↓ 5%
Level 5	↑ 17%	↑ 17%	↑ 0%		↓ 10%
Level 4	↑ 0%	↑ 0%	↑ 0%		↓ 10%
Level 3	↑ 15%	↑ 15%	< 25		↓ 10%
Level 2	↑ 0%	< 35	↓ 0%		↓ 0%
Level 1	< 75	↓ 0%	↓ 50%		↓ 15%

Legend:
< **xx** arrivals (persons/5-minutes)
↑ up destination probability (%) ↓ down destination probability (%).
Figure 16.6 Describing traffic (after Peters, 2000)

Either method requires a great deal of data entry, which is where the spreadsheet techniques {E} of copy, cut, paste and fill are invaluable. The availability of input data templates {P} for uppeak and down peak traffic patterns simplifies the task.

The stair factor {E} allows the arrivals at any floor to be reduced in deceasing ratios if they intend to travel one, two or more floors. This could be allowed for {P} by reducing the arrival rate or reducing the floor population.

Simulation is a powerful tool and like all tools can be misapplied. The main area of care in simulation is to ensure the model to be simulated is correct in all respects. A simulation based on incorrect input data looks as convincing as the real thing ! Once all the input data has been entered it should be checked. To do this a print out {P} or paged/scrolled display should be available.

16.4.2 Simulation Phase

The structure of a typical simulation module could be as shown in Figure 16.7. There will be a control sub module for each of the available traffic control algorithms. A simulation can be arranged for one hour of activity {P} divided into 12, 5-minute periods, or defined for any length and number of periods {E}.

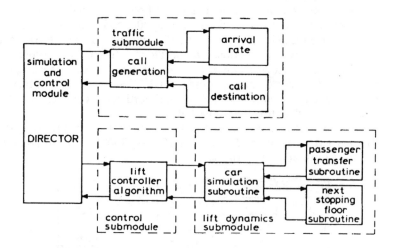

Figure 16.7 The structure of a control and simulation module

The previous history of a group of lifts will always influence the future performance. There is always a difficulty to initialise a simulation in order to achieve a realistic result. Exactly where should the lifts be in the building, what is their status, and how many passengers do they contain, and how many calls are there ? This is solved by running a simulation for 5 minutes at the initial demand levels, before data collection starts.

A simulation will take some time to execute, dependent on its complexity, and the interest of the designer needs to be maintained. This can be achieved by presenting an activity display. One method {P} is to show the arrival envelope at a selected floor, together with an indication of the number of passengers queuing at that floor. The length of the queue does give some indication of the viability of the design. An example is shown as Figure 16.8 for a gaussian shaped profile and at Simulation Case Study CS18.1 for a stepped profile.

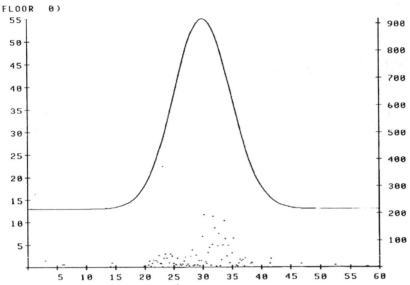

Figure 16.8 Display presented during the simulation phase of a PC-LSD design
The solid line with the right-hand vertical axis indicates passenger arrival rate.
The dots with the left-hand vertical axis represent the queue length.

Another method {E} is to display a mimic of the lift installation and to show the lifts moving, carrying out their door operations, etc. Also other numerical information can be displayed. This display is very fast moving and attractive to architects, but gives little design information. An example is shown at Figure 16.9.

16.4.3 Output Phase

During the simulation, data are acquired. The traffic designer must be able to appreciate the design results easily, in order to iterate rapidly to reach a satisfactory design conclusion. In 1972 there were no precedents for the form of the output information to be provided to the designer. A large number of presentations were developed. The discussion that follows selects those that have proven to be useful {P} over the past 30 years and relates them to modern techniques {E}. Many of the graphs display results averaged over 5 minute periods. This will introduce "end effects" from time to time, eg: a passenger is picked up in one 5 minute period and alights in another. Other graphical displays are presented for one hour of activity.

The importance of checking that the input data are correct was mentioned in Section 16.4.1.3. To make an initial check, an examination of the spatial movements of the lifts will show whether they are serving the defined floors. For example, a floor with no attraction or in an express zone should normally receive no stops. Displays of this type are shown in Figures 4.5, 4.8 and 4.9, and in all of the Simulation Case Studies.

An appreciation of the level of car occupancy and the likely interval at the main terminal is important, if the study is for uppeak, or down peak, or for a traffic condition, where all lifts visit the main terminal floor, such as during a lunch time period (this can be determined by examining a spatial plot). Examples of this type of display are shown in CS18.3 and CS19.3.

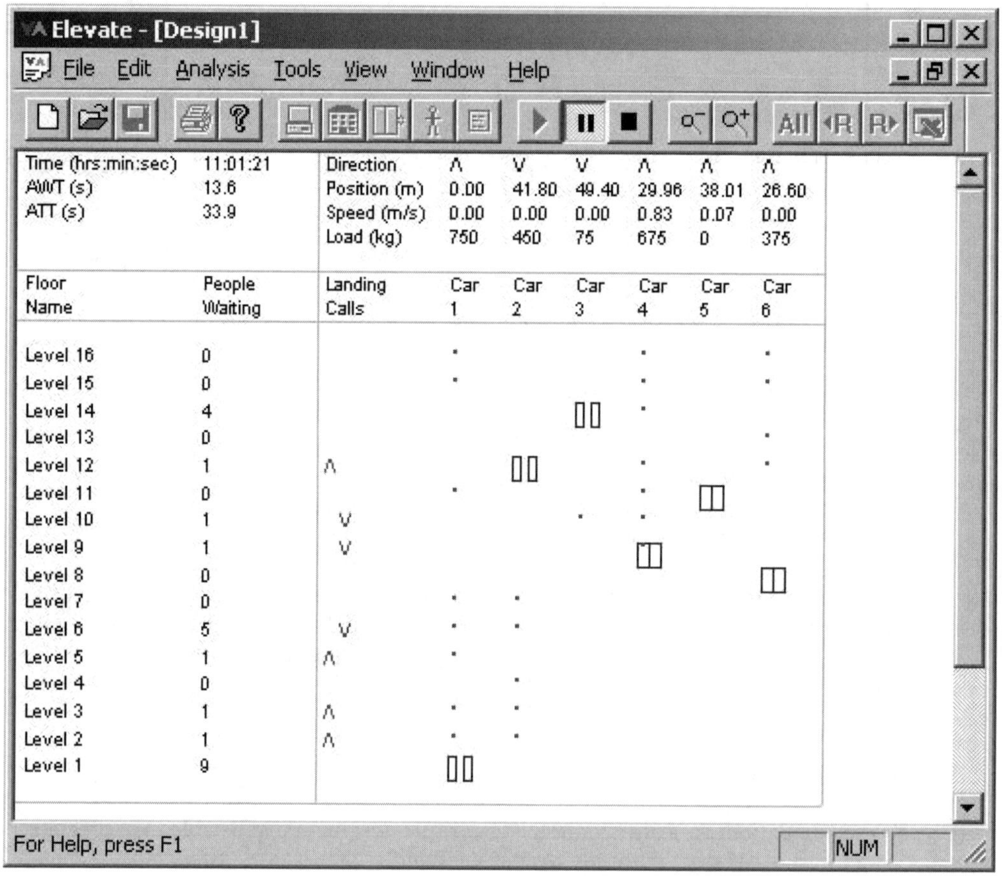

⩗A Elevate – [Design1]

File Edit Analysis Tools View Window Help

Time (hrs:min:sec)	11:01:21	Direction	Λ	V	V	Λ	Λ	Λ
AWT (s)	13.6	Position (m)	0.00	41.80	49.40	29.96	38.01	26.60
ATT (s)	33.9	Speed (m/s)	0.00	0.00	0.00	0.83	0.07	0.00
		Load (kg)	750	450	75	675	0	375

Floor Name	People Waiting	Landing Calls	Car 1	Car 2	Car 3	Car 4	Car 5	Car 6
Level 16	0			·			·	·
Level 15	0			·			·	·
Level 14	4					🞔🞔	·	
Level 13	0							·
Level 12	1	Λ		🞔🞔			·	·
Level 11	0			·			·	🞔
Level 10	1	V				·	·	
Level 9	1	V				🞔		
Level 8	0							🞔
Level 7	0		·	·				
Level 6	5	V	·	·				
Level 5	1	Λ	·					
Level 4	0			·				
Level 3	1	Λ	·	·				
Level 2	1	Λ	·	·				
Level 1	9		🞔🞔					

For Help, press F1   NUM

Figure 16.9 Simulation phase of ELEVATE

The next important parameter is passenger waiting time and, to a less extent, passenger journey time. Figure CS18.4 is an example of both these parameters for a single floor, in this case the main terminal. Here it is possible to view both parameters as average values (solid line) and maximum values (dotted line) for every 5 minute period. As Simulation Case Study CS18 is for an uppeak traffic condition, the main terminal floor is of most interest. In the interfloor and lunch time traffic conditions the activity at all floors is of interest. To cope with this passenger waiting time (and journey time, not shown) plots are required. Figures CS19.4, CS20.2 and CS21.4 give examples of multi-floor presentation of passenger waiting times. The bars represent the average values and the hats (^) the maximum values. This type of display assists the detection of problem floors and allows the traffic control algorithm to be tuned.

The bar/hat displays do not reveal the distribution of waiting/journey times. These can be shown as percentile plots. Figure CS21.5 shows a plot of the number of calls answered in 5 s time bands and a percentile of these calls. The figure is for one hour of activity for all floors in the building. If an examination of the activity at a specific floor is required, consider Figure CS21.6, which looks at Floor 7 for one hour. If an examination of the activity at a specific floor is required for a specific time period, consider Figure CS18.5, which shows the activity at Floor 0 for Period 7 of the simulation, or Figure CS19.6, which shows activity at all floors for Periods 6 and 7 of the simulation. The importance of these plots is that any floor or all floors can be examined for one period or several periods.

If plots of calls answered are required, then graphs such as Figure 16.10 can be displayed for single periods or several periods. This figure allows the identification of floors with long wait calls. Figure 16.11 shows the percentile plot as a companion to Figure 16.10. This type of figure will show, by its irregular shapes, where there might be a problem, eg: look at Figure CS19.5.

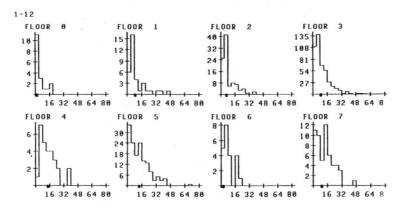

Figure 16.10 Distribution of waiting times (PC-LSD)

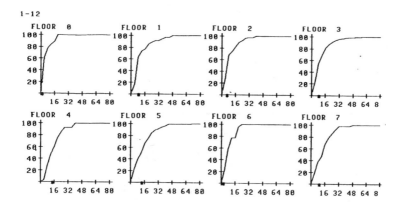

Figure 16.11 Distribution of percentile values (PC-LSD)

A more modern plot {E} derived from spreadsheet ideas is shown at Figure 16.12, where the percentile plot is continuously presented. This plot can be obtained for the duration of the simulation, which might be 5 minutes or longer.

Designers do like to quote actual numbers. These can be obtained from listings. Examples of such listings are shown in the Simulation Case Studies {P} and Table 16.1 {E}. Figure 16.12 is the output from a spreadsheet application program and allows the standard tools available in the spreadsheet to plot the data in various ways. The form of these plots will only be limited by spreadsheet tools available and the imagination of the user. The disadvantage is that the range of output presentations is not standardised.

16.5 CONCLUSIONS

Simulation allows a greater understanding of complex situations, but is not necessary for straightforward designs. Because simulation can follow each passenger from arrival to departure, a vast amount of data can be collected. Graphical figures give a close feel to what is happening and confirmation of the underlying exact values can be obtained from tabulated print outs. The ability to plot graphs for any period of time for any floor or all floors is important to the appreciation of the design. It can be concluded from the examination of the results presented here that digital modelling can be very realistic.

Day and Barney (1993) proposed a check list to evaluate computer programs used for traffic design, which Peters refined in 2000. A summary of the questions is given in Table 16.2 with a broad comparison between PC-LSD and ELEVATE.

A general comparison of the features available in two simulation programs is given in Tables 16.3a−16.3e.

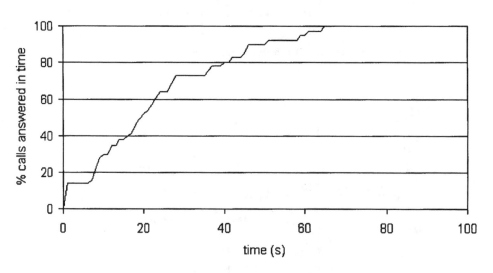

Figure 16.12 Percentile output (ELEVATE)

Table 16.1 Extract from ELEVATE output listing showing input data and principal results

ELEVATE Version 4.0 © Peters Research Ltd. 2001

JOB DATA
Job Title	Elevate Manual
Job No	M4b
Calculation Title	High Rise Zoning
Made By	rdp
File	Example 4b.elv
Date	13-Apr-02

ANALYSIS DATA
Analysis Type	Simulation
Measurement system	Metric
Dispatcher Algorithm	Up peak 1
Time slice between simulation calculations (s)	0.1
No of time slices between screen updates	10
Random number seed for passenger generator	1

BUILDING DATA
Floor Name	Floor Height (m)
Car Park	3.6
Level 1	5.0
Level 2	3.8
Level 3	3.8
Level 4	3.8
Level 5	3.8
Level 6	3.8
Level 7	3.8
Level 8	3.8
Level 9	3.8
Level 10	

ELEVATOR DATA
No of Elevators	3
Capacity (kg)	1000
Door Pre-opening Time(s)	0
Door Open Time (s)	1.8
Door Close Time (s)	2.9
Speed (m/s)	2.5
Acceleration (m/s²)	0.8
Jerk (m/s³)	1.6
Motor Start Delay (s)	0.5
Home Floor	Level 1

PASSENGER DATA
Loading Time (s)	1.2
Unloading Time (s)	1.2
Passenger Mass (kg)	75
Capacity Factor (%)	80
Stair Factor (%)	0
Start Time (hrs:mins)	09:00
End Time (hrs:mins)	09:15
Arrival Rate as	% building population in 5 mins

Floor Name	No of people	Area (m²)	Area/person	Arrival Rate
Car Park	0	1		
Level 1	0	14		
Level 2	0	0		
Level 3	0	0		
Level 4	0	0		
Level 5	0	0		
Level 6	0	0		
Level 7	0	0		
Level 8	0	0		
Level 9	50	0		
Level 10	50	0		

RESULTS   (3 No. 1000 kg elevators @ 2.50 m/s)

Average Waiting Time (s) 60.2 Average Transit Time (s) 70.9   Average Journey Time (s) 131
Longest Waiting Time (s) 138 Longest Transit Time (s) 125.3   Longest Journey Time (s) 262.8

Table 16.2 Comparison of programs: Questions and answers

Question	PC-LSD	ELEVATE
Will the program run on user's computer (eg: operating system) ?	Maybe	YES
Any special program requirements (eg: spreadsheet, compiler) ?	NO	YES
Does program include calculation and simulation ?	NO	YES
Does the program use an iterative process ?	NO	YES
What traffic flows can it analyse (eg: uppeak, down peak, etc.) ?	ALL	ALL
Can the program analyse single and double deck ?	NO	YES
What are the inputs and outputs ?	Full range	Full range
Who are the authors ?	Barney	Peters
Are results correct ?	n/a	n/a
What are the initial and recurrent costs ?	n/a	n/a
Is there technical support ?	NO	YES
Is the program copy protected ?	NO	YES
Is user training available ?	Some	YES
Is the source code available for review ?	YES	NO
Are there plans for upgrades, etc. ?	NO	YES

Table 16.3a System parameters

Feature	PC-LSD	ELEVATE
Job title	Yes	Yes
Measuring system	Metric	Imperial/metric
Operating system	DOS	Windows
Calculation programs	No (1)	Yes
Number of users	20	500
Select time slice	No = $t_s/10$	Yes
Select seed	No	Yes
Select simulation time	No = 60 minute	Yes
Display during simulation	Yes	Yes

Table 16.3b Building parameters

Feature	PC-LSD	ELEVATE
Number of floors	25	100
Floor identification	By number	By number or name
Enter individual interfloor distances	No	Yes
Total travel	Yes	Yes
Express zones	Yes	Yes
Express zone travel	Yes	Yes
Limit floors served	No	Yes

Notes to tables, shown (*n*), are listed after Table 16.3e.

Table 16.3c Passenger parameters

Feature	PC-LSD	ELEVATE
Arrival templates	Yes	No
Gaussian arrival pattern	Yes	No
Stepped arrival pattern	Yes	Yes
Sloped arrival pattern	Yes	No
Define arrivals by floor	Yes	Yes (4)
Define arrivals by period	Yes	Yes (4)
Simple definition of destinations	Yes	Yes
Advanced definition of destinations	No	Yes (4)
Floor populations	Yes	Yes
Passenger transfer times	Yes	Yes
Passenger weight	No	Yes
Stair factor	No	Yes
Mix traffic types	No	Yes

Table 16.3d Lift system parameters

Feature	PC-LSD	ELEVATE
Number of lifts	8	12 (2)
Rated capacity (persons)	1 − 24	Any
Lifts of different capacities	No	Yes
Rated speed (m/s)	Yes	Yes
Lifts of different speeds	No	Yes
Entrance width (mm)	1 − 5000	n/a
Door open/closing time (s)	Yes	Any
Advance door opening	Yes	Yes
Acceleration	Yes	Yes
Jerk (3)	No	Yes
Enter individual flight times	Yes	No
Traffic control algorithms	Collective control Timed collective Nearest car control Fixed bidirectional sector control Fixed timed sector control Dynamic sector control ETA control Stochastic control Hall call allocation control	Collective Dynamic sectoring Hall call allocation ETA Bespoke versions of ELEVATE contain proprietary algorithms

Table 16.3e Output facilities

Feature	PC-LSD	ELEVATE
Summary of input data	Yes – printed	No
Printed output	Yes (5)	Yes (6)
Export printed output	Yes, word processor	Yes, spreadsheet (7)
Output graphs	Car spatial graph Car load graph Interval graph Waiting time graphs Journey time graphs Percentile graphs No. of calls graphs	Car spatial graphs Percentile graphs Spreadsheet graphics allow user defined presentations

(1) Available separately
(2) Sensible limit is 8
(3) Usually not known
(4) {E} extensive facilities
(5) Includes: Number of passengers in lifts and associated *RTT*
(6) Includes: List of passengers and their service times
(7) {E} allows spreadsheet graphs

SIMULATION CASE STUDIES

Contents

See the next page for an introduction to these case studies.

INTRODUCTION TO THE SIMULATION CASE STUDIES

The four traffic situations, uppeak, down peak, interfloor and lunch time traffic will be considered by simulation in these case studies. In Chapter 9, a 31 storey trader building was the example for the selection and design of the lift systems used to illustrate the classical calculation method. The Upper Zone is the most demanding for the traffic design and as the lifts serve the upper zone and also the "awkward" requirements of a facility floor positioned part the way up the Low Zone. The design of such a system by calculation methods is not convincing. A summary of the target building is given here for convenience.

The building contains trader floors served by escalators, offices associated with the traders, in Low Zone, a prestigious Mid Zone used as offices and a less prestigious Upper Zone used as a call centre. See Table 9.1 and Figure 9.1 below.

Table 9.1 Building data

Floor	Activity	Population
1–3	Traders	600/floor
4–6	Office	160/floor
7	Facilities	80/floor
8–9	Office	160/floor
10–20	Office	153/floor
21–30	Call Centre	160/floor

Elevation. Floor 1: 11.7 m. Floor 10: 49.5 m. Floor 21: 98.8 m.

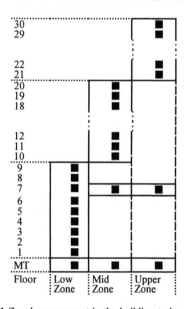

Figure 9.1 Zonal arrangement in the building to be considered

A facilities floor is situated at Floor 7 in the Low Zone and provides restaurant, snack bar, reprographics, etc. for the whole building. Stops are possible to this floor by the Upper Zone lifts. It will be assumed that about 5% of the passengers arriving during the 5 minute uppeak period will stop at the facilities floor. There are no transfer floors except at Floor 7 (or ground).

CASE STUDY EIGHTEEN

Uppeak Simulation for Upper Zone

The Upper Zone (Floors 21 – 30, 10 possible stops) is served by eight lifts from the main terminal (Floor 0). All lifts can stop at the facilities floor (Floor 7). In the calculation method, it was assumed this floor was positioned at Floor 20, in order to allow a calculation to be performed. It was assumed that this would not affect the outcome greatly. In the simulation, Floor 7 is positioned properly and the first possible stop in the Upper Zone is at Floor 21.

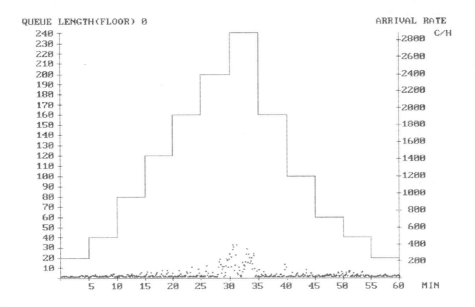

Figure CS18.1 Uppeak arrival pattern

Figure CS18.1 shows the arrival profile at the main terminal floor (Floor 0). In the calculation procedure the population of the zone was taken as 1600 persons (160 per floor) and a demand of 80 persons assumed to Floor 7, ie: a total of 1680 persons. At a 15% demand this is 252 persons/5-minutes.

The arrival profile of Figure CS18.1 comprises a series of 12 steps representing the numbers of arrivals over the 12 periods of 5 minutes. The values for each step might have been obtained by observers counting arrivals or could have been estimated by the traffic designer. The overall period of the simulation is 60 minutes (one hour) of activity. In broad terms the steps represent the call sequence:

$20 - 40 - 80 - 120 - 160 - 200 - 240 - 160 - 100 - 60 - 40 - 20$ calls per 5 minutes.

These step values can be seen in Figure CS18.1.

The total number of calls represented by this sequence is thus 1240 persons, which is some 74% of the building population (1240/1680). It will be seen later that the random number generator in the simulation has produced different values for the total arrivals (1248) and the peak 5 minutes (260). Because random number generators are random (*sic*), it is important to check that the desired values are generated or values quite close to them, otherwise the simulation will not be accurate. The right-hand axis scales the arrival rate in calls/hour (C/H). The arrivals peak at Period 7 (30−35 minutes), which could be the 5 minute period before the starting time of the building occupants, eg: 08:55−09:00. There is a slow build up of arrivals up to the peak 5 minutes and then a more rapid fall after the peak.

The left-hand axis scales the number of passengers queuing at the main terminal. The dots along the bottom axis represent the queues. As can be seen, these reach some 35 persons at the peak period. This is approximately equivalent to two car loads and the last passenger might have to wait for the third car to arrive before boarding. The number of passengers who are queuing gives the designer a "feel" for the usefulness of the design. If the queues became excessive the design could be modified or abandoned, before any further work is carried out.

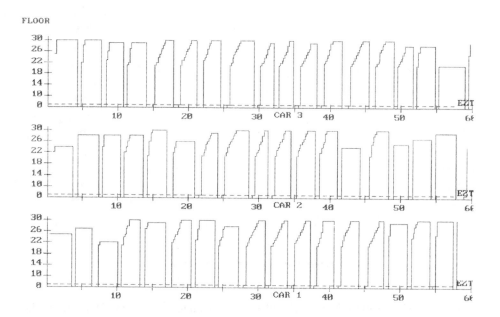

Figure CS18.2 Spatial plot of three lifts during uppeak

Figure CS18.2 shows a spatial plot of how three of the eight lifts (Cars 1−3) move between floors during this uppeak simulation. This graph is used to confirm that the simulation is set up correctly as the expected movements can be confirmed. Again one hour of time is represented. The left-hand axis indicates the floor identifications. Note that a dotted line is inserted (labelled EZT) at the facilities floor (Floor 7) level to indicate a non-stop trip from the main terminal (Floor 0). Careful examination of this dotted line will reveal some lifts stop at Floor 7. There are then no stops as the lifts travel through the Mid Zone of this building until Floor 21 is reached, which is the lowest floor of the Upper Zone. The lifts then stop as required at all floors up to Floor 30, the top floor of the zone.

In the central portion of the plot, at the 30-minute point, it will be seen that the lifts display a staircase pattern, as they call at most of the upper floors. The highest floor (30) is not always served. Note the express trip from the highest reversal floor back to the main terminal floor. The plot is idealised and does not show the actual time/distance profile between floors, only indicating the times the lift is at each floor. If graphs for all eight lifts were examined, the average interval, average highest reversal floor and average number of stops could be determined.

Thus Figure CS18.2 confirms that the lifts stop correctly.

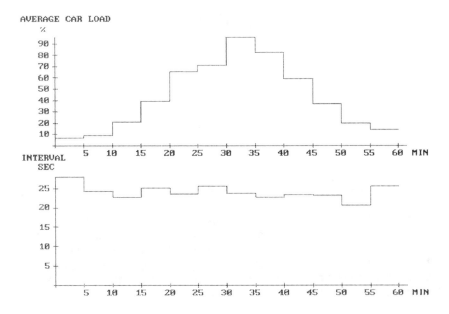

Figure CS18.3 Car load and interval during uppeak

Figure CS18.3 shows the percentage load in the car and the interval at the main terminal floor (Floor 0). The lift has been declared in the simulation to have a maximum occupancy of 20 persons, although it has an EN81 rated capacity of 26 persons (see Table 7.2). In Period 7 it can be seen that some cars are loaded to some 95%. At 95% this means an average loading of 19 persons. The print-out in Table CS18.3 indicates that 36 cars leave the main terminal with 20 passengers. This loading will not present a problem to the intending passengers as the cars can accommodate 20 persons, although it will be somewhat crowded. Also passengers arriving close to their official starting time will often crowd a car in order not to be late on duty. However, this density level would not be acceptable in a prestigious building.

The calculation method suggested that the average car loading would be 19 passengers. There is no significant difference here.

The interval during Period 7 is some 24 s, which is again acceptable. If the interval is too small there may be more than one car loading at the lobby at any time. This often leads to inefficient passenger loading and cars leaving with small numbers of passengers. In this case, 20 passengers can load in 13 s (6 × 1.0 + 14/2 × 1.0), see Section 5.9 and Example 5.6.

The calculation method suggested that the average interval would be 23 s. There is no significant difference here.

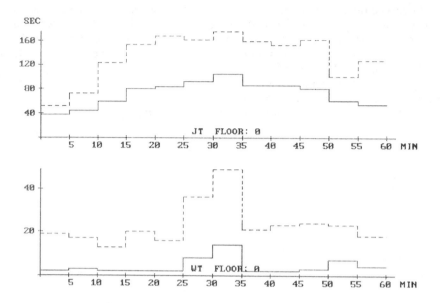

Figure CS18.4 Waiting times and journey times during uppeak

The two graphs in Figure CS18.4 indicate the passenger journey times (upper graph) and waiting times (lower graph) for passengers arriving at the main terminal (Floor 0). Both graphs show, by means of a solid line, the average values and, by means of the dotted line, the maximum values. Looking at the passenger waiting time graph, it can be seen that the worst performance is during Period 7, as would be expected. The average waiting time is 14 s and the maximum waiting time is 49 s.

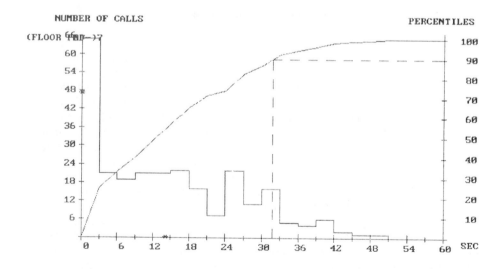

Figure CS18.5 Percentiles and number of calls at Floor 0, for Period 7 during uppeak

Figure CS18.5 (for the worst period, Period 7) shows the number of calls (left-hand axis) against time bands of 3 s (bottom axis). The * on the left axis indicates the number

(48) of calls answered with no waiting time. The * on the bottom axis indicates the average waiting time (14 s).

The right-hand axis is a percentile scale and shows the number of calls answered as a percentage. This graph is derived from the number of calls graph. The dashed line indicates the ninety percentile (31 s). Table 6.5 defines an excellent system for 5 minutes of activity, as one where 65% of all calls are answered in 30 s and 92% in 60 s. This is then an excellent system, as 90% are answered in 31 s. Note that this period also contains the worst maximum waiting time of 49 s.

As a note of interest, if Figure CS18.5 were to be produced for a one hour period (often shown on data logger reports) the average waiting time becomes 6 s, the maximum remains at 49 s and the ninety percentile becomes 20 s (see Table CS18.2). The average and ninety percentile are considerably improved when averaged over one hour of activity and do not give any guidance to the performance during the peak 5 minutes.

Table CS18.1 is a print out of the design data concerning the building, the lifts and the passengers. Note that the building, lift and passenger parameters are the same as those used in the calculations in Chapter 9. The attraction of Floor 7 has been made half that of the other floors as indicated by the bias being set at −50. The supervisory control system used is one based on an estimated time of arrival (ETA) algorithm.

Table CS18.2 gives the performance data. The figures given here have been used to derive many of the graphical representations above. Note the total arrivals over the hour of simulated activity is 1248 persons, with 260 persons arriving during the peak 5 minute period. This corresponds well with the calculation parameters.

Table CS18.3 gives data concerning the car trips (number of passengers carried, round trip times, etc.) and the average interval and car load. All data are presented for 5 minute periods.

Examination of the individual car trips shows that just prior to the onset of the peak period that some cars are departing from the main terminal with very low numbers of passengers. Thus if this system were to be employed it might be necessary to increase the lobby dwell time during the uppeak hour to prevent this.

Examination of the spatial plot (Figure CS18.2) also shows lifts apparently parking high in the building. Thus again a stronger parking algorithm should be employed to bring cars to the main terminal.

IN CONCLUSION

In the peak period the average passenger waiting time is 14 s, which is excellent. The maximum waiting time is 49 s and the ninety percentile is only 20 s, which again is excellent. The car loading is higher than desired at 95.4% (19.1 passengers), but this probably would be acceptable if the 2500 kg lifts were installed.

The values for the average interval and the car loading are different to those obtained by calculation. Table 9.4 indicates an interval of 23.0 s with 19.3 passengers in the lift.

There is a difficulty here for the designer. It is unlikely that 2500 kg lifts would be acceptable. This would mean using 2000 kg lifts, which are short on capacity. Here the traffic designer might draw some comfort from the likelihood of absentees reducing the demand on the lifts during uppeak.

Table CS18.1 Print-out for uppeak traffic simulation of Upper Zone: Input data

```
BUILDING DATA:

        GROUP SERVES                                 EXPRESS ZONE
        EXPRESS STOP                                    7 FLOOR
        EXPRESS JUMP                                36.90 METRES
        EXPRESS ZONE TERMINAL                          21 FLOOR
        NUMBER OF LEVELS SERVED ABOVE EZT               9 LEVELS
        DISTANCE FROM M.T. TO EZT                   98.80 METRES

LIFT DATA:

        NUMBER OF CARS IN GROUP                         8 CARS
        CAPACITY                                       20 PERSONS
        TERMINAL SPEED                               6.00 M/S
        ACCELERATION                                 1.00 M/S/S
        DOOR CLOSING TIME                            2.60 SECONDS
        DOOR OPENING TIME                            1.30
        WIDTH OF ENTRANCE                           1100. MILLIMETRES
        SUPERVISORY SYSTEM                            ETA SYSTEM
        JUMP(LEVELS):          1     2     3     4     5     6     7     8
        FLIGHT TIME(S):      5.1   6.8   8.2   9.3  10.3  11.2  12.0  12.8

PASSENGER DATA:

        ARRIVAL PATTERN                               UPP PATTERN
        PATTERN PROFILE                              STEP PROFILE
        MAIN TERMINAL                                   0 FLOOR
        PASSENGER TRANSFER TIME                      1.00 SECONDS
        RATES OF ARRIVAL AT M.T. IN CALLS/5 MINUTES:
        PR 1  PR 2  PR 3  PR 4  PR 5  PR 6  PR 7  PR 8  PR 9  PR10  PR11  PR12
        20.   40.   80.  120.  160.  200.  240.  160.  100.   60.   40.   20.

        BIAS FOR CALL DESTINATION:
        FLOOR:     0     7    21    22    23    24    25    26    27    28    29    30
        BIAS :     0   -50     0     0     0     0     0     0     0     0     0     0
```

Table CS18.2 Print out for uppeak traffic simulation of Upper Zone: System performance

FLOOR	PRD 1	PRD 2	PRD 3	PRD 4	PRD 5	PRD 6	PRD 7	PRD 8	PRD 9	PRD10	PRD11	PRD12	HOURLY
AVERAGE WAITING TIMES													
0	2	3	2	2	2	8	14	2	2	3	7	4	6
MAXIMUM WAITING TIMES													
0	19	17	13	20	16	36	49	21	23	24	23	18	49
NINETY PERCENTILES													
0	7	12	9	10	9	26	31	9	11	15	21	16	20
AVERAGE JOURNEY TIMES													
0	38	45	60	81	84	93	104	86	86	81	61	55	86
MAXIMUM JOURNEY TIMES													
0	52	73	124	154	168	162	176	159	153	162	100	127	176
NUMBER OF PICKED UP CALLS													
0	17	37	76	124	170	182	260	157	98	64	44	19	1248

Table CS18.3 Print-out for uppeak traffic simulation of Upper Zone: Other information

```
INFORMATION CONCERNING THE CAR TRIPS

CAR 1
ROUND TRIP TIMES & PASSENGERS/TRIP:
217  164  187  180  206  197  187  191  208  180  180  184  195  181  183  188  186
  1    1    2    9    6   14    3    5   20   20   20    6   14   19    2    8    3
NUMBER OF TRIPS(IN ONE HOUR): 17
AVERAGE NUMBER OF PASSENGERS/TRIP: 9.0
AVERAGE ROUND TRIP TIME: 189.1

CAR 2
ROUND TRIP TIMES & PASSENGERS/TRIP:
202  198  183  185  195  206  186  250  159  179  192  180  177  187  150  181  201
  2    1    2    5    5    3   20   20   20   20   20   10    1    6    1    1    1
NUMBER OF TRIPS(IN ONE HOUR): 17
AVERAGE NUMBER OF PASSENGERS/TRIP: 8.1
AVERAGE ROUND TRIP TIME: 188.9

CAR 3
ROUND TRIP TIMES & PASSENGERS/TRIP:
227  194  176  186  204  185  182  254  163  158  183  197  199  185  154  184  240
  2    3    4    5   20   20   10   20   20   20   20   13   20   17    8    3    1
NUMBER OF TRIPS(IN ONE HOUR): 17
AVERAGE NUMBER OF PASSENGERS/TRIP: 12.1
AVERAGE ROUND TRIP TIME: 192.4

CAR 4
ROUND TRIP TIMES & PASSENGERS/TRIP:
231  199  180  178  197  203  188  164  206  206  194  172  201  170  160  180  232
  1    1    2    4    2    6    5    7   20   20    5   20    1    3    2    3    1
NUMBER OF TRIPS(IN ONE HOUR): 17
AVERAGE NUMBER OF PASSENGERS/TRIP: 6.1
AVERAGE ROUND TRIP TIME: 191.8

CAR 5
ROUND TRIP TIMES & PASSENGERS/TRIP:
242  187  187  200  210  189  195  241  171  163  194  206  226  179  161  195  219
  1    2    3    5   16   20   20   20   20   10    3    6    1    6    3    3
NUMBER OF TRIPS(IN ONE HOUR): 17
AVERAGE NUMBER OF PASSENGERS/TRIP: 9.4
AVERAGE ROUND TRIP TIME: 197.9

CAR 6
ROUND TRIP TIMES & PASSENGERS/TRIP:
227  188  187  188  206  180  195  256  153  170  192  190  234  167  167  209  201
  2    3    2    2    5    8    6   20   20   20   14   20    3    2    1    3    1
NUMBER OF TRIPS(IN ONE HOUR): 17
AVERAGE NUMBER OF PASSENGERS/TRIP: 7.8
AVERAGE ROUND TRIP TIME: 194.7

CAR 7
ROUND TRIP TIMES & PASSENGERS/TRIP:
226  197  177  203  206  183  182  234  180  166  176  195  214   52  124  168  187
  1    2   10    4   20   20   11   20   20   10   13   13    1    7    5   11
NUMBER OF TRIPS(IN ONE HOUR): 17
AVERAGE NUMBER OF PASSENGERS/TRIP: 11.1
AVERAGE ROUND TRIP TIME: 180.6

CAR 8
ROUND TRIP TIMES & PASSENGERS/TRIP:
166  192  176  199  209  173  187  233  180  166  180  186  221  174  162  200
  1    3    2    9    1   20   20   16   20   20   19   11    4    7    1    1
NUMBER OF TRIPS(IN ONE HOUR): 16
AVERAGE NUMBER OF PASSENGERS/TRIP: 9.7
AVERAGE ROUND TRIP TIME: 187.8

NUMBER OF PASSENGERS: 1245

AVERAGE LOAD OF CARS(%) & INTERVAL(SECONDS):
   7.0    9.1   20.8   39.6   65.5   71.5   95.4   82.9   59.1   36.9   20.0   14.1
  28.0   24.3   22.7   25.2   23.5   25.6   23.8   22.7   23.4   23.2   20.7   25.6
```

Down Peak Simulation for Upper Zone

Normally a properly designed lift system able to serve the requirements of the uppeak traffic condition will inherently have a 50% or more larger handling capacity during down peak than during uppeak traffic. This, coupled with the desire of most building occupants to leave at the end of the day, causes the down peak traffic condition to last for ten minutes or longer.

The Upper Zone has an uppeak handling capacity of 252 persons/5-minutes and a 50% increase means that the down peak traffic demand is 378 persons/5-minutes, but for two 5 minute periods. The building zone has ten floors in the zone plus the opportunity to collect passengers from the facilities floor at Floor 7. To take a worst case scenario it will be assumed that all 11 floors contribute passengers. Thus the arrival rate equates to 36 persons arriving every 5 minutes at Floors 21 − 30 and half that number (say) at Floor 7. Figure CS19.1 displays the arrival profile and the queues at Floor 24.

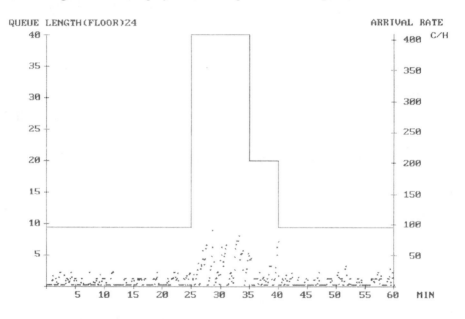

Figure CS19.1 Down peak arrival pattern at Floor 24

Figure CS19.1 shows the arrival pattern with the solid line and the queueing by the dots. The queue reaches some 9 persons occasionally, which is acceptable. Examination of Table CS19.2 shows the randomness of the arrivals (picked up calls) at each floor. This is similar to the behaviour of the real system. The total arrivals over the one hour is 1665 persons, which is nearly all the building population leaving in the hour. This is likely.

The spatial plot, Figure CS19.2, confirms that the lifts call at the correct floors and exhibit a reverse staircase movement pattern compared to the uppeak situation. It is noticeable that the number of stops is smaller. Occasionally the lifts can be seen parking, whenever the demand is low.

As all lifts during the peak period (Periods 6−7) call at the main terminal, it is meaningful to examine Figure CS19.3, which shows the car loading and the interval. The average car loading is 86% at an interval of 14 s. The loading is slightly high, but acceptable at the end of the day. The 14 s interval is, as expected, shorter than the 24 s uppeak interval. This is close to the two thirds value foreseen by the analysis in Chapter 13. The handling capacity is thus increased during down peak.

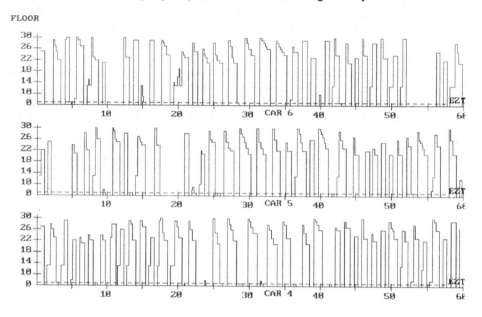

Figure CS19.2 Spatial plot of lift movements during down peak for lifts 4−6

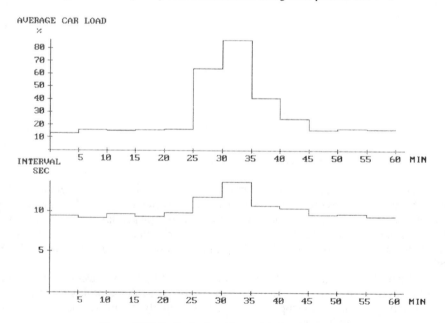

Figure CS19.3 Car load and interval during down peak

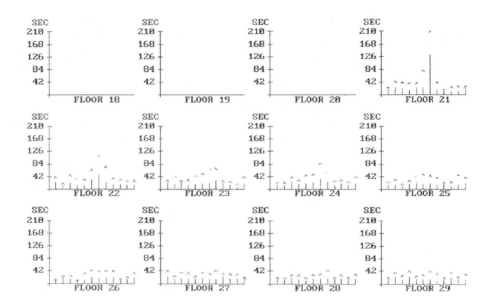

Figure CS19.4 Waiting time graphs for Floors 18−29 during down peak

Figure CS19.4 gives an indication of the average and maximum passenger waiting times at Floors 18−29. This figure is a more compact version of Figure CS18.4 as it represents the average values by the height of the solid bar and the maximum values by a "hat" (^). Note that Floors 18−20 show no activity as they are not served by the Upper Zone group. This type of display enables an analysis of relative performance across a number of floors. As an example note that the performance during Period 7 at Floor 21 is very poor. This may be because this floor is the lowest floor during down peak traffic and that many cars are full when they arrive there. The passenger waiting times at Floor 21 should be examined closely to identify the problem during Period 7.

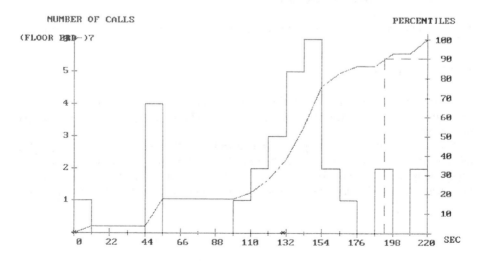

Figure CS19.5 Percentile graph for Floor 21 during Period 7 of down peak

Figure CS19.5 is a percentile graph for Floor 21 and shows a very irregular percentile shape. The number of calls (left-hand axis) shows some very long wait calls, ie: two passengers waiting over 209 s. A traffic designer noting this deficiency would then alter some parameters in an attempt to improve the situation. Suitable changes might be to park cars at Floor 21 or to declare it a priority floor. After each change a simulation would be run to evaluate the effectiveness.

Figure CS19.6 shows the percentile and number of calls graphs for Periods 6 and 7 during down peak for all the floors in the zone. The average waiting time is some 25 s and the ninety percentile is just over 50 s, both figures being very satisfactory.

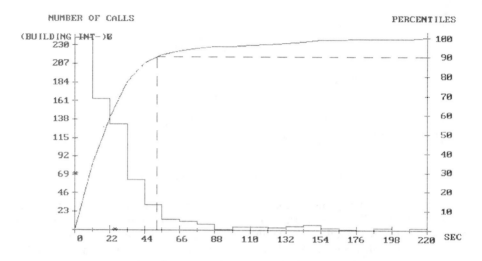

Figure CS19.6 Percentile graph for all floors during Periods 6 and 7 of down peak

Table CS19.1 gives the input data for the simulation.
Table CS19.2 gives the system performance.
Table CS19.3 gives the other data.

These tabulations of the actual figures have been used to derive many of the figures.

IN CONCLUSION

The average waiting time is 25 s, during Periods 6 and 7. A calculation using Equation (13.5) gives an average waiting time of 25.9 s. The estimate is thus close to that obtained by simulation.

Service to Floor 21 is very poor and that to Floor 22 is poor. The reason may be that cars are full by the time they reach these floors. Attention would need to be given to improving this problem. However, the ninety percentile is only 40 s which is only twice that achieved during uppeak. The system is generally performing satisfactorily.

Table CS19.1 Print-out for down peak traffic simulation of Upper Zone: Input data

```
BUILDING DATA:

     GROUP SERVES                              EXPRESS ZONE
     EXPRESS STOP                                 7 FLOOR
     EXPRESS JUMP                             36.90 METRES
     EXPRESS ZONE TERMINAL                       21 FLOOR
     NUMBER OF LEVELS SERVED ABOVE EZT            9 LEVELS
     DISTANCE FROM M.T. TO EZT                98.80 METRES

LIFT DATA:

     NUMBER OF CARS IN GROUP                      8 CARS
     CAPACITY                                    20 PERSONS
     TERMINAL SPEED                            6.00 M/S
     ACCELERATION                              1.00 M/S/S
     DOOR CLOSING TIME                         2.60 SECONDS
     DOOR OPENING TIME                         1.30
     WIDTH OF ENTRANCE                         1100. MILLIMETRES
     SUPERVISORY SYSTEM                         ETA SYSTEM
     JUMP(LEVELS):        1     2     3     4     5     6     7     8
     FLIGHT TIME(S):     5.1   6.8   8.2   9.3  10.3  11.2  12.0  12.8

PASSENGER DATA:

     ARRIVAL DEFINITION:
     ARRIVAL PATTERN                            DEF PATTERN
     PATTERN PROFILE                           STEP PROFILE
     MAIN TERMINAL                                0 FLOOR
     PASSENGER TRANSFER TIME                   1.00 SECONDS
     RATES OF ARRIVAL: RATES IN CALLS/5 MINUTES:
     FLOOR  PR 1 PR 2 PR 3 PR 4 PR 5 PR 6 PR 7 PR 8 PR 9 PR10 PR11 PR12 POPUL.  BIAS
        0   0.0  0.0  0.0  0.0  0.0  0.0  0.0  0.0  0.0  0.0  0.0  0.0     0    100
        7   4.0  4.0  4.0  4.0  4.0 17.0 17.0  8.5  4.0  4.0  4.0  4.0    80      0
       21   8.0  8.0  8.0  8.0  8.0 34.0 34.0 17.0  8.0  8.0  8.0  8.0   160      0
       22   8.0  8.0  8.0  8.0  8.0 34.0 34.0 17.0  8.0  8.0  8.0  8.0   160      0
       23   8.0  8.0  8.0  8.0  8.0 34.0 34.0 17.0  8.0  8.0  8.0  8.0   160      0
       24   8.0  8.0  8.0  8.0  8.0 34.0 34.0 17.0  8.0  8.0  8.0  8.0   160      0
       25   8.0  8.0  8.0  8.0  8.0 34.0 34.0 17.0  8.0  8.0  8.0  8.0   160      0
       26   8.0  8.0  8.0  8.0  8.0 34.0 34.0 17.0  8.0  8.0  8.0  8.0   160      0
       27   8.0  8.0  8.0  8.0  8.0 34.0 34.0 17.0  8.0  8.0  8.0  8.0   160      0
       28   8.0  8.0  8.0  8.0  8.0 34.0 34.0 17.0  8.0  8.0  8.0  8.0   160      0
       29   8.0  8.0  8.0  8.0  8.0 34.0 34.0 17.0  8.0  8.0  8.0  8.0   160      0
       30   8.0  8.0  8.0  8.0  8.0 34.0 34.0 17.0  8.0  8.0  8.0  8.0   160      0
```

Table CS19.2 Print-out for down peak traffic simulation of Upper Zone: System performance

FLOOR	PRD 1	PRD 2	PRD 3	PRD 4	PRD 5	PRD 6	PRD 7	PRD 8	PRD 9	PRD10	PRD11	PRD12	HOURLY
AVERAGE WAITING TIMES													
7	12	12	22	13	21	16	36	15	9	0	8	10	18
21	23	25	23	14	23	23	130	19	11	10	13	20	42
22	22	11	19	13	13	29	47	21	15	16	12	12	25
23	19	17	21	14	15	17	19	28	19	12	7	18	19
24	13	10	23	18	17	21	33	23	11	18	14	18	21
25	13	14	10	17	19	16	18	14	17	9	16	20	16
26	9	21	14	7	12	21	17	18	21	7	13	19	17
27	22	16	17	15	14	16	15	13	22	16	19	16	17
28	14	12	19	14	10	14	15	19	15	15	13	16	15
29	16	23	12	22	16	14	14	10	12	19	15	14	15
30	10	21	21	14	15	13	18	11	16	14	14	12	15
BUILD	16	17	17	16	16	19	31	19	16	14	14	17	20
MAXIMUM WAITING TIMES													
7	16	19	22	16	37	56	86	39	14	0	15	22	86
21	28	46	44	42	39	82	210	43	19	28	30	29	210
22	42	21	47	37	33	67	114	78	38	34	28	30	114
23	30	44	30	32	46	53	70	71	28	27	20	42	71
24	24	23	42	32	47	51	87	59	28	32	27	42	87
25	27	34	18	30	44	51	47	40	24	25	49	40	51
26	14	29	28	12	36	46	42	45	44	15	24	36	46
27	44	28	39	37	26	38	34	47	35	33	33	23	47
28	23	21	30	30	22	32	34	46	24	30	22	33	46
29	22	39	29	44	27	32	42	36	22	39	25	34	44
30	21	28	28	34	25	34	47	27	28	25	32	15	47
BUILD	44	46	47	44	47	82	210	78	44	39	49	42	210
NINETY PERCENTILES													
7	19	20	31	20	35	35	81	35	18	0	16	24	36
21	30	46	48	31	41	69	193	38	19	20	24	31	140
22	39	19	32	34	36	60	105	76	32	34	28	26	63
23	27	42	30	25	46	45	46	63	31	25	17	39	43
24	26	23	37	30	32	42	72	43	27	27	21	32	43
25	28	35	19	27	46	37	34	29	30	25	48	36	32
26	17	31	26	17	31	40	35	45	46	19	28	35	36
27	47	27	32	31	30	34	30	40	37	35	35	30	32
28	26	20	30	29	24	29	32	40	30	30	27	35	30
29	28	40	21	43	29	26	31	33	21	38	29	34	31
30	19	31	30	32	27	28	41	25	28	26	26	20	31
BUILD	29	31	31	31	32	41	76	42	30	29	27	32	39
AVERAGE JOURNEY TIMES													
7	26	31	41	31	40	35	58	34	27	0	27	29	37
21	53	54	52	45	49	60	170	73	44	41	41	45	75
22	52	41	51	45	47	69	83	60	53	51	43	46	61
23	56	53	52	45	47	61	66	67	57	46	51	55	60
24	51	49	57	54	55	68	78	67	59	50	49	53	63
25	46	53	48	55	66	65	74	63	64	52	59	64	63
26	58	63	49	70	62	78	77	67	64	46	62	64	69
27	63	65	64	59	58	70	72	81	66	64	70	63	69
28	67	67	63	65	59	84	89	83	82	62	65	65	77
29	59	73	61	80	70	93	94	72	83	70	85	65	82
30	56	87	75	72	69	93	95	102	85	75	61	70	84
BUILD	55	58	57	57	58	72	87	71	66	57	55	58	69
MAXIMUM JOURNEY TIMES													
7	26	36	41	34	56	79	109	59	32	0	33	42	109
21	58	75	72	70	68	131	244	185	77	60	58	58	244
22	72	50	78	67	64	100	154	114	78	75	54	62	154
23	72	78	61	65	78	101	117	111	88	57	64	85	117
24	68	64	86	69	89	105	140	106	82	63	59	73	140
25	59	75	61	73	89	105	112	86	95	65	92	84	112
26	69	78	70	75	86	114	108	91	82	77	94	92	114
27	100	88	85	84	87	97	110	113	95	81	78	99	113
28	86	99	78	77	77	106	110	97	112	81	84	90	112
29	87	86	80	114	98	121	129	112	126	79	109	99	129
30	88	118	83	119	84	121	120	136	135	106	87	81	136
BUILD	100	118	86	119	98	131	244	185	135	106	109	99	244
NUMBER OF PICKED UP CALLS													
7	2	5	1	4	8	18	12	8	6	0	5	8	77
21	2	8	6	11	5	27	29	21	6	14	8	5	142
22	8	11	10	9	7	37	33	21	10	9	12	6	173
23	5	11	4	7	8	26	36	26	7	7	9	8	154
24	11	9	6	7	10	31	29	24	10	5	11	10	163
25	9	8	9	5	8	32	37	16	5	7	6	7	149
26	4	6	11	4	15	39	31	26	8	5	8	15	172
27	7	10	11	14	5	33	35	16	12	8	8	5	164
28	6	9	6	7	8	33	23	16	9	8	5	8	138
29	4	7	10	10	9	31	48	19	13	5	7	9	172
30	8	10	6	10	5	32	37	14	11	11	12	5	161
BUILD	66	94	80	88	88	339	350	207	97	79	91	86	1665

Table CS19.3 Print-out for down peak traffic simulation of Upper Zone: Other data
(Full details given for Lifts 1−3, basic details for Lifts 4−8.)

```
INFORMATION CONCERNING THE CAR TRIPS

CAR 1
ROUND TRIP TIMES & PASSENGERS/TRIP:
 64  63 134  58  54 110  72  68  91  84  66  76  48 119  44  50 139  85 127 105
  1   2   4   1   1   5   3   2   5   4   1   2   2   8   2   1   9   5  20  15
ROUND TRIP TIMES &  PASSENGERS/TRIP:
 29  33 127 121 111  77  30  79 100  82 108  99  51  77  78  99  56  42  29 103 87
  1   3  20  20   6   6   1   6   4   2   5   5   1   3   4   5   1   2   1   7  4
NUMBER OF TRIPS(IN ONE HOUR): 41
AVERAGE NUMBER OF PASSENGERS/TRIP: 4.9
AVERAGE ROUND TRIP TIME:   79.9

CAR 2
ROUND TRIP TIMES & PASSENGERS/TRIP:
 62  54  82  73  29 127  81  88  58  69 101  77  77  72  95  39   6
  1   1   2   3   1  11   5   5   3   3   6   4   5   3   6   1   3
ROUND TRIP TIMES & PASSENGERS/TRIP:
102  77 137 137 126 112  84 103  72  67  70  90 114  92  98  77 102
  3   5  20  20  20  20   8   7   4   8   4   5   9   3   3   3  10
ROUND TRIP TIMES & PASSENGERS/TRIP:
 90 101  67  29
  5   2   1   0
NUMBER OF TRIPS(IN ONE HOUR): 38
AVERAGE NUMBER OF PASSENGERS/TRIP: 5.9
AVERAGE ROUND TRIP TIME:   84.2

CAR 3
ROUND TRIP TIMES & PASSENGERS/TRIP:
 61  61 123 108 111  73 128  61  70  58  67  57  29 103  80  59 111
  1   2   7   4   6   2   2   2   1   1   2   2   1   6   3   4   6
ROUND TRIP TIMES & PASSENGERS/TRIP:
112  18  31 116 128 124  92 107 164  73  77 106  57  90  68  53  87
 19   0   2  20  20  20   5  11  20   2   3   4   2   7   1   2   5
ROUND TRIP TIMES & PASSENGERS/TRIP:
139  34  90  30
  8   3   5   1
NUMBER OF TRIPS(IN ONE HOUR): 38
AVERAGE NUMBER OF PASSENGERS/TRIP: 5.6
AVERAGE ROUND TRIP TIME:   83.1

CAR 4
NUMBER OF TRIPS(IN ONE HOUR): 42
AVERAGE NUMBER OF PASSENGERS/TRIP: 5.2
AVERAGE ROUND TRIP TIME:   75.7

CAR 5
NUMBER OF TRIPS(IN ONE HOUR): 37
AVERAGE NUMBER OF PASSENGERS/TRIP: 5.3
AVERAGE ROUND TRIP TIME:   77.5

CAR 6
NUMBER OF TRIPS(IN ONE HOUR): 35
AVERAGE NUMBER OF PASSENGERS/TRIP: 5.7
AVERAGE ROUND TRIP TIME:   84.1

CAR 7
NUMBER OF TRIPS(IN ONE HOUR): 38
AVERAGE NUMBER OF PASSENGERS/TRIP: 5.4
AVERAGE ROUND TRIP TIME:   83.3

CAR 8
NUMBER OF TRIPS(IN ONE HOUR): 36
AVERAGE NUMBER OF PASSENGERS/TRIP: 5.7
AVERAGE ROUND TRIP TIME:   82.5

NUMBER OF PASSENGERS: 1658

AVERAGE LOAD OF CARS(%) & INTERVAL(SECONDS):
 13.3    16.0    15.4    16.1    16.7    63.9    86.4    40.4    25.0    15.8    17.0    16.9
  9.5     9.1     9.7     9.4     9.8    11.8    13.7    10.8    10.4     9.6     9.7     9.4
```

Interfloor simulation for Upper Zone

The interfloor traffic simulation considers one third (or thereabouts) of all the buildings' occupants using the lifts every hour. It is not necessary for the demand from every floor to be the same, but the demand range should be restricted so that the interfloor traffic demand can be considered to be balanced interfloor traffic.

The interfloor simulation documented in the print-out below (Figure CS20.1) applies about 528 calls into the system over the period of one hour. This closely corresponds to the calculated assumption of 560 calls per hour (one third of 1680). Table CS20.2 indicates from the "number of picked up calls" that 510 calls were generated in the simulation. It should also be noted that the randomised range of calls per floor was 33 to 60 calls per hour and the randomised range of calls generated every 5 minute period was 30 to 56 calls per 5-minutes. This model emulates a real situation very well.

Table CS20.1 indicates that the calls defined to be made at the main terminal (Floor 0) and the facilities floor (Floor 7) are half those made elsewhere in the zone. However, these two floors have been set to have twice the attraction of the other floors (this is achieved by changing the bias values).

Graphically a plot of the arrival profile (such as Figure CS18.1) would give little information as the arrival rate is constant over the whole hour and queueing is not expected to be significant. The spatial plot (Figure CS20.1) serves only to confirm the random movements of the cars. It should be noted, however, that only a small number of visits are made to the main terminal and facilities floors.

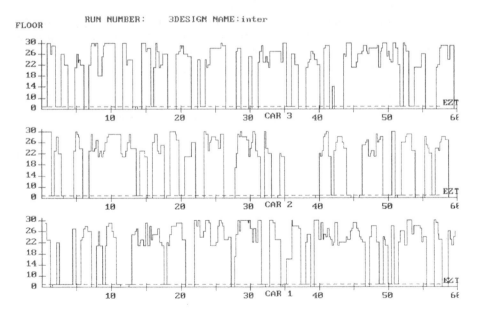

Figure CS20.1 Spatial plot during interfloor traffic for Lifts 1−3

Note that Figure CS20.1 only shows the movements of three of the eight lifts.

The most interesting graph is Figure CS20.2, which shows the passenger waiting times on a floor by floor basis (Floors 21 – 29). Floor 21 can be seen to offer the worst performance, although Floor 22 has a long maximum waiting time during Period 10. Graphs make it easier to see these variations, which can be easily missed in tables. Again some system tuning is needed to balance the performances between floors.

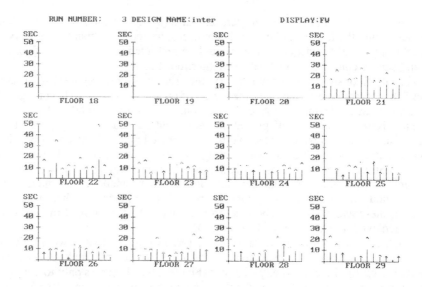

Figure CS20.2 Passenger waiting times for Floors 18 – 29 during interfloor traffic

Figure CS20.3 presents a percentile curve for all floors over one hour. The average for the hour is some 7 s and the ninety percentile is 15 s. There is one very long wait call at 50 s. Typically this would happen in a real system.

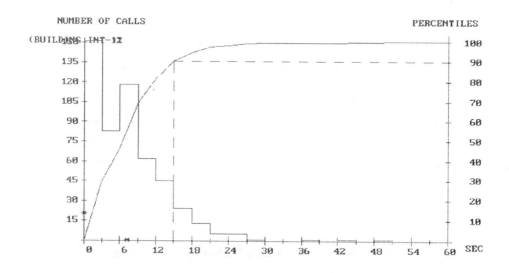

Figure CS20.3 Percentile and number of calls for all floors for one hour of interfloor activity

IN CONCLUSION

The average waiting time for the one hour period for all floors is quite small at 7 s and the maximum waiting time is only 50 s. Using Equation (13.5) the calculated average passenger waiting time is 11.2 s, which is larger than the simulated value. The most useful parameter indicating performance is the ninety percentile. This is 14 s for all floors over the one hour. This is an excellent performance.

Table CS20.1 Print-out for interfloor traffic simulation of the Upper Zone: Input data

```
BUILDING DATA:

        GROUP SERVES                             EXPRESS ZONE
        EXPRESS STOP                               7 FLOOR
        EXPRESS JUMP                           36.90 METRES
        EXPRESS ZONE TERMINAL                     21 FLOOR
        NUMBER OF LEVELS SERVED ABOVE EZT          9 LEVELS
        DISTANCE FROM M.T. TO EZT              98.80 METRES

LIFT DATA:

        NUMBER OF CARS IN GROUP                    8 CARS
        CAPACITY                                  20 PERSONS
        TERMINAL SPEED                          6.00 M/S
        ACCELERATION                            1.00 M/S/S
        DOOR CLOSING TIME                       2.60 SECONDS
        DOOR OPENING TIME                       1.30
        WIDTH OF ENTRANCE                       1100. MILIMETRES
        SUPERVISORY SYSTEM                      ETA SYSTEM
        JUMP(LEVELS):        1     2     3     4     5     6     7     8
        FLIGHT TIME(S):    5.1   6.8   8.2   9.3  10.3  11.2  12.0  12.8

PASSENGER DATA:

        ARRIVAL DEFINITION:
        ARRIVAL PATTERN                         DEF PATTERN
        PATTERN PROFILE                         STEP PROFILE
        MAIN TERMINAL                             0 FLOOR
        PASSENGER TRANSFER TIME                 1.00 SECONDS
        RATES OF ARRIVAL:  RATES IN CALLS/5 MINUTES:
        FLOOR  PR 1 PR 2 PR 3 PR 4 PR 5 PR 6 PR 7 PR 8 PR 9 PR10 PR11 PR12  POPUL.  BIAS
            0  2.0  2.0  2.0  2.0  2.0  2.0  2.0  2.0  2.0  2.0  2.0  2.0     80     14
            7  2.0  2.0  2.0  2.0  2.0  2.0  2.0  2.0  2.0  2.0  2.0  2.0     80     14
           21  4.0  4.0  4.0  4.0  4.0  4.0  4.0  4.0  4.0  4.0  4.0  4.0    160      7
           22  4.0  4.0  4.0  4.0  4.0  4.0  4.0  4.0  4.0  4.0  4.0  4.0    160      7
           23  4.0  4.0  4.0  4.0  4.0  4.0  4.0  4.0  4.0  4.0  4.0  4.0    160      7
           24  4.0  4.0  4.0  4.0  4.0  4.0  4.0  4.0  4.0  4.0  4.0  4.0    160      7
           25  4.0  4.0  4.0  4.0  4.0  4.0  4.0  4.0  4.0  4.0  4.0  4.0    160      7
           26  4.0  4.0  4.0  4.0  4.0  4.0  4.0  4.0  4.0  4.0  4.0  4.0    160      7
           27  4.0  4.0  4.0  4.0  4.0  4.0  4.0  4.0  4.0  4.0  4.0  4.0    160      7
           28  4.0  4.0  4.0  4.0  4.0  4.0  4.0  4.0  4.0  4.0  4.0  4.0    160      7
           29  4.0  4.0  4.0  4.0  4.0  4.0  4.0  4.0  4.0  4.0  4.0  4.0    160      7
           30  4.0  4.0  4.0  4.0  4.0  4.0  4.0  4.0  4.0  4.0  4.0  4.0    160      7
```

Table CS20.2 Print-out for interfloor traffic simulation of the Upper Zone: Performance data (continued overleaf)

FLOOR	PRD 1	PRD 2	PRD 3	PRD 4	PRD 5	PRD 6	PRD 7	PRD 8	PRD 9	PRD10	PRD11	PRD12	HOURLY
AVERAGE WAITING TIMES													
0	0	4	0	0	0	0	3	0	0	0	2	0	2
7	4	5	5	8	0	6	5	0	4	8	0	0	4
21	11	8	7	9	7	21	20	7	10	12	8	12	10
22	9	5	14	3	7	9	8	8	8	17	4	2	9
23	9	9	4	7	6	14	5	10	7	10	8	7	8
24	10	8	7	9	7	9	8	7	10	6	7	9	8
25	0	9	5	7	6	12	7	17	6	12	7	4	8
26	7	8	7	5	1	10	11	8	5	7	5	2	7
27	2	4	7	10	6	2	3	6	7	8	11	9	7
28	8	8	0	3	4	8	0	11	16	5	9	7	7
29	8	7	8	0	3	3	11	7	4	5	1	4	6
30	2	6	4	1	11	6	7	6	5	4	7	6	6
BUILD	7	8	7	6	5	9	8	8	7	9	6	6	7
MAXIMUM WAITING TIMES													
0	0	15	0	0	0	0	15	0	0	0	7	0	15
7	8	8	8	8	0	19	8	0	10	8	0	0	19
21	18	26	7	18	19	28	42	16	16	24	14	18	42
22	18	8	36	10	13	13	20	11	12	50	13	5	50
23	16	18	7	13	8	21	14	16	12	13	8	9	21
24	11	14	14	9	14	25	8	9	14	12	10	16	25
25	0	11	5	14	13	18	8	17	8	14	13	7	18
26	7	10	11	9	2	15	13	11	8	12	8	5	15
27	5	11	11	22	8	5	8	8	12	25	22	10	25
28	15	9	0	7	7	10	0	23	16	14	15	16	23
29	24	17	8	0	5	5	23	15	6	5	1	5	24
30	5	12	19	2	16	17	14	16	19	7	16	14	19
BUILD	24	26	36	22	19	28	42	23	19	50	22	18	50
NINETY PERCENTILES													
0	0	16	2	2	2	2	16	2	2	2	7	2	2
7	8	8	8	8	2	20	8	2	10	8	2	2	8
21	19	24	8	18	19	29	44	16	16	25	12	20	19
22	20	8	37	11	14	14	19	11	13	49	14	5	19
23	17	18	8	13	8	23	14	17	12	14	8	11	15
24	11	13	12	11	13	24	8	11	14	13	11	17	14
25	0	11	5	13	13	20	8	17	8	14	13	8	14
26	8	11	11	11	2	16	14	11	8	13	8	5	12
27	5	11	10	22	8	5	8	8	13	24	23	11	11
28	17	11	2	8	8	11	0	23	17	13	16	15	15
29	25	16	8	0	5	5	22	16	8	5	2	5	15
30	4	13	19	2	17	16	13	16	19	8	17	14	16
BUILD	17	14	13	13	13	19	15	16	14	14	14	14	14

Table CS20.2 Print-out for interfloor traffic simulation of the Upper Zone: Performance data (continued)

FLOOR	PRD 1	PRD 2	PRD 3	PRD 4	PRD 5	PRD 6	PRD 7	PRD 8	PRD 9	PRD10	PRD11	PRD12	HOURLY
AVERAGE JOURNEY TIMES													
0	0	32	34	29	37	26	32	19	36	36	31	30	32
7	29	29	26	33	22	33	30	26	26	31	24	23	28
21	31	28	0	29	21	43	47	26	33	35	26	25	30
22	34	34	37	21	22	25	33	33	26	41	18	18	31
23	30	26	17	25	23	27	25	22	22	28	27	19	25
24	23	34	30	19	27	33	35	26	33	30	25	29	30
25	0	27	20	24	31	28	23	48	20	26	27	30	27
26	33	26	24	18	25	27	24	26	30	20	18	23	25
27	32	27	21	25	18	28	24	33	24	30	27	18	26
28	48	35	0	27	20	24	0	54	36	27	30	23	29
29	30	34	23	0	25	37	29	38	28	20	14	34	31
30	27	26	29	34	36	17	35	26	31	31	38	21	30
BUILD	31	30	28	27	25	28	31	30	29	31	28	26	29
MAXIMUM JOURNEY TIMES													
0	0	44	35	34	37	33	43	19	43	36	33	37	44
7	36	34	33	33	25	36	37	28	33	33	24	29	37
21	39	50	0	35	30	50	70	39	50	50	35	28	70
22	40	53	76	33	27	30	39	38	36	79	27	21	79
23	34	41	19	39	39	33	32	31	30	41	39	22	41
24	24	55	46	19	39	55	37	48	56	38	29	51	56
25	0	42	20	32	46	33	39	48	22	26	36	48	48
26	33	39	45	21	35	39	27	36	44	28	18	36	45
27	32	42	23	49	20	34	35	35	45	39	49	19	49
28	64	47	0	49	32	43	0	69	56	42	50	43	69
29	45	72	23	0	49	39	39	83	41	20	14	42	83
30	31	37	53	53	58	28	47	36	37	53	58	28	58
BUILD	64	72	76	53	58	55	70	83	56	79	58	51	83
NUMBER OF PICKED UP CALLS													
0	0	4	1	5	2	2	6	1	3	1	5	3	33
7	5	3	3	1	2	3	4	2	4	3	2	2	34
21	4	9	2	9	4	2	3	5	7	5	7	3	60
22	3	4	6	3	2	2	5	4	4	6	3	2	44
23	2	8	2	5	3	3	3	3	7	5	3	2	46
24	2	5	7	1	5	8	1	5	5	4	2	6	51
25	0	3	1	4	5	3	3	1	3	2	4	3	32
26	1	6	5	3	2	5	2	6	3	5	2	2	42
27	2	3	5	5	3	2	3	2	4	7	3	2	41
28	2	2	1	6	7	3	0	3	1	4	4	7	40
29	5	5	1	0	5	2	4	5	3	1	1	6	38
30	4	4	5	3	3	6	5	5	4	2	5	3	49
BUILD	30	56	39	45	43	41	39	42	48	45	41	41	510

Mid Day Simulation for Upper Zone

CS21.1 DEFINITION OF DEMAND

Nowadays the mid day demand is often considered to be the most severe traffic pattern, as a result of significant coincident demands between the various office floors and the main terminal floor and "attractive" other floors, such as restaurant and health club floors. The mid day traffic pattern is very complex with movements to/from the main terminal, to/from attractions (facilities floors, restaurants, shops, etc.) and to/from any other floor to another floor.

The building being considered has a facilities floor at Floor 7, which is intended to be significantly attractive to the building occupants. There is no standard "benchmark" that can be applied in this case although several have been suggested as indicated in Chapter 15. Simulation, not calculation, is the only convincing method of analysis. In order to analyse this particular building, a number of assumptions will be made. These are based on experience and are to be regarded as a sufficient test of the installation. Of course, other assumptions can be made based on other experiences, or to meet a different set of requirements. This is a great advantage of design by simulation methods.

Two levels of activity will be simulated, the likely demand and a level of demand 50% larger. The higher demand simulation will indicate whether the lift system is able to provide an acceptable performance. The likely traffic level assumes about one third of each floor population use the lifts over each hour. The remaining one third of any floor population do not use the lifts at all during the two hour lunch period. The higher traffic level will be set at 50% larger than the likely traffic demand, ie: 50% of the one half of each floor's population use the lifts every hour. To produce this level of activity in reality would require some occupants to use the lifts more than once, or for the activity pattern not to be balanced over the two hour lunch period. The Upper Zone will again be considered as it is the most difficult case.

CS21.2 ASSUMPTIONS

The mid day period will be considered to be of a two hour duration. The Facilities floor will be assumed to experience four cycles of 30 minute activity with the average length of stay by persons assumed to be 30 minutes. The simulations should be considered to represent the "middle hour" of the two hour simulation period.

Considering the assumptions made above about one third (53 persons) of each floor population leave each floor in the Upper Zone every hour. It will be assumed that about two thirds (36 persons) travel to the facilities floor and the rest (17 persons) travel to the main terminal. The number of persons travelling back to the Upper Zone from the facilities floor over one hour will be 360 persons and from the main terminal floor will be 170 persons.

The pattern of passenger arrivals at any Upper Zone floor (and who subsequently depart) will be greater in the first periods and tail off in the later periods. Assuming two "waves" of six, 5 minute periods, the ratio of departures from any office floor over the six periods is assumed as: $6-4-2-1-1-1$. This "wave" is repeated for the second half hour of the simulation period of one hour. Similarly the ratio of departures from the facilities floor and the main terminal floor over six, 5 minute periods is

$1-1-1-1-3-3$, again repeated in the second half hour. However the return "wave" is stronger at the end of each half hour period, leading to skewed arrival and departure profiles.

Table CS21.1 Ratio of departures from Floors 21−30, Floor 7 and the main terminal floor

Period	1	2	3	4	5	6	7	8	9	10	11	12	13	14	15	16	17	18	19	20	21	22	23	24
Departures Floors 21−30	6	4	2	1	1	1	6	4	2	1	1	1	6	4	2	1	1	1	6	4	2	1	1	1
Departures Floor 7 and MT	1	1	1	1	3	3	1	1	1	1	3	3	1	1	1	1	3	3	1	1	1	1	3	3

The simulation will emulate some interfloor traffic, but this does not need to be defined. There will also be persons travelling from the office floors to the facilities floor, then later travelling to the main terminal and later still returning to their office floor. Other persons may travel to the main terminal, then later travel to the facility floor and later still return to their office floor. It is assumed that the stops that the various cars make to the facilities and main terminal floors to pick up and set down passengers arising from the main demand will serve these transferring passengers.

To obtain these patterns of arrivals, the passenger input data has to be "tuned" to obtain the correct numbers of picked up calls. Thus examination of Rates of Arrival in Table CS21.2 shows some unusual numbers. However, examination of Number of Picked Up Calls later in Table CS21.3 shows a realistic and randomised pattern. When examining this part of Table CS21.3 remember that 530 persons leave Floors 21−30 over one hour and 530 persons return, making a total number of calls picked up in one hour of approximately 1060 persons.

CS21.3 LIKELY TRAFFIC DEMAND

This section considers the lower of the two mid day scenarios: the likely demand. Figure CS21.1 illustrates the arrival profile at Floor 24, which represents the arrival profile at Floors 21-30. The queues do not exceed three persons.

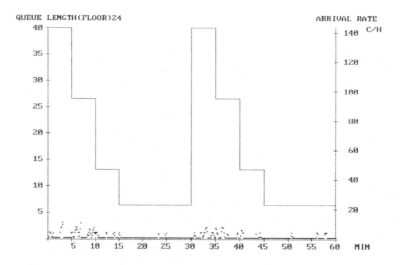

Figure CS21.1 Floor 24 arrival pattern: likely demand

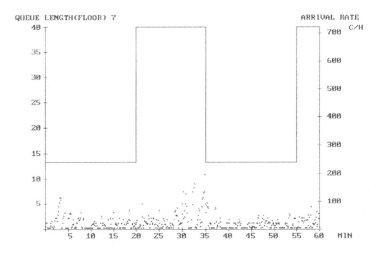

Figure CS21.2 Floor 7 arrival pattern: likely demand

Figure CS21.2 shows the arrival profile for Floor 7. The profile shape for Floor 0 (main terminal) is identical, but with a different arrival rate (see Table CS21.2). Queues up to 11 persons can be seen to accumulate at Floor 7. Note the arrival profiles for Floor 0 and Floor 7 are skewed.

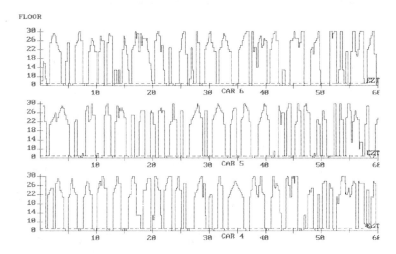

Figure CS21.3 Spatial plot for Lifts 4−6: likely demand

The movement of the three lifts (Lifts 4, 5 and 6) shown in Figure CS21.3 are random in appearance. Closer examination shows an inclination to visit the main terminal (Floor 0) and the facilities floor (Floor 7), both of which have been declared as priority floors. There is also some indication of parking at Floors 21 and 24. Several abortive attempts to move are readily evident for Lift 6, where it has moved away from the lower floors and then reversed direction and returned.

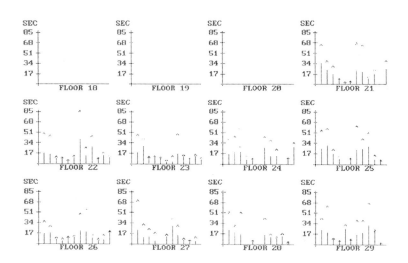

Figure CS21.4 Waiting time Floors 21−29: likely demand

Figure CS21.4 shows the average (bar) and maximum (hat^) passenger waiting times for Floors 18−29. The other floors can be similarly displayed. This display enables a rapid evaluation to be made of the numerical values presented in Table CS21.3. All floors appear generally to be well served. There are a few long wait calls. With this type of traffic demand it is better to look at ninety percentiles.

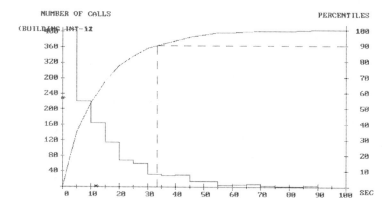

Figure CS21.5 Percentiles and number of calls for all floors for one hour of mid day activity: likely demand

The percentile curve shown in Figure CS21.5 and Table CS21.3 indicate an average passenger waiting time of 13 s and a ninety percentile of 33 s for the whole hour and all floors. Table CS21.3 shows a few floors with ninety percentiles times over 45 s (the excellent criterion of Table 6.5). Overall these times are very acceptable. The values for Floor 7, shown in Figure CS21.6 are better than the building average.

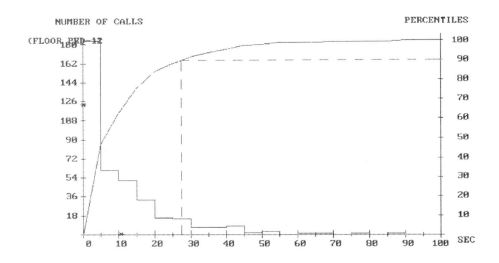

Figure CS21.6 Percentiles and number of calls for Floor 7 for one hour of mid day activity: likely demand

CS21.4 HIGHER TRAFFIC DEMAND

This section considers the higher of the two mid day scenarios: the higher demand. Figure CS21.7 illustrates the arrival profile at Floor 7. The queues increase to some 14 persons, slightly larger than during the likely demand case.

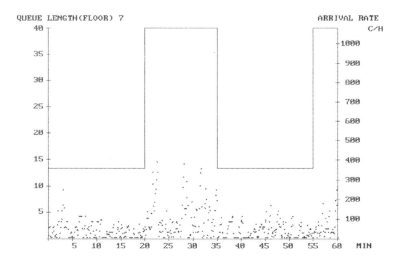

Figure CS21.7 Floor 7 arrival profile: higher demand

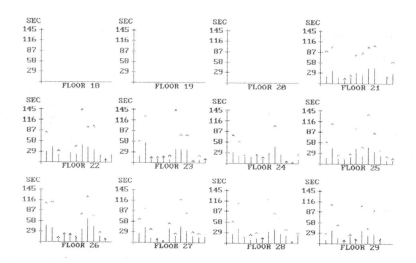

Figure CS21.8 Waiting times Floors 21–28: higher demand

The waiting times shown in Figure CS21.8 are generally 50% larger than for the likely demand case, when examining all floors and one hour values. Some individual values are higher. This is confirmed by examination of the print-out in Table CS21.5. Here it can be seen that the one hour, all floors average waiting time is 20 s, compared to 13 s for the likely demand.

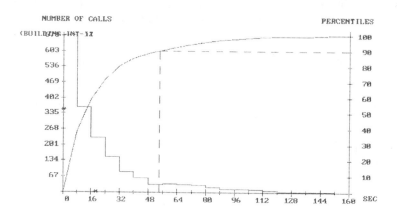

Figure CS21.9 Percentiles and number of calls for all floors for one hour: higher demand

From Figure CS21.9 the ninety percentile, for one hour, for all floors, is 53 s compared to 33 s for the likely demand case.

If the evaluation criteria for the ninety percentile values given in Table 6.5 are accepted, the performance of the lift system for the higher demand can only be graded "fair" for all floors for a one hour period. However, at some of the upper floors the performance would be graded "unacceptable". In the likely demand case, the ninety percentiles over a one hour period are "excellent".

If a 15 minute evaluation window is applied to both demand levels, the likely demand case is still "excellent", but the higher demand case is still "unacceptable".

The activity at Floor 7 for a one hour period is illustrated in Figure CS21.10. Here the average waiting time is 12 s and the ninety percentile is 31 s. These figures are better than the overall building values, indicating good service to the facilities floor.

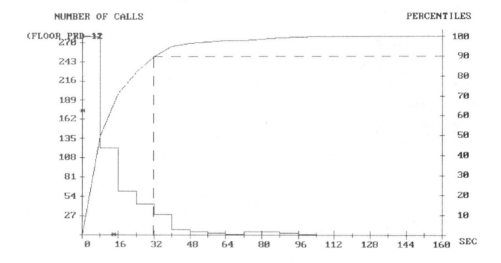

Figure CS21.10 Percentiles and number of calls for Floor 7 for one hour: higher demand

CS21.5 IN CONCLUSION

This case study has shown that the complex mid day traffic demand can be analysed. The likely demand level was found to provide excellent performance. However, by increasing the demand by 50% (see increased population figures in Table CS21.4 compared to Table CS21.2) the installation gives a poorer service. In real systems this level of demand will occur as surges of people leave, or travel, to the facilities floor and main terminal.

For one hour of activity, in the likely demand case, the average passenger waiting time was 13 s and in the higher demand case it was 20 s. However, if the most demanding period (Period 7) is examined, the figures are 18 s and 24 s. Using Equation (15.1) and the data from Table 9.4, the mid day round trip time is 230 s, the mid day interval is 28.7 s and the average passenger waiting time from Equation (15.3) is 24.4 s. The calculated value and the simulated value for the higher demand case are almost identical.

Table CS21.2 Print-out for mid day traffic simulation of the Upper Zone: likely condition — Input data

```
BUILDING DATA:

        GROUP SERVES                              Z EXPRESS ZONE
        EXPRESS STOP                              7 FLOOR
        EXPRESS JUMP                          36.90 METRES
        EXPRESS ZONE TERMINAL                    21 FLOOR
        NUMBER OF LEVELS SERVED ABOVE EZT         9 LEVELS
        DISTANCE FROM M.T. TO EZT             98.80 METRES

LIFT DATA:

        NUMBER OF CARS IN GROUP                   8 CARS
        CAPACITY                                 20 PERSONS
        TERMINAL SPEED                         6.00 M/S
        ACCELERATION                           1.00 M/S/S
        DOOR CLOSING TIME                      2.60 SECONDS
        DOOR OPENING TIME                      1.30
        WIDTH OF ENTRANCE                      1100. MILIMETRES
        SUPERVISORY SYSTEM                       ETA SYSTEM
        JUMP(LEVELS):        1     2     3     4     5     6     7     8
        FLIGHT TIME(S):     5.1   6.8   8.2   9.3  10.3  11.2  12.0  12.8

PASSENGER DATA:

        ARRIVAL PATTERN                        DEF PATTERN
        PATTERN PROFILE                       STEP PROFILE
        MAIN TERMINAL                            0 FLOOR
        PASSENGER TRANSFER TIME               1.00 SECONDS
        RATES OF ARRIVAL:   RATES IN CALLS/5 MINUTES:
        FLOOR  PR 1 PR 2 PR 3 PR 4 PR 5 PR 6 PR 7 PR 8 PR 9 PR10 PR11 PR12  POPUL.  BIAS
           0   10.0 10.0 10.0 10.0 30.0 30.0 30.0 10.0 10.0 10.0 10.0 30.0    200    12
           7   20.0 20.0 20.0 20.0 60.0 60.0 60.0 20.0 20.0 20.0 20.0 60.0    400    24
          21   12.0  8.0  4.0  2.0  2.0  2.0 12.0  8.0  4.0  2.0  2.0  2.0     60     6
          22   12.0  8.0  4.0  2.0  2.0  2.0 12.0  8.0  4.0  2.0  2.0  2.0     60     6
          23   12.0  8.0  4.0  2.0  2.0  2.0 12.0  8.0  4.0  2.0  2.0  2.0     60     6
          24   12.0  8.0  4.0  2.0  2.0  2.0 12.0  8.0  4.0  2.0  2.0  2.0     60     6
          25   12.0  8.0  4.0  2.0  2.0  2.0 12.0  8.0  4.0  2.0  2.0  2.0     60     6
          26   12.0  8.0  4.0  2.0  2.0  2.0 12.0  8.0  4.0  2.0  2.0  2.0     60     6
          27   12.0  8.0  4.0  2.0  2.0  2.0 12.0  8.0  4.0  2.0  2.0  2.0     60     6
          28   12.0  8.0  4.0  2.0  2.0  2.0 12.0  8.0  4.0  2.0  2.0  2.0     60     6
          29   12.0  8.0  4.0  2.0  2.0  2.0 12.0  8.0  4.0  2.0  2.0  2.0     60     6
          30   12.0  8.0  4.0  2.0  2.0  2.0 12.0  8.0  4.0  2.0  2.0  2.0     60     6
```

Table CS21.3 Print-out for mid day traffic simulation of the Upper Zone: likely condition – Performance data (continued overleaf)

FLOOR	PRD 1	PRD 2	PRD 3	PRD 4	PRD 5	PRD 6	PRD 7	PRD 8	PRD 9	PRD10	PRD11	PRD12	HOURLY
AVERAGE WAITING TIMES													
0	5	10	4	2	3	3	10	15	4	2	3	7	6
7	20	15	7	8	5	5	16	14	10	8	13	6	10
21	34	23	17	10	5	6	22	20	10	17	0	25	21
22	17	15	8	11	5	9	39	14	27	10	14	10	19
23	18	28	11	14	10	4	4	16	13	9	12	7	15
24	15	20	19	7	10	0	26	13	13	0	11	28	17
25	25	30	16	8	0	11	23	24	28	7	7	0	21
26	16	17	7	4	12	9	20	19	9	5	6	21	15
27	22	10	11	5	0	11	29	14	7	7	4	0	15
28	22	18	16	0	6	0	15	11	10	12	5	0	16
29	22	19	9	7	25	9	15	18	30	24	3	0	20
30	21	18	6	8	10	8	21	24	12	18	11	15	16
BUILD	20	19	10	7	6	5	18	17	14	8	9	8	13
MAXIMUM WAITING TIMES													
0	16	19	11	18	17	17	30	30	14	14	9	22	30
7	51	42	20	29	28	25	86	79	37	25	33	36	86
21	66	39	30	11	5	6	68	65	14	25	0	40	68
22	50	47	12	11	8	14	87	28	46	10	19	14	87
23	48	42	12	14	12	8	14	49	16	10	16	11	49
24	42	46	30	14	10	0	44	39	25	0	11	35	46
25	55	58	25	14	0	11	63	42	52	16	8	0	63
26	37	30	11	9	12	13	50	58	16	8	15	21	58
27	72	33	25	19	0	16	42	31	10	7	12	0	72
28	53	31	52	0	7	0	44	17	12	18	5	0	53
29	44	63	11	7	33	9	40	40	68	24	3	0	68
30	42	32	11	20	29	8	53	47	47	18	16	17	53
BUILD	72	63	52	29	33	25	87	79	68	25	33	40	87
NINETY PERCENTILES													
0	14	16	9	9	14	13	28	24	12	10	8	19	18
7	44	42	19	16	16	17	43	41	29	19	24	15	27
21	66	37	32	13	9	9	62	44	14	27	0	43	44
22	50	34	13	14	9	13	84	26	48	14	19	14	48
23	31	42	14	14	13	8	11	47	18	13	18	14	35
24	40	34	33	13	14	0	42	35	27	0	14	38	35
25	55	56	27	13	0	14	61	41	52	17	9	0	48
26	37	30	13	8	14	13	47	55	17	9	17	24	35
27	64	32	26	17	0	18	42	32	13	9	13	0	35
28	29	30	52	0	9	0	40	18	14	18	9	0	29
29	41	44	13	9	33	9	40	37	67	24	4	0	43
30	34	31	11	22	28	9	50	45	46	19	19	18	37
BUILD	43	41	22	15	16	15	43	39	38	19	21	19	33

Table CS21.3 Print-out for mid day traffic simulation of the Upper Zone: likely condition — Performance data (continued)

FLOOR	PRD 1	PRD 2	PRD 3	PRD 4	PRD 5	PRD 6	PRD 7	PRD 8	PRD 9	PRD10	PRD11	PRD12	HOURLY
AVERAGE JOURNEY TIMES													
0	53	40	40	48	43	57	72	80	50	37	33	51	55
7	62	64	43	39	47	62	70	76	43	37	42	54	56
21	73	55	37	31	17	28	54	51	27	36	0	54	49
22	51	43	25	21	26	28	73	45	47	39	26	28	46
23	45	53	37	34	36	12	35	47	40	27	31	29	42
24	39	53	44	40	20	35	69	37	37	0	24	45	44
25	62	58	39	34	0	42	47	72	54	31	24	43	52
26	40	50	31	26	38	23	67	52	56	25	28	46	47
27	58	54	45	28	0	29	58	74	44	27	18	0	51
28	66	65	36	62	21	0	51	41	22	37	32	0	50
29	60	75	52	30	36	58	71	45	70	28	15	0	62
30	50	74	47	30	42	36	71	79	58	36	42	48	58
BUILD	55	60	41	37	43	57	65	64	50	35	34	51	53
MAXIMUM JOURNEY TIMES													
0	73	51	61	90	86	127	163	140	109	62	47	105	163
7	117	113	61	70	86	132	165	185	61	62	72	103	185
21	113	64	58	36	17	28	101	95	32	47	0	62	113
22	92	57	28	21	33	28	134	61	55	39	36	37	134
23	102	75	47	36	39	12	78	94	59	33	32	41	102
24	58	74	61	43	20	35	91	84	49	0	24	59	91
25	126	104	59	44	0	42	81	159	77	59	32	43	159
26	80	93	39	36	53	28	123	117	132	33	41	64	132
27	101	86	70	72	0	29	122	138	76	41	28	0	138
28	128	145	69	62	24	0	86	87	23	56	32	0	145
29	100	195	58	36	47	58	137	88	142	41	15	0	195
30	135	133	100	45	63	36	118	187	109	36	59	49	187
BUILD	135	195	100	90	86	132	165	187	142	62	72	105	195
NUMBER OF PICKED UP CALLS													
0	10	6	10	10	31	31	33	15	9	9	12	37	213
7	21	17	24	22	57	51	61	26	13	30	14	54	390
21	7	4	5	2	1	1	14	10	4	4	0	2	54
22	8	10	3	1	5	2	11	6	3	1	3	4	57
23	16	8	3	2	3	2	12	11	2	2	2	3	66
24	8	11	3	2	2	0	5	9	4	0	1	2	47
25	9	7	5	4	0	1	7	7	4	4	4	0	52
26	10	8	3	4	1	2	14	9	4	5	5	1	66
27	10	5	7	4	0	2	8	5	4	1	3	0	49
28	10	8	5	0	2	0	8	5	3	2	1	0	44
29	12	20	2	2	2	1	8	14	9	1	1	0	72
30	10	6	6	4	3	1	9	8	7	1	3	2	60
BUILD	131	110	76	57	107	94	190	125	66	60	49	105	1170

INFORMATION CONCERNING THE CAR TRIPS IS NOT RELEVANT

Table CS21.4 Print-out for mid day traffic simulation of the Upper Zone: higher demand — Input data

```
BUILDING DATA:

        GROUP SERVES                            Z EXPRESS ZONE
        EXPRESS STOP                            7 FLOOR
        EXPRESS ZONE TERMINAL                  21 FLOOR
        EXPRESS JUMP                        36.90 METRES
        NUMBER OF LEVELS SERVED ABOVE EZT       9 LEVELS
        DISTANCE FROM M.T. TO EZT           98.80 METRES

LIFT DATA:

        NUMBER OF CARS IN GROUP                 8 CARS
        CAPACITY                               20 PERSONS
        TERMINAL SPEED                       6.00 M/S
        ACCELERATION                         1.00 M/S/S
        DOOR CLOSING TIME                    2.60 SECONDS
        DOOR OPENING TIME                    1.30
        WIDTH OF ENTRANCE                    1100. MILIMETRES
        SUPERVISORY SYSTEM                     ETA SYSTEM
        JUMP(LEVELS):        1     2     3     4     5     6     7     8
        FLIGHT TIME(S):    5.1   6.8   8.2   9.3  10.3  11.2  12.0  12.8

ARRIVAL DEFINITION:

        ARRIVAL PATTERN                       DEF PATTERN
        PATTERN PROFILE                      STEP PROFILE
        MAIN TERMINAL                           0 FLOOR
        PASSENGER TRANSFER TIME              1.00 SECONDS
        RATES OF ARRIVAL:    RATES IN CALLS/5 MINUTES:
        FLOOR  PR 1 PR 2 PR 3 PR 4 PR 5 PR 6 PR 7 PR 8 PR 9 PR10 PR11 PR12  POPUL.  BIAS
           0   15.0 15.0 15.0 15.0 45.0 45.0 45.0 15.0 15.0 15.0 15.0 45.0   300    12
           7   30.0 30.0 30.0 30.0 90.0 90.0 90.0 30.0 30.0 30.0 30.0 90.0   600    24
          21   18.0 12.0  6.0  3.0  3.0  3.0 18.0 12.0  6.0  3.0  3.0  3.0    60     6
          22   18.0 12.0  6.0  3.0  3.0  3.0 18.0 12.0  6.0  3.0  3.0  3.0    60     6
          23   18.0 12.0  6.0  3.0  3.0  3.0 18.0 12.0  6.0  3.0  3.0  3.0    60     6
          24   18.0 12.0  6.0  3.0  3.0  3.0 18.0 12.0  6.0  3.0  3.0  3.0    60     6
          25   18.0 12.0  6.0  3.0  3.0  3.0 18.0 12.0  6.0  3.0  3.0  3.0    60     6
          26   18.0 12.0  6.0  3.0  3.0  3.0 18.0 12.0  6.0  3.0  3.0  3.0    60     6
          27   18.0 12.0  6.0  3.0  3.0  3.0 18.0 12.0  6.0  3.0  3.0  3.0    60     6
          28   18.0 12.0  6.0  3.0  3.0  3.0 18.0 12.0  6.0  3.0  3.0  3.0    60     6
          29   18.0 12.0  6.0  3.0  3.0  3.0 18.0 12.0  6.0  3.0  3.0  3.0    60     6
          30   18.0 12.0  6.0  3.0  3.0  3.0 18.0 12.0  6.0  3.0  3.0  3.0    60     6
```

Table CS21.5 Print-out for mid day traffic simulation of the Upper Zone: higher demand — Performance data (continued overleaf)

FLOOR	PR 1	PR 2	PR 3	PR 4	PR 5	PR 6	PR 7	PR 8	PR 9	PR10	PR11	PR12	HOURLY
AVERAGE WAITING TIMES													
0	14	18	14	4	5	8	12	13	13	9	7	8	10
7	15	21	10	8	12	10	14	32	15	13	9	7	12
21	21	35	17	10	17	29	26	43	42	0	14	28	28
22	29	41	21	0	26	21	47	42	33	16	7	21	34
23	18	55	12	10	15	9	38	37	35	5	7	11	29
24	31	21	18	17	26	11	30	45	25	5	5	11	27
25	15	43	13	12	18	42	21	45	34	21	9	8	26
26	46	37	8	20	22	14	34	65	42	16	7	0	36
27	23	41	10	9	5	37	16	43	30	16	13	15	27
28	23	39	15	8	11	13	33	37	23	20	5	23	28
29	9	36	15	10	19	9	34	22	25	11	0	0	24
30	26	35	8	8	15	1	32	25	42	11	13	10	26
BUILD	22	36	13	9	11	12	24	36	27	13	9	9	20
MAXIMUM WAITING TIMES													
0	34	41	32	23	18	34	47	42	35	29	20	28	47
7	53	57	38	33	98	84	94	81	68	53	36	40	98
21	92	103	43	15	25	81	86	103	107	0	20	63	107
22	84	124	34	0	56	46	147	98	101	22	10	22	147
23	61	130	16	17	19	22	145	78	79	8	22	12	145
24	78	61	24	22	26	19	137	106	75	7	5	25	137
25	61	112	25	19	29	100	48	137	86	41	22	16	137
26	109	114	13	24	22	14	79	128	76	29	10	0	128
27	67	95	38	9	5	54	25	120	82	30	25	22	120
28	61	112	19	24	30	15	124	84	71	39	5	30	124
29	28	109	53	10	26	13	107	44	92	14	0	0	109
30	76	109	40	14	25	1	98	78	99	19	22	25	109
BUILD	109	130	53	33	98	100	147	137	107	53	36	63	147
NINETY PERCENTILES													
0	35	40	31	17	15	22	34	29	30	25	19	24	26
7	35	47	28	19	31	29	33	79	31	32	27	24	31
21	50	60	42	14	30	85	82	92	108	0	22	61	63
22	81	81	35	0	61	44	138	91	96	22	14	23	83
23	53	109	21	20	22	20	79	72	76	14	19	15	74
24	73	47	28	22	31	20	114	107	75	7	7	28	72
25	51	107	25	19	27	101	48	116	71	45	19	18	67
26	104	101	14	27	23	15	74	111	75	29	14	0	93
27	44	77	34	15	7	54	29	120	80	28	29	22	64
28	46	88	21	28	26	15	100	81	67	36	7	30	79
29	23	95	49	15	28	14	69	44	46	15	7	0	57
30	68	73	22	14	27	7	65	55	87	21	22	28	67
BUILD	53	83	31	21	24	29	55	80	74	29	22	25	53

Table CS21.5 Print-out for mid day traffic simulation of the Upper Zone: higher demand — Performance data (continued)

FLOOR	PR 1	PR 2	PR 3	PR 4	PR 5	PR 6	PR 7	PR 8	PR 9	PR10	PR11	PR12	HOURLY
AVERAGE JOURNEY TIMES													
0	78	82	66	39	48	80	90	93	57	58	48	72	73
7	65	93	51	50	67	65	90	100	54	60	48	63	69
21	63	76	46	34	40	49	73	84	65	0	34	49	65
22	67	74	43	0	57	47	84	72	72	38	27	36	66
23	49	105	31	35	31	36	85	77	82	56	24	27	67
24	71	81	57	36	42	67	76	80	67	18	17	50	69
25	45	80	39	37	48	70	68	80	90	46	30	29	61
26	69	84	40	41	39	0	65	114	78	47	21	0	71
27	73	91	50	20	20	50	38	91	61	42	27	24	61
28	84	84	56	39	48	55	61	108	63	35	14	37	73
29	68	101	63	60	76	30	77	109	83	50	13	0	81
30	86	107	57	42	43	25	102	111	78	35	32	39	82
BUILD	68	89	52	42	58	68	84	94	70	53	38	60	70
MAXIMUM JOURNEY TIMES													
0	152	142	122	75	86	195	193	193	87	117	83	147	195
7	131	169	87	114	128	148	196	183	131	102	100	126	196
21	139	133	72	71	49	104	112	139	153	0	44	85	153
22	148	154	85	0	103	75	176	164	158	61	47	39	176
23	95	176	40	53	44	46	182	148	148	89	42	29	182
24	124	141	91	38	42	79	180	176	154	21	17	83	180
25	92	198	49	66	75	110	103	176	165	89	42	58	198
26	105	201	88	49	39	0	160	173	140	65	24	0	201
27	168	152	106	24	20	67	47	167	127	61	39	31	168
28	198	226	72	56	103	83	114	237	149	50	14	44	237
29	178	173	174	60	109	30	160	254	160	75	13	0	254
30	198	224	140	46	56	32	159	239	170	43	50	53	239
BUILD	198	226	174	114	128	195	196	254	170	117	100	147	254
NUMBER OF PICKED UP CALLS													
0	10	9	12	15	44	54	39	25	12	15	12	53	300
7	35	20	34	25	91	88	84	11	20	39	26	86	559
21	17	14	7	4	2	3	13	14	4	0	4	3	85
22	15	18	6	0	3	4	17	15	9	6	5	2	100
23	13	13	5	8	2	4	11	19	8	2	5	2	92
24	16	11	4	2	1	4	17	11	12	2	1	4	85
25	15	15	8	6	5	3	9	14	12	3	6	7	103
26	9	13	5	5	1	1	7	10	5	3	4	0	63
27	14	13	7	2	1	4	5	9	9	4	3	5	76
28	12	19	3	4	7	2	19	8	11	4	1	2	92
29	13	20	8	1	4	2	13	9	12	2	1	0	85
30	14	28	12	3	6	1	18	11	10	6	3	4	116
BUILD	183	193	111	75	167	170	252	156	124	86	71	168	1756

INFORMATION CONCERNING THE CAR TRIPS IS NOT RELEVANT

APPENDICES

Contents

APPENDIX ONE

Equations of Motion of a Lift Drive

A1.1 SIMPLE KINEMATICS[1]

When considering kinematics during school physics, the following equations are developed:

$$s = ut + \tfrac{1}{2}at^2 \tag{A1.1}$$

$$v = u + at \tag{A1.2}$$

$$v^2 = u^2 + 2as \tag{A1.3}$$

These formulae make the assumption (for simplicity) that the values of acceleration are instantaneously attained. Equations of motion describing real systems, such as the movement of a lift car in a shaft, cannot attain instantaneous values of acceleration, owing to such factors as the time delay for drive motor currents to reach working values, mechanical stiction, brake release delays, etc., termed "start delays". A body can only attain its maximum value of acceleration at a specific rate of change of acceleration. This physical effect is called jerk.

A1.2 LIFT KINEMATICS

The movement of a lift car from rest at one floor to rest at another floor requires four periods of varying acceleration, as listed in Table A.1. These are illustrated in Figure A.1.

Table A.1 Periods of acceleration of a lift car

Period	Jerk	Acceleration
1	Positive constant	Positive increasing
2	Negative constant	Positive decreasing
3	Negative constant	Negative[*] increasing
4	Positive constant	Negative[*] decreasing

[*] Deceleration

---

[1] This Appendix is not claimed as original, but sets down in one place information on lift dynamics: it arose as a result of a number of events. In November 1987 the Author sent Mr G.Doek of Wolter & Dros-Evli, The Netherlands, a "Traffic Design Card", which included suggested lift floor to floor flight times, etc. Doek queried some of the values as being too optimistic and sent the Author a copy of a paper by H.D.Motz of Solingen, West Germany. At the same time (November 1987) the late Joris Schroeder of Schindler, Switzerland, published a paper on drive dynamics. In 1989, whilst preparing the book "Elevator Electric Drives" with Messrs Loher of Ruhstorf, the Motz paper was translated into English. This, together with the work of Roschier and Kaakinen, of Kone, Finland, published in 1980, was the basis of the computer program published in "Elevator Electric Drives". During the writing of the CIBSE Guide D: 1993, Mr P.Day, the Volume Chairman, consulted Mr M.Kaakinen who confirmed the details and pointed out one error. The Motz paper has been translated into Italian and appears in Elevatori (1/1991). Thus an interesting international, industrial, consulting, estate management, academic collaboration has thrown light on a complex subject.

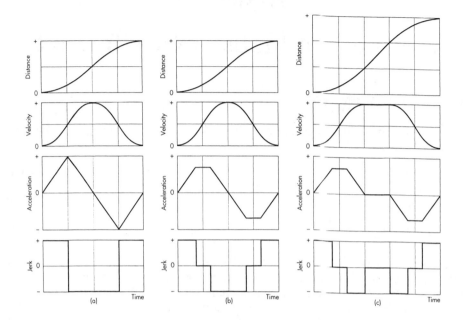

(a) rated speed reached before rated acceleration
(b) rated acceleration reached, rated speed not reached
(c) both rated acceleration and rated speed reached.
Figure A1.1 Lift dynamics

It can be seen in all cases that:

(a) maximum velocity[1] is reached when the acceleration is zero
(b) the distance to be travelled occurs at the end of the four periods
(c) there is a symmetry to the process.

In Figure A.1.1(a) the rated speed is reached before the rated acceleration is attained, which is an unlikely event for a well designed lift drive system. Figure A.1.1(b) shows that the rated acceleration is reached, but not the rated speed, which is typical of single floor run for a high speed lift. The rated acceleration and the rated speed are reached in Figure A.1.1(c), which is typical of a low speed lift or a high speed lift making a multiple floor run.

A1.3 LIFT FORMULAE

The derivation of the formulae for the motion of a lift system are more complex than for the school physics case, where the jerk (J) was considered to be infinite. Motz (1976) provides details of the derivations, Roschier & Kaakinen (1980) contribute summary tables, and Schroeder (1987) and Barney and Loher (1990) suggest computer programs based on the equations. The main formulae used are given below:

---

[1] Velocity and speed can be considered synonymous in this case.

Distance (SA) in order to reach rated acceleration (A), in m/s^2:

$$SA = 2A^3/J^2 \qquad (A1.4)$$

Time (TA) elapsed in order to reach rated acceleration (A):

$$TA = 4A/J \qquad (A1.5)$$

Speed (VA) at rated acceleration (A):

$$VA = (A \times SA/2)^{1/2} \qquad (A1.6)$$

Distance (SVM) to reach rated speed (VM):

$$SVM = VM^2/A + A \times VM/J \qquad (A1.7)$$

Time (TVM) to reach rated speed (VM):

$$TVM = SVM/VM + VM/A + A/J \qquad (A1.8)$$

Maximum speed achieved during a single floor jump V_1:

$$V_1 = \sqrt{\left[\frac{A^2}{2J}\right]^2 + Ad_f} - \left[\frac{A^2}{2J}\right] \qquad (A1.9)$$

If V_1 is less than VM then the flight time $t_f(1)$ is given by:

$$t_f(1) = A/J + \sqrt{(A/J)^2 + (4\,d_f/A)} \qquad (A1.10)$$

A1.4 MULTI-FLOOR JUMPS

Figure A1.2 illustrates a practical example of a lift making single and multi-floor jumps.

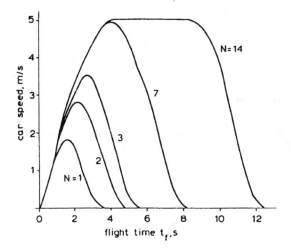

Figure A1.2 Flight times $t_f(N)$ associated with a rated speed (V) of 5 m/s
The interfloor distance (d_f) is 3 m and the acceleration (A) is 1.5 m/s^2

A1.5 LIFT EXAMPLE

Data: $A = 1.0$ m/s^2; $J = 2.0$ m/s^3; $VA = 2.5$ m/s; $d_f = 4.0$ m.

Distance (SA) to reach rated acceleration (A) using Equation (A1.4):

$SA = 2 \times 1.0/4.0 = 0.5$ m

Time (TA) to reach rated acceleration (A) using Equation (A1.5):

$TA = 4 \times 1.0/2.0 = 2.0$ s

Speed (VA) to reach rated acceleration (A) using Equation (A1.6):

$VA = (1.0 \times 0.5/2)^{\frac{1}{2}} = 0.5$ m/s

Distance (SVM) to reach rated speed (VM) using Equation (A1.7):

$SVM = 6.25/1.0 + 1.0 \times 2.5/2.0 = 7.5$ m

Time (TVM) to reach rated speed (VM) using Equation (A1.8):

$TVM = 7.5/2.5 + 2.5/1.0 + 1.0/2.0 = 6.0$ s

Maximum speed during single floor jump using Equation (A1.9):

$$V_1 = \sqrt{(1 \times 4) + \left[\frac{1^2}{2 \times 2}\right]^2} - \frac{1^2}{2 \times 2} = 1.77 \text{ m/s}$$

Flight time $t_f(1)$ using Equation (A1.10):

$$t_f(1) = 1/2 + \sqrt{1/2^2 + (4 \times 4/1)} = 4.53 \text{ s}$$

The times obtained from the formulae above are ideal theoretical times. To obtain practical times add from 0.2 s to 0.5 s (dependent on drive quality) to allow for delays.

A1.6 COMPUTER PROGRAM

In practice a computer program would be used to calculate the dynamics and the result would be presented as Table A2.4.6 in Appendix 2.

SIMPLE:
S̲uite of I̲terative balance M̲ethod and other P̲rograms for L̲ift and E̲levator design

A2.1 INTRODUCTION

In the preceding chapters and case studies, computer programs have been used to determine various traffic designs. There are six programs and these are available at:

www.sponpress.com/supportmaterial

To run the *GOSIMPLE* program suite, select and click the *GOSIMPLE* icon and follow the instructions. To get help, select and click the HELP icon.

All programs offer EDITING switches indicated by curly parentheses {*edit*}.
Hard copies of the screens can be obtained on a local printer by using the clipboard.

No responsibility is taken for your use of these programs.

Send errors and omissions with your full details to:
Gina Barney Associates, PO Box 7, SEDBERGH, Cumbria, LA10 5GE, UK
Fax: +44(0)15396-20578.   Email: None.   Website: www.liftconsulting.org

A2.2 INPUT DESIGN PARAMETERS

All input numerical values should be rounded to two and preferably one place of decimals. Example floor height 3.467 m becomes 3.5 m.

Design name:	Up to 25 characters free format.
Run No.:	Up to 12 characters free format.
Date:	Up to 9 characters free format.

[To avoid transatlantic confusion use 3-letter abbreviation for the month, eg: JAN.]

{N}umber of floors   = Any number >1.   Typical range up to 40.
Number of {P}assengers   = Any number >1.   Typical range up to 33.
Number of {L}ifts   = Any number >1.   Typical range 1 to 8.
   [For double deck calculations - the number of floors served by one deck in uppeak.]
Desired interval (s)   = Any number >0.   Typical range 15 to 60.
Arrival rate (per/5-min)   = Any number >0.
Building {POP}ulation   = Any number >0.
Interfloor {D}istance (m)   = Any number >0.
   [For double deck lifts it is twice single interfloor distance.]
   [Hint: 1 foot is 300 mm; multiply feet by 0.3 to obtain m.]
Base{M}ent floors   = Any number >1.   Typical range 1 to 5.
% {B}asement {P}as'gers   = Any number >1.   Typical range 5 to 25.
{E}xpress {J}ump (m)   = Any number   Additional travel.
   [Allows for analysis of high rise zones.]
{P}assenger {T}ransfer time (s) = Any number.   Typical range 0.8 – 3.0.
Speed (m/s) {v}   = Any number >0.   Typical range 0.25 to 20 m/s.
   [Hint: divide feet/minute by 200 to obtain m/s.]
{AC}celeration (m/s²)   = Any number >0.   Typical range 0.5 to 1.5.
{J}erk (m/s³)   = Any number >0.   Typical range 0.8 to 2.2.
{S}tart {D}elay (s)   = Any number.   Typical range 0 to 1.0.
Door {C}lose time (s)   = Any number.   Typical range 2.5 to 4.0.
Door {O}pen time (s)   = Any number.   Typical range 1.8 to 4.0.
Door {AD}vance time (s)   = Any number.   Typical range 0.3 to 0.8.

A2.3 OUTPUT RESULTS

Average high reversal floor, number.
Average probable number of stops, number.
Average travel time: Time to travel to the average floor in the building, seconds.
DNP/UPP ratio: ratio of handling capacities, number.
Flight time, seconds.
Handling capacity: Uppeak handling capacity, persons per 5-minutes.
HM: Lowest reversal floor, number.
Interval: Uppeak interval, seconds.
Number of lifts, number.
Passengers: passengers in the lift, number.
Performance time, seconds.
%POP: Percentage of the building population handled in the peak 5-minutes, number.
RTT: Round trip time, seconds.
SM: Number of stops below main terminal, number.
tfB: single floor flight time below main terminal, seconds.

A2.4 THE SIX PROGRAMS

All programs have five fields (except the dynamics program).

> The first is the program header.
> The second is the project identification.
> The third is the data entry field.
> The fourth is the results field.
> The last is the editing field.

A2.4.1 Iterative Balance Method (IBM)

This program determines the size of an installation, which will meet a specified passenger arrival rate and a specified lift system interval.

```
*******************************************************************
Lift traffic calculations                      Iterative Balance Method
*******************************************************************
Design name: Case Study Twelve        Run No.: 6         Date: 2002
...................................................................
Desired {I}nterval (s)      =  25    {AR}rival rate (5-min) =  166
{N}umber of floors          =   5    Building {POP}ulation  = 1275
Interfloor {D}istance (m)   = 4.2    {E}xpress {J}ump (m)   = 38.8
{P}ass. {T}ransfer time (s) = 1.0    Speed (m/s) {v}        = 4.0
{AC}eleration (m/s2)   = 1.2    {J}erk (m/s3) = 1.5 {S}tart {D}elay (s) = 0.3
Door times (s):{C}lose = 2.6    {O}pen      = 1.8 {AD}vance          = 0.5
...................................................................
Flight time (s) = 4.63                Performance time (s) = 8.83
...................................................................
Average high reversal floor = 5.0    Average number of stops =  4.8
Passengers =  14.3  Interval =  25.9 Number of lifts = 4 %POP =  13.0
Average travel time =  65.3
...................................................................
Average high reversal floor = 4.9    Average number of stops =  4.5
Passengers =  10.3  Interval =  18.6 Number of lifts = 5 %POP =  13.0
Average travel time =  54.5
*******************************************************************
```

A2.4.2 Lift Traffic Design (LTD)

This program determines the performance of a lift installation given a defining data set.

```
*******************************************************************
Lift traffic calculations                        Lift Traffic Design
*******************************************************************
Design name: Case Study Fourteen      Run No.: 1         Date: 2002
...................................................................
{N}umber of floors          =  16    Number of {P}assengers =  13.5
Number of {L}ifts           =   6    Building {POP}ulation  = 1125
Interfloor {D}istance (m)   =   4    {E}xpress {J}ump (m)   = 0
{P}ass. {T}ransfer time (s) = 1.0    Speed (m/s) {v}        = 4.0
{AC}eleration (m/s2)   = 1    {J}erk (m/s3) = 1.5 {S}tart {D}elay (s) = 0.4
Door times (s): {C}lose = 2.8   {O}pen     = 2.0 {AD}vance          = 0.5
...................................................................
Flight time (s) = 4.72                Performance time (s) = 9.42
...................................................................
Average high reversal floor = 15.3   Average number of stops =  9.3
RTT =  144.4  Interval =  24.1  Handling capacity = 168.2  %POP =  15.0
Average travel time =  79.1
*******************************************************************
```

A2.4.3 Lift Traffic Design with Basements (LTDB)

This program determines the performance of a lift installation serving basements.

```
*****************************************************************
Lift traffic calculations                 Lift traffic design with basements
*****************************************************************
Design name: Case Study Ten          Run No.: 1          Date: 2002
.................................................................
{N}umber of floors          = 4     Number of {P}assengers =  9.1
Base{M}ent floors           = 3     %{B}asement {P}as'gers = 56
Number of {L}ifts           = 2     Building {POP}ulation  =  440
Interfloor {D}istance (m)   = 3.8   {B}asement {D}istance (m)   =  2.7
{P}ass. {T}ransfer time (s) = 1.0   Speed (m/s) {v}        = 1.6
{AC}eleration (m/s2)  = 0.8  {J}erk (m/s3) = 1.2 {S}tart {D}elay (s) = 0.5
Door times (s): {C}lose = 2.6  {O}pen      = 2.0 {AD}vance        = 0.4
.................................................................
Flight time (s) = 5.08               Performance time (s) = 9.78
.................................................................
Average high reversal floor =  3.9   Average number of stops =  3.7
HM =   2.9    SM =      2.6    tfB =   4.40
RTT =   95.9  Interval =   47.9  Handling capacity =  57.0  %POP =  12.9
*****************************************************************
```

A2.4.4 Lift Traffic Double Deck (LTDD)

This program determines the performance of a lift installation served by double deck lifts.

```
*****************************************************************
Lift traffic calculations              Lift Traffic Design for Double Decks
*****************************************************************
Design name: Case Study Fourteen     Run No.: 2          Date: 2002
.................................................................
{N}umber of floors          = 8     Number of {P}assengers = 13.5
Number of {L}ifts           = 6     Building {POP}ulation  = 1181
Interfloor {D}istance (m)   = 8     {E}xpress {J}ump (m)   = 0
{P}ass. {T}ransfer time (s) = 1.0   Speed (m/s) {v}        = 4.0
{AC}eleration (m/s2)  = 1.0  {J}erk (m/s3) = 1.5 {S}tart {D}elay (s) = 0.4
Door times (s): {C}lose = 2.8  {O}pen      = 2.0 {AD}vance        = 0.5
.................................................................
Flight time (s) = 6.36               Performance time (s) =11.06
.................................................................
Average high reversal floor =  7.8   Average number of stops =  6.7
RTT =  142.3  Interval =   23.7  Handling capacity = 170.8  %POP =  14.5
Average travel time =   70.2
*****************************************************************
```

A2.4.5 Lift Traffic Down Peak design (DNP)

This program indicates the performance of a lift installation during down peak traffic.

```
*****************************************************************
Lift traffic calculations           by Dr Gina Barney          Down Peak Estimate
*****************************************************************
Design name: Down peak               Run No.: 1          Date: 2002
.................................................................
{N}umber of floors          = 16    Number of {P}assengers = 12.8
Number of {L}ifts           = 4     Building {POP}ulation  = 647
Number of {SEC}tors         = 4
Interfloor {D}istance (m)   = 4.0   {E}xpress {J}ump (m)   = 0
{P}ass. {T}ransfer time (s) = 1.0   Speed (m/s) {v}        = 2.5
{AC}eleration (m/s2)  = 1.0  {J}erk (m/s3) = 1.5 {S}tart {D}elay (s) = 0.4
Door times (s): {C}lose = 3.0  {O}pen      = 2.0 {AD}vance        = 0.12
.................................................................
Flight time (s) = 4.72               Performance time(s) =10.00
.................................................................
Average high reversal floor =  8.0   Average number of stops =  4.5
RTT =    97.4  Interval =   24.3  Handling capacity = 157.7  %POP =  24.4
Floor interval =   97.4              DNP/UPP ratio = 1.63
*****************************************************************
```

A2.4.6 Calculation of lift system dynamics (DYN3)

This program calculates the time to travel a specified number of floors and the speed then attained (see Appendix 1).

```
**************************************************************************
Lift traffic calculations                        Calculation of lift dynamics
**************************************************************************
Max rated speed (m/s) = 2.5    Acceleration (m/s2) = 1.0    Jerk (m/s3) = 2.0
Floor height (m)      = 4.0    Number of floors    = 6
.........................................................................
          Floor:         Distance:      Flight time:    Speed:
          No.              (m)              (s)          (m/s)
           1             4.00             4.53           1.77
           2             8.00             6.20           2.50
           3            12.00             7.80           2.50
           4            16.00             9.40           2.50
           5            20.00            11.00           2.50
           6            24.00            12.60           2.50
.........................................................................
To reach rated acceleration:   0.50            2.00           0.50
To reach contract speed:       7.50            6.00           2.50
**************************************************************************
```

Measuring Performance

A3.1 MEASURING PERFORMANCE

If it is necessary to determine the performance of a lift, whether it is just commissioned or has been in service for some time, then some important parameters must be measured. The important times to find are the single floor transit time, the performance time and an estimate of the passenger transfer time. What should be measured ? Table A3.1 indicates the principal parameters.

Table A3.1 What to measure

Building:	Lift ID:		Date:	Surveyed by:
Item	Unit	Value	Comment	
Interfloor distance	m		Measure one stair riser, multiply by number of steps	
Rated speed up	m/s		Use tachometer	
Rated speed down	m/s		Use tachometer	
Car depth	mm		Check with rated capacity Table 7.2	
Car width	mm		Check with rated capacity Table 7.2	
Car shape			Note ease of movement in/out	
Rated capacity	kg/Per		Check with rated capacity Table 7.2	
Door width	mm		Clear opening width	
Door type			Side, centre, swing, single panel, two panel, etc.	
Door closing time	s		See Definition 4.15	
Door opening time	s		See Definition 4.16	
Single floor flight time up	s		See Definition 4.20	
Single floor flight time down	s		See Definition 4.20	
Performance time up	s		See Definition 4.23	
Performance time down	s		See Definition 4.23	
Cycle time up	s		See Definition 4.24	
Cycle time down	s		See Definition 4.24	
Car call dwell time	s		See Definition 4.18	
Landing call dwell time	s		See Definition 4.19	

To make these measurements, a stop watch, a tachometer and a tape measure are required.
Table A3.1 could be used as a template for data documentation.

References and Principal Citations

References are cited in alphabetical (first) author order.
Principal citations are to sections.

Alexandris, N.A. (1976): Expected number of stops and highest reversal floor in lift performance, Control Systems Centre report 314, UMIST. - 7.6.3, 12.2.

Alexandris, N.A. (1977): Statistical models in lift systems, PhD thesis, UMIST. - 5.6.1, 12.4.

Alexandris, N.A., Barney, G.C. and Harris, C.J. (1979): Multi-car lift system analysis and design, Appl. Math. Modelling, 3, August. - 12.4, 14.1.

Al-Sharif, L.R. (1992): New concepts in lift traffic analysis: the inverse S-P method, Elevator Technology 4, IAEE Publications. - 12.5.

Al-Sharif, L.R. (1993): Bunching in lift systems, Elevator Technology 5, IAEE Publications. - 12.6.

Al-Sharif, L.R. (1996): Escalator handling capacity, Elevatori, 4/96 and 5/96. - 1.3.4.

ANSI/BOMA Z65.1 (1996): Methods of measuring floor area in office buildings, ANSI-USA. - 6.4.4.

Approved document M (1999): Building Regulations: 1991; Access and facilities for disabled people, Approved Document M, Stationery Office. - 1.5.

Barney, G.C. and Dos Santos, S.M., (1975): Improved trafic design methods for lift systems, Bldg. Sci., 10. - 6.3, 6.10.1.

Barney, G.C. and Dos Santos, S.M. (1977): Lift traffic analysis design and control, Peter Pereginus. - 6.6.3, 11.5.1, 12.4, 13.2, 14.4.

Barney, G.C. and Dos Santos, S.M. (1985): Elevator traffic analysis design and control, Peter Pereginus. - 11.2.4, 11.4.3, 14.1, CS16.1.

Barney, G.C. (1988): Questions and answers, Elevator World, November. - 8.3.1.2.

Barney, G.C. and Loher, A.G. (1990): Elevator electric drives, Ellis Horwood. - A1.3.

Barney, G.C. (1992): Uppeak revisited, Elevator Technology 4, IAEE Publications. - 12.8.1.

Barney, G.C. (1993): Guide D: Transportation systems in buildings, Section 3, CIBSE. - 8.3.3.

Barney, G.C. (1996): Load and capacity, Elevatori, 1/96. - 7.9.5.

Barney, G.C. and Pearce, C. (2000): Calculating basement service, Elevatori, November/December. - 8.3.3.

Barney, G.C. (2000a): Engineering: Load and capacity, Elevator World, March. - 7.9.5, 11.4.7, 12.10.

Barney, G.C. (2000b): Uppeak handling capacity improved — What gives ?, Elevator World, May. - 11.4.7.

Barney, G.C. and Pearce, C. (2001): Calculating basement service, Elevator World, January. - 8.3.3.

Barney, G.C. and Imrak, E. (2001): Application of neural networks to lift traffic control, Elevator World, May. - 11.4.4, 11.4.5.

Barney, G.C. Cooper, D.A. and Inglis, J. (2001): Elevator and escalator micropedia, GBA Publications. - 6.11, 12.7.1.

Bassett Jones (1923): The probable number of stops made by an elevator, GE Rev., 26, (8). - 5.3, 6.3, 12.5.

Bates, V.Q. (1993): What's in a word ?, Elevator Technology 5, IAEE Publications. - 12.7.1.

Bedford, R.J. (1966): Lift traffic recording and analysis, GEC J., 33. - 14.2, 14.6.

Beebe, J.R. (1980): Lift management, PhD, UMIST. - 11.2.2.

Beddington, N. (1982): Design of shopping centres, Butterworth. - 2.

British Council (1997): Best practice in the specification for offices, British Council for Offices. - CS12.1.

Browne, J.J. and Kelly, J.J. (1968): Simulation of elevator system for the worlds tallest buildings, Transp. Sci, 2, (1), 1968. - 8.2.1.

BS CP407 (1972): Electric, hydraulic and hand-powered lift, British Standards Institution. - CS9.3, CS13.2.

BS 5655: Part 6: (2002): Code of practice for the selection and planning of new lifts, British Standards Institution. - 5.7.2, 10.3.4.

BS 5655: Part 7: (1983): Specification for manual control devices, indicators and additional fixings, British Standards Institution. - 10.3.4.

BS 8300: (2001): Design of buildings and their approaches to meet the needs of disabled people, British Standards Institution. - 1.5.

BS EN115 (1995): Safety rules for the construction and installation of escalators and passenger conveyors, British Standards Institution. - 3.2.

BS EN81-1/2 (1998): Safety rules for the installation of lifts, British Standards Institution. - 5.3, 5.4, 7.9.1, 7.9.2, 8.4.2.

BS ISO4190 (1999/2001): Lift (US: elevator) installation − Part 1: Class I, II, III, and VI lifts, ISO. - 5.3, 5.4, 6.8.2, 12.10.

Chan, W.L. and So, A.T.P. (1996): Dynamic zoning for intelligent supervisory control, Int. J. Elev. Eng., 1. - 11.4.4.

Christensen, I. (1988): Why do users press call buttons unnecessarily ? Elevator Technology II, IAEE Publications. - 11.4.5.

CIBSE Guide D: 2000: Transportation systems in buildings, CIBSE. - 1.4, 8.2.3.

Closs, G.D. (1970): The computer control of passenger traffic in large lift system, PhD thesis, UMIST. - 10.2.3, 11.2.1, 11.4.4, 12.8.4, CS17.2.3.

Closs, G.D. (1972): Lift control conservation or progress, Electon. Pwr., 18, Institution of Electrical Engineers. - 10.2.3, 12.8.4.

Day, P and Barney G.C. (1993): Section 11, Transportation systems in buildings, CIBSE. - 16.7.

Day, P. (2001a): Passenger comfort − Are you travelling comfortably ?, Elevator World, April. - 7.9.1, 7.9.2, 12.10.

Day, P. (2001b): Lift passenger comfort have we got it right ?, Elevatori, September. - 7.9.1, 7.9.2, 12.10.

Disability Discrimination Act (1995): Stationary Office, UK. - 1.5.

Dober, R.P. (1969): Environmental design, Van Nostrad Reinhold. - 2.

Dos Santos, S.M. (1972): Lift simulation, MSc dissertation, UMIST. - 16.3.4.

Elevator World Source (1999-2000): Elevator World. - 8.3.2.1.

Fletcher, P.T. (1954): The planning of lift installations in commercial buildings, RIBA Journal, 61. - 10.1.1, 10.2.2.

Fortune, J.W. (1986): Top/down sky lobby design, Elevator Technology, Ellis Horwood, 1986. - 8.2.1.

Fortune, J.W. (1990): Top/down lift design – a case study, Elevator Technology 3, IAEE Publications. - 8.2.1, 8.3.2.5.

Fortune, J.W. (1992): Elevatoring Frank Lloyd Wright's mile high building, Elevator Technology 4, IAEE Publications. - 8.3.2.1, 8.3.2.7.

Fortune, J.W. (1995): Modern double deck elevator applications and theory, Elevator Technology 6, IAEE Publications. - 8.2.1, 8.2.2, 8.3.5.

Fortune, J.W. (1996): Modern double deck elevator applications and theory, Elevator World, August 1996. - 8.2.1, 8.2.2.

Fortune, J.W. (1997): Mega-High Rise elevatoring, Elevator World, December. - 8.3.2.1.

Forwood, B. and Gero, J.S. (1971): Computer simulated lift design analysis for office buildings, Achit. Sci. Rev., 14. - 12.4.

Fruin, J.J. (1971): Pedestrian planning and design, Metropolitan Association of Urban Designers and Environmental Planners. - 1.1, 2.1, 3.1, 7.9.5.

Gaver, D.P. and Powell, B.A. (1971): Variability in round trip times for an elevator car during uppeak, Transpn. Res. 5 (4). - 7.4.1, 7.6.2.

GLC 25 (unknown): Traffic generation: users guide and review of studies, Greater London Council. - 6.5.

Godwin, A.M. (1993): Unique design for a high rise office building, Elevator Technology 5, IAEE Publications. - CS11.

Godwin, M. (1986): Bush House: lifts for the world, Elevator Technology, Ellis Horwood. - 11.4.3.

Halpern, J.B. (1992): Variance analysis a new way of evaluating elevator dispatching systems, Elevator World, September. - 11.4.3.

Halpern, J.B. (1993): Variance analysis of hall call response time, Elevator Technology 5, IAEE Publications. - 11.4.3.

Halpern, J.B. (1995): Statistical analysis ... of call response times, Elevator Technology 6, IAEE Publications. - 11.4.3.

Hall, E.T. (1966): The hidden dimension, Doubleday. - 1.1.

Hikita, S. Amano, M. and Ando, H. (2001): The latest elevator group control system, Elevator Technology 11, Elevcon Books. - 11.4.5.

Hirbod, S. (1975): Simulation extensions for lift supervisory control, MSc, UMIST. - 11.2.3, 16.2.

Ho, M. and Robertson, B. (1994): Elevator group supervisory control using fuzzy logic, Canadian Conference on Elevator and Computer Engineering, 2. - 11.4.4.

HTM (1995): Health Technical Memorandum 2024, design considerations, lifts, HMSO. - 8.4.4.

Hunt, S. (1975): Control of high speed lifts - continuous pattern system, GEC J., 32 (3). - 7.4.1.

ISO 4190, Part 6 (1984): Planning and selection of passenger lifts to be installed in residential buildings, ISO. - 5.7.2, 5.9.

ISO 4190-1 (1999): Lift installation – Part 1: Class I, II, III and VI lifts, ISO. - 7.9.6, 8.3.6, 12.10.

ISO 4190-2 (2001) Lift installation – Part 2: Class IV lifts, ISO. - 7.9.6.

ISO/TR 11071-2 (1996): Comparision of world wide lift safety standards – Part 2: Hydraulic lifts, ISO. - 7.9.5.

Jackson, C. (1970): Analytical techniques: simulation case study, Architects Journal Information Library, 8. - 16.1.

Jappsen, H.M. (2000): Elevators for the Commerzbank building in Frankfurt, Elevator Technology 10, Elevcon Books. - CS11.4.

Jones, B.W. (1971): Simple analytical method for evaluating lift performance during uppeak, Transpn. Res., 5 (4). - 7.4.1, 8.3.6.

Kavounas, G.T. (1989): Elevatoring analysis with double deck elevators, Elevator World, November 1989. - 8.2.2.

Kavounas, G.T. (1992a): Lowest call express floor, a third order consideration for round trip calculations, Elevator World, January. - 12.3.

Kavounas, G.T. (1992b): Bunching: Private communication. - 12.6.

Kavounas, G.T. (1993a): Lowest call express floor: a full theoretical treatment, Part 1, Elevatori, January/February. - 12.3.

Kavounas, G.T. (1993b): Lowest call express floor: a full theoretical treatment, Part 2, Elevatori, March/April. - 12.3.

Kavounas, G.T. (1993c): Designing elevator systems also for the average waiting time, Elevator World, January. - 12.7.2.

Kavounas, G.T. (1993d): Pedestrian traffic calculations among lineraly connected vertical modes, Elevator Technology 5, IAEE Publications. - 1.1.

Lauer, R.J. (1984): Group supervisory control, Elevator World, November. - CS16.4.

Lim, S.H. (1983): A computer based lift control algorithm, PhD, UMIST. - 11.2.4, 11.4.3, 14.4, 14.6, 14.7, CS16.2, CS16.3, 16.2.

Mayo, A.J. (1966): A study of escalators and associated flow system, MSc, Imperial College, London. - 1.3.4.

Miravete, A. (1999): Genetics and intense vertical traffic, Elevator World, July. - 11.4.4.

Morley, C.G. (1962): Installations mechanical: lifts escalators paternosters, Architects Journal, 136. - 7.5.

Motz, H.D. (1976): On the kinematics of the ideal motion of lifts (in German), Foerden und haben, 26. - A1.

Motz, H.D. (1991): On the kinematics of the ideal motion of lifts (in English and Italian), Elevatori, January. - A1.

Moussallati, M.Z. (1974): Lift performance under balanced interfloor traffic conditions, MSc dissertation, UMIST. - 16.2.

Nahon, J. (1990): Traffic analysis for small scale projects, Elevator World, February. - 8.3.3.

NEII-1: 2000 Building transportation standards and guidelines, NEII. - 6.11, 12.7.1.

Papoulis, A. (1965): Probability random variables and stochastic process, McGraw Hill. - 7.7.2.

Pearce, C. (1995): Effects of basements on lift traffic analysis – a review, Elevator Technology 6, IAEE Publications. - 8.3.4.

Pearce, C. (1996): The lobby – a trivial element in lift design, Elevator Technology 7, IAEE Publications. - 8.3.5.

Peters, R.D. (1990): Lift traffic analysis: formulae for the general case, CIBSE -BSERT, 11, 2. - 4.5.2, 16.2.

Peters, R.D. (1995): General analysis double decker lift calculation, Elevator Technology 6, IAEE Publications. - 8.2.2.

Peters, R.D., Mehta, P. and Haddon, J. (1996): Lift passenger traffic patterns: applications, current knowledge and measurement, Elevator Technology 7, IAEE Publications. - 1.4.4.

Peters, R.D. (1997a): Vertical transportation planning in buildings, Eng.D. thesis, Brunell University. - 8.3.3, 12.9, 14.1.

Peters, R.D. (1997b): Private communication. - 7.4.1.

Peters, R.D. and Mehta, P. (1998): Green lift control strategies, Int. J. Elev. Eng., 2. - 11.4.2.

Peters, R.D. (2000): Guide D: Transportation systems in buildings, Section 4, CIBSE. - 8.2.2, 16.4.1.3, 16.5.

Peters, R.D. and Sung, A.C.M. (2000): Beyond the uppeak, Elevator Technology 10, Elevcon Books. - 15.2.1, 15.2.3, 15.3.

Peters, R.D. (2001): Getting started with Elevate, Peters Research. - 12.7.1.

Petigny, B. (1972): Le calcul des ascenceurs, Transpn. Res, 6. - 7.4.1, 7.6.3.

Phillips, R.S. (1973): Electric lifts, Pitman. - 5.9, 7.4.1, 7.5, 7.6, 8.3.6.

Port, L.W. (1961): Australian patent specification 255218, 1961. - 11.4.6.

Port, L.W. (1968): The Port elevator system, University of Sydney, June. - 11.4.6.

Powell, B.A. (1992): Important issues in uppeak traffic handling, Elevator Technology 4, IAEE Publications. - 12.8.3.

Powell, B.A. (2001): Artificial neural networks in elevator dispatching, Elevator Technology 10, Elevcon Books, International Association of Elevator Engineers. - 11.4.7.

Powell, B.A. (2002): Elevator planning and analysis on the web, Int. J. Elev. Eng., 4. - 15.3.

prEN81-70: (1999): Accessibility to lifts for persons including persons with disability, CEN. - 1.5.

prEN81-72 (2002): Safety rules for the construction and installation of lifts − Part 72: Firefighters lifts, CEN (to be ratified). - 8.2.3.

Prowse, R.W. Thomson, T. and Howells, D. (1992): Design and control of lift systems using expert systems and traffic sensing, Elevator Technology 4, IAEE Publications. - 11.4.4.

Qun, Z. Ding, S., Yu, C. and Xiaofeng, L. (2001): Elevator group control system modelling based on object orientated Petri Net, Elevator World, August. - 11.4.4.

Roschier, N.R. and Kaakinen, M.J. (1980): New formulae for elevator round trip time calculation, Elevator World, August. - 7.4.1, A1.

Schroeder, J. (1955): Personenaufzuege (passenger lifts), Foerden und Heben, 1 (in German). - 5.4, 6.3.

Schroeder, J. (1984): Down peak handling capacity of elevator groups, Proc. Int. Lift Sym., Amsterdam. - 13.3.1, 13.3.2.

Schroeder, J. (1987): Elevator trip profiles, Elevator World, November. - A1.

Schroeder, J. (1989a): Sky lobby elevatoring, Elevator World, January 1989. - 8.2.1, 8.3.2.5.

Schroeder, J. (1989b): Value elevatoring by computer, Profi Press. - 8.2.1.

Schroeder, J. (1990a): Elevator traffic: elevatoring, simulation, data recording, etc., Elevator World, June. - 12.6, 12.8.5.

Schroeder, J. (1990b): Elevatoring for modern supervisory systems, Elevator Technology 3, IAEE Publications. - 12.8.4.

Schroeder, J. (1990c): Advanced dispatching, destination hall calls, Elevator Technology 3, IAEE Publications. - 11.4.5, 12.8.5.

Schroeder, J. (1990d): Elevatoring calculation, probable stops and reversal floor, Elevator Technology 3, IAEE Publications. - 12.8.5.

Siikonen, M-L. (1998): Double deck elevators: savings in time and space, Elevator World, July. - 8.2.2, 15.2.1, 15.3.

Siikonen, M-L. (2000): On traffic planning methodology, Elevator Technology 10, Elevcon Books. - 13.6.

Siikonen, M-L. Susi, T. and Hakonen, H. (2001): Passenger traffic flow simulation in tall buildings, Elevator World, August. - 11.4.4.

Siikonen, M-L. (2002): Private communication. - 14.8, 15.2.1, 15.2.2, 15.3.

Smith, R. and Peters, R. (2002): ETD algorithm with destination dispatch and booster option, Elevator Technology 12, ELevcon Books. - 11.2.3, 11.4.5.

So, A.T.P. and Suen, W.S.M. (2002a): Assessment of real time lift traffic performance, CIBSE-BSER&T (to be published). - 11.4.2.

So, A.T.P. and Suen, W.S.M. (2002b): A new statistical formula for estimating average travel time under uppeak conditions, Elevatori, July. - 12.7.3.

Strakosch, G.R. (1967): Elevators and escalators, 1st edition, Wiley. - 5.9, 7.4.1, 7.5, 7.6, 8.3.3, 8.3.6, 10.1.1, 10.5.1, 12.4, 13.1.2, 13.1.5, 13.2, 13.3, 14.1, CS9.

Strakosch, G.R. (1983): Vertical transportation, 2nd edition, Wiley. - 7.9.1, 7.9.5, 8.2.2, 12.4, 13.1.2.

Strakosch, G.R. (1998): Vertical transportation handbook, 3rd edition, Wiley. - 5.9, 6.7.1, 7.9.1, 12.4.

Swindells, W. (1975): Software for the computer control of lift systems, M.Phil. thesis, UMIST. - 16.2.

Tregenza, P.R. (1971): The prediction of passenger lift performance in multi-storey buildings, PhD thesis, University of Nottingham. - 1.1.

Tregenza, P.R. (1972): The prediction of passenger lift performance, Archit. Sci. Rev., 15. - 1.1, 5.6.2, 6.10.1, 7.4.1, 7.5, 7.6.2, 12.2, 12.4, 14.1.

Tregenza, P.R. (1976): The design of interior circulation: people and buildings, Crosby Lockwood. - 1.1.

Tutt, D. and Adler, P. (1990): New metric handbook, planning and design data, Butterworth. - 8.4.8, CS11.

US-A17.1 (1984): Safety code for elevators and escalators, ASME. - 7.9.2.

Zimmermann, F.O. (1973): Elevator traffic analysis, Elevator World, May. - 12.4. 13.1.2, 13.2, 13.3.4.

Index